Low-Energy Electrons

Low-Energy Electrons

Fundamentals and Applications

edited by

Oddur Ingólfsson

PAN STANFORD PUBLISHING

Published by

Pan Stanford Publishing Pte. Ltd.
Penthouse Level, Suntec Tower 3
8 Temasek Boulevard
Singapore 038988

Email: editorial@panstanford.com
Web: www.panstanford.com

British Library Cataloguing-in-Publication Data
A catalogue record for this book is available from the British Library.

Low-Energy Electrons: Fundamentals and Applications

Cover image: Taken from the public domain (https://pixnio.com/nature-landscapes/night/aurora-borealis-astronomy-atmosphere-phenomenon-planet-majestic-sky-night) and modified by Reza Tafrishi

ISBN 978-981-4800-00-6 (Hardcover)
ISBN 978-0-429-05882-0 (eBook)

To my late father,
Ingólfur G. Sigurðsson, in appreciation of
his support and inspiration throughout
the past few decades

Contents

Preface

Low-energy electrons are ubiquitous in nature and play an important role in many potential and current industrial processes; accordingly, the importance of understanding electron–molecule interactions has been highlighted in recent decades. In this book, the fundamental characteristics of low-energy electron–molecule interactions are described, and their role in different fields of science and technology, including plasma processing, nanotechnology and health care, as well as astro- and atmospheric physics and chemistry, is reviewed and discussed.

The book starts with three chapters offering an account of the fundamentals of low-energy electron–molecule interactions, followed by a series of independent chapters devoted to specific fields where such interactions are of importance. It is not only directed at researchers active within this field but may also serve as a graduate study course material where selected chapters can be used either individually or in combination as the basis to highlight and study specific aspects of low-energy electron–molecule interactions. Specific topics may then be deepened through further studies of selected references cited in context to the respective topic being studied. For example, the first three chapters may be used to teach the fundamentals of electron–molecule interactions. Alternatively, the second chapter in combination with Chapter 6 can be used to address the role of dissociative electron attachment in radiation damage. Likewise, the first two or the first three chapters may be used in combination with Chapter 4 as course material to study plasma physics and chemistry. Hence the book may be used in a many ways, defining either a narrow or a broad-based course.

The first three chapters deal with fundamental aspects of low-energy electron–molecule interactions. The first chapter describes the fundamentals of electron-scattering processes, with an emphasis on definition of critical quantities; differential-, integral- and total cross sections; and a description of how these may be derived from experiments. The second chapter deals with the fundamental aspects of low-energy electron-induced dissociation processes, with an emphasis on dissociative electron attachment and a section devoted to the experimental approaches developed to study such dissociation processes. The third chapter deals with theoretical aspects of low-energy electron–molecule interactions, specifically offering an account of the Schwinger multichannel method, the discrete momentum representation method and the *R*-matrix method.

These three chapters provide the basis for the rest of the book and are followed by five independent chapters reviewing the significance of low-energy electrons in different fields. The first of these, Chapter 4, focuses on electron processes in plasma systems, reviewing critical parameters in plasma processes and the underlying electron–molecule interactions. Chapter 5 deals with the role of low-energy electrons in nanotechnology, more specifically the role of secondary and backscattered electrons in focused electron beam–induced deposition and etching and describes the surface science approaches used to explore the respective electron-induced surface chemistry. Chapter 6 focuses on the role of low-energy electrons in radiation damage, specifically the potential DNA damage caused by secondary electrons formed when living organisms are exposed to high-energy radiation, and includes a section on the physicochemical mechanism of radio sensitization in cancer therapy. Chapter 7 then turns to reviewing the role of low-energy electrons in atmospheric processes, including auroral emissions, lightning, ozone depletion and the production of negative Ions in the ionosphere. Finally, Chapter 8 deals with electrons in space, starting with the Big Bang and the birth of the molecular universe before turning to electrons in the interstellar medium and the potential role of electron interactions in the formation of stars, finally offering a discussion of the role of electrons in planetary atmospheres.

The book is authored by 16 active researchers within their respective fields, is packed with illustrative examples, both on the fundamental and the application side, features about 130 figures and lists over 800 references.

Oddur Ingólfsson
2019

Chapter 1

Molecular Processes and Techniques for Measuring Their Scattering Cross Sections

Michael J. Brunger
College of Science and Engineering, Flinders University, GPO Box 2100, Adelaide, SA 5001, Australia
michael.brunger@flinders.edu.au

1.1 Prologue and Definitions

Molecules, both diatomic and polyatomic, possess many internal degrees of motion, which when quantized lead to rotational energy levels, vibrational energy levels, discrete electronic-state energy levels, and the continuum. Therefore, when a molecule interacts with some incident projectile, which might be electromagnetic radiation or an electron or a positron (for instance), those internal degrees of freedom may be excited, with the probability for such an excitation ocurring being known as the cross section. For example, if the incident electromagnetic radiation was in the microwave part of the spectrum, the rotational levels of a heteronuclear diatomic molecule could be excited and from the resulting microwave

Low-Energy Electrons: Fundamentals and Applications
Edited by Oddur Ingólfsson

ISBN 978-981-4800-00-6 (Hardback), 978-0-429-05882-0 (eBook)
www.panstanford.com

spectrum important physical quantities such as the rotational constant (B) and bond length (R) could, in principle, be determined for that molecule. On the other hand if the incident radiation was in the visible or near-ultraviolet part of the spectrum, discrete electronic-state excitations might be probed [1, 2], a technique known as photoabsorption spectroscopy. Finally, if X-rays are the source radiation then enough energy can be imparted to the target molecule so that one or more of the bound electrons are moved into the continuum. In this case we are dealing with the process of photoionization, with photoelectron spectroscopy [3] being a powerful tool by which one might investigate molecular structure and correlation. A similar story also occurs when electrons are the incident projectile, insofar as rotational, vibrational, discrete electronic-state excitation and ionization may again occur. In this case, however, the incident projectile is charged, so the Coulomb interaction between that electron and the bound electrons and nuclei of the target molecule now plays a significant role in the scattering process. For instance, the process of electron exchange, which gives rise to the excitation of optically forbidden electronic-state transitions, can now happen. One particularly important process, apparently ubiquitous to electrons as the incident projectile, is when the incoming electron is temporarily captured by the molecule. This is generally referred to as being a resonant or resonance process [4]. After a certain lifetime the electron may then autodetach, leaving the target molecule in, for example, one or more excited vibrational levels (including the ground vibrational level) and/or electronic states, depending on the incident energy of those electrons. Note that the population of those vibrational levels and/or electronic states is typically much greater via this resonance process than via direct excitation. A competing, and very important, resonance effect is when the captured electron leads to molecular rearrangement. This process is better known as dissociative electron attachment [5, 6] and is described in significant detail in a later chapter.

This chapter will concentrate on measurements of total scattering, differential (in angle), integral and momentum transfer, and dissociative electron attachment cross sections (whether absolute or relative) for electron–molecule scattering, with some reasons for why a knowledge of those cross sections is fundamentally important being given shortly. The differential cross section (DCS) is typically measured in a crossed-beam configuration (see Sections 1.2 and 1.3) for well-defined values of both the incident and scattered electron energy, E_0 and E_r, for an interaction of the form

$$e^-(E_0) + AB_i \rightarrow e^-(E_r) + AB_j, \tag{1.1}$$

where AB_i and AB_j represent the initial and final states of some generic diatomic or polyatomic molecule, and for the moment we ignore dissociation. If the energy resolution of the incident beam and the nature of the molecular target are such that the initial and final states are well defined, the DCS at a scattering angle θ is given by

$$\mathrm{DCS}(E_0, \theta)_{i\rightarrow f} = \frac{d\sigma}{d\Omega}(E_0, \theta)_{i\rightarrow f} = \frac{k_r}{k_0}\left|f_{i\rightarrow f}(E_0, \theta)\right|^2, \tag{1.2}$$

where k_0 and k_r are the initial and final momenta given by (in atomic units)

$$k_{0,r} = \sqrt{\frac{E_{0,r}}{E_\mathrm{R}}}, \tag{1.3}$$

with $f_{i\rightarrow f}$ being the scattering amplitude for the interaction and E_R the Rydberg energy. As we shall see later, in general, state-of-the-art energy resolution of modern crossed-beam spectrometers ($\approx$10 meV) is only sufficient to resolve rotational excitation in H_2. Also, most quantitative experiments do not involve spin-polarized beams, or targets which have been "dressed" in a single fine or hyperfine level, let alone a single rotational level, resulting from the (usually) elevated temperature of the molecular beam. Thus, the measured quantities typically involve an average over the possible initial states of the molecular target and a sum over the final states.

The integral cross section (ICS) for the excitation process $i \rightarrow f$, $Q_{i\rightarrow f}$ is usually obtained (see Section 1.4) by integration of the DCS over all scattering angles:

$$Q_{i\rightarrow f} = 2\pi \int_0^\pi \left(\frac{d\sigma}{d\Omega}\right)_{i\rightarrow f} \sin\theta d\theta, \tag{1.4}$$

and the total cross section (see Section 1.5), TCS $\equiv \sigma_t$, is the sum of the integral cross sections for all energetically available processes n:

$$\sigma_t(E_0) = \sum_n Q_n(E_0). \tag{1.5}$$

Another commonly measured quantity, particularly from electron swarm experiments [7], is the momentum transfer cross section $Q_m^{i\to f}$, which, in terms of the differential cross section for any given process $i \to f$, is defined as

$$Q_m^{i\to f}(E_0) = 2\pi \int_0^{\pi} \left(\frac{d\sigma}{d\Omega}\right)_{i\to f} \sin\theta(1-\cos\theta)d\theta. \tag{1.6}$$

It is worth noting that in the analysis of swarm experiments using the two-term Boltzmann equation [7], the momentum transfer cross section which arises is, in fact, not the elastic momentum transfer cross section but rather the following quantity (which is referred to as the total momentum transfer cross section):

$$Q_m(E_0) = Q_m^{\text{elastic}}(E_0) + \sum_k Q_m^k(E_0) + \sum_k Q_m^{-k}(E_0), \tag{1.7}$$

where $Q_m^{\text{elastic}}(E_0)$ is the elastic momentum transfer cross section and Q_m^k and Q_m^{-k} are the momentum transfer cross sections for the kth inelastic and superelastic scattering processes, respectively. This is a particularly important distinction when the inelastic scattering cross sections are significant, when compared to the elastic, as, for example, in the case of gases with Ramsauer minima, such as methane, or for gases where resonance effects are large, such as N_2.

While the importance of electron-driven processes in several applications, both technological and medical, are addressed in detail later in this volume, it seems appropriate at this time to make some brief introductory remarks on this theme and present three relevant exemplars. Microscopic electron collision processes with atoms, molecules, and ions are fundamental to a complete understanding of the macroscopic behavior of all gas discharges and low-temperature plasmas. In most cases, collisions between low- to intermediate-energy electrons and atoms and molecules, are the precursors for the production of critical species in low-temperature plasmas and plasma-processing gas discharges [8]. Precise measurements of atomic and molecular collisions can provide an understanding of low-energy processes such as vibrational excitation, dissociative attachment, electronic excitation, and ionization, all of which play an important role in the production of excited species, free radicals, and negative and positive ions. Long-lived metastable species can also be important, due to their large amounts of internal energy and their typically large interaction cross sections with charged particles and other neutral species, in processes such as Penning ionization. In many cases, due to the selection rules that govern the metastability of these states, they can only be efficiently excited by electron impact through the exchange interaction. Of particular interest (although probaby more so 5–10 years ago) to plasma processing are electron-driven processes involving perfluorocarbons (PFCs). These have invited a lot of interest both due to the industrial application of these molecules [9] as well as growing environmental concerns [10, 11]. In particular, low-energy, that is, between ~0–20 eV, electron scattering cross sections—total, elastic, rotational, vibrational, dissociation, and momentum transfer—are of critical importance in the modeling of low-temperature plasmas, plasma etch discharges, and plasma-enhanced chemical vapour deposition. These collision cross sections are the basis for the derivation of electron transport properties, mobilities, and energy—and spatial—distribution functions [12, 13]. From a fundamental point of view, when these PFCs are compared to their corresponding pure hydrocarbon counterparts, the effect of fluorination on electron scattering cross sections has attracted quite a lot of theoretical and experimental attention over the past couple of decades [14]. However, the use of PFCs offers an interesting case as their wide application is offset by environmental concerns. They are important as greenhouse gas molecules, most of them are chemically inert, they have large capacities for trapping heat, and they have long atmospheric lifetimes (e.g., up to 50,000 years for C_3F_6) [15]. Thus, from an environmental point of view, the interaction properties of these molecules need to be correctly known and well characterized in order to achieve a better compromise between their individual usefulness and the possible environmental ramifications.

Another, more natural, role of electron-driven processes is their effect on plasma processes in cometary and planetary atmospheres. Electrons from our sun, often accelerated by magnetospheric processes, produce low-density plasmas in the upper atmospheres of planets and their satellites. The secondary electrons can produce further ionization, dissociation, and excitation, leading to enhancement of chemical reactions and light emission. Similar processes are driven by photoelectrons produced by sunlight in upper atmospheres during daytime. Sunlight and solar electrons also drive the same processes in the atmospheres of comets. Thus for both an understanding of planetary atmospheres and prediction of emissions for comparison with remote observations, it is necessary to address the role and importance of electron-driven processes, and the cross sections that characterize those processes, toward gaining a quantitative understanding of the phenomena in question. Campbell et al. [16, 17], for example, have thus conducted a number of studies of plasma processes in the upper atmospheres of Earth and several other planets, in the atmosphere of Titan, and in the coma of comet Hale–Bopp. In all cases it was necessary to determine an appropriate model of the electron flux spectrum, considering thermal, photo-, and auroral electrons. While some calculations (such as electron cooling) required just a formula to determine the results from the electron parameters and electron impact excitation cross sections, most of the applications required a statistical equilibrum [16, 17] calculation to incorporate the many gain and loss processes for each level of each excited state. In addition, due to the long lifetimes of some species, it was necessary to calculate equilibrium populations for fast processes by statistical equilibrium and then import them into a time-step [16, 17] calculation that considered the slower processes. Most of those calculations were performed in order to assess the importance of using more accurate electron impact cross sections. Accurate cross sections were found by them to be critical in the case of the contribution of vibrational excitation of N_2 in electron heating of the ionopshere of Titan and in calculations of NO infrared emissions in Earth's ionosphere [16]. They were also important in calculations of electron cooling by CO_2 in the atmosphere of Mars and significant at low electron temperatures in calculations of CO infrared emissions in the atmospheres of Mars and Venus [16]. A major outcome of some of their investigations was to prove that electron impact excitations were a factor at all. In the case of the fourth positive emissions from the comet Hale–Bopp, which were previously ascribed entirely to fluorescence, they showed [16] that electron impact could account for 40% of the emissions, thus reducing by 40% the calculated abundance of CO in the comet. Thus, once again, we find that for a quantitative understanding of (in this case) atmospheric phenomena, detailed and accurate cross-section sets for many processes are essential.

The final exemplar I discuss, for the importance of having accurate and complete electron scattering cross section databases, deals with studying charged particle tracks in matter at the nanoscale, with a particular focus on radiation damage in matter and especially in the human body. In particular, biomedical uses of radiation are demanding an increased level of detail [18–21] in order to describe the interaction processes initiating radiation damage. For example, reduced volume irradiation techniques such as brachytherapy, in which radiation emitters are placed close to the target tumour, not only require traditional dosimetric methods to prescribe the treatment but also require the evaluation of possible side effects in the surrounding sensitive areas. The important role of low-energy, sub-ionizing, secondary electrons for inducing radiation damage in biomolecular systems has been extensively studied in the past decade [22, 23]. According to these studies, realistic radiation interaction models should include Low-energy electron interactions up to their final thermalization in the medium. In addition, due to the efficiency of positron emission tomography (PET) scanners in detecting premature cancers and cancer metastasis, positron-emitting sources are increasingly being used in nuclear medicine [24]. The resolution of these systems has been recently improved by new medical imaging techniques and developments. However, positron sources used in medicine are beta emitters, which show a wide energy distribution from very low energies up to the megaelectron volt (MeV) domain, with typical average energies of hundreds of kiloelectron volts (keV). Since positron annihilation mainly occurs at relatively low energies, below 100 eV, modeling positron tracks from emission energies down to final annihilation, in biologically relevant media, can be useful to improve PET resolution and in designing more efficient radiation sources [25]. This type of model represents a real challenge for fundamental research, as it will need both differential and integral interaction cross sections for all processes taking place across the whole energy range. Note also that as the incident positrons thermalize through interactions with atoms and molecules, secondary electrons are increasingly likely to be produced. As a consequence, complete cross-section databases for electron impact interactions are also needed if one is to build a total picture of positron transport and its consequences for radiation damage. Recently, this need of describing radiation effects at the molecular level has led to the development of the concept of nanodosimetry, that is, a procedure to quantify radiation-induced molecular damage in nanovolumes [21, 26]. Due to the variety of interactions expected and their respective different biological implications, the absorbed dose alone is probably no longer a proper parameter to describe effects at the nanoscale. Rather, a detailed description of the interaction processes occurring in a nanosized target and their implications in terms of radiation damage (number of dissociative events, type of radicals generated, etc.) is really required. As this nanoregion of interest could be relatively far (microns or even millimetres) from the original track of the incident high-energy primary particles, an accurate description of single tracks of secondary particles would seem to be essential for that purpose. While there are several groups pursuing this line of research, three of note are, respectively, based in Madrid [20, 21, 27], Belgrade [28], and Townsville [29, 30].

There are now several quite large and extensive databases that describe scattering phenomena in the applications discussed before. I note that this statement holds true for both electrons and positrons, although the electron database is currently much more extensive. Some of those cross-section compilations include the work of Anzai et al. [31], Pitchford et al. [32] and Brunger et al. [33].

Having provided a few definitions, relating to the important cross sections in describing electron scattering phenomena, and having briefly advanced some rationales for why a knowledge of those cross sections is important, let us now move on to addressing the techniques that are currently being employed in an attempt to accurately measure those data.

1.2 Introduction

In the remainder of this chapter we focus on describing techniques currently being used to measure absolute elastic (where no energy is transferred from the incident electron to the target molecule) cross

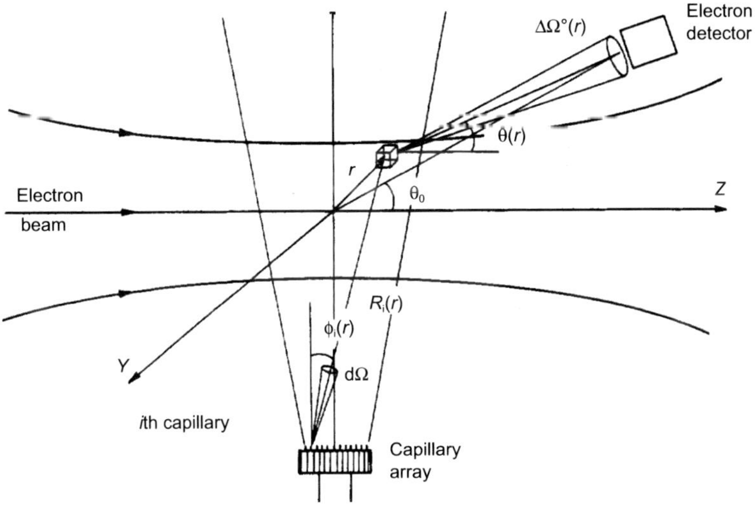

Figure 1.1 Geometry used in a typical crossed-beam experiment. From Ref. [34] © IOP Publishing. Reproduced with permission. All rights reserved.

sections and absolute discrete inelastic (rotational, vibrational, and electronic-state excitation in molecules) cross sections. Important molecular scattering processes such as ionization and attachment are dealt with in other chapters of this volume, and so they are not described further here.

A schematic diagram for a typical crossed electron beam–target beam scattering configuration is given in Fig. 1.1 [34]. Notice the finite spatial extent of both electron and target molecular beams. The relationship, for a given scattering process n, between the DCS and the number N_n of scattered electrons detected per second with nominal energy $\overline{E}_n = \overline{E}_0 - \Delta E$, where $\overline{E}_0$ is the nominal electron impact energy and ΔE is the energy loss corresponding to process n, at the nominal angle $\overline{\Omega}(\overline{\theta}, \overline{\phi})$, can be written as

$$N_n(\overline{E}_0, \overline{E}_n, \overline{\Omega}) = \int_r \int_{E_0} \int_{E_n} \int_{\Omega} \rho(\underline{r}) f(E_0, \underline{r}) \xi(E_n, \underline{r}) \frac{d^3\sigma(E_0, E_n, \Omega)}{dE_n d\Omega} d\Omega dE_0 dE_n d^3r, \tag{1.8}$$

where $\rho(\underline{r})$ is the density distribution of the target molecules, $f(E_0, \underline{r})$ is the energy and spatial distribution of the incident electron flux, $\xi(E_n, \underline{r})$ is the response function of the detector for electrons of energy E_n scattered by a target molecule at position $\underline{r}$, and $d^3\sigma/dE_n d\Omega$ is the DCS.

In Eq. 1.8 we assume that the background contribution to the scattering signal has been removed. This can be achieved by taking, as the true scattering signal N_n, it to be the difference between the signals with the target molecular beam on and off. Note that for this procedure to be valid the scattering conditions in the beam on and off modes must be identical. To ensure this latter criterion is met, one usually introduces the proper leakage of gas into the background that was eliminated by shutting off the target molecular beam.

A detailed knowledge of the instrumental function, and the target density and electron beam flux distributions, is required to extract the cross section from Eq. 1.8. In general, however, some simplifying assumptions can be made. For excitation of a discrete final state n it is usually possible to integrate over the energy loss profile, assuming constant system response over the natural line width. Furthermore, if the cross section does not change significantly over the energy and angular resolution of the experiment, it can be removed from within the integral so that Eq. 1.8 simplifies to give

$$\begin{aligned} N_n(\overline{E}_0, \overline{E}_n, \overline{\Omega}) &= \kappa(\underline{r}, E_0, \overline{E}_n, \Omega) \cdot \overline{\frac{d\sigma(E_0, \Omega)}{d\Omega}} \\ &= \kappa(\underline{r}, E_0, \overline{E}_n, \Omega)\mathrm{DCS}_n(\overline{E}_0, \overline{\Omega}), \end{aligned} \tag{1.9}$$

where

$$\kappa(\underline{r}, E_0, \overline{E}_n, \Omega) = \int_r \int_{E_0} \int_\Omega \rho(\underline{r}) f(E_0, \underline{r}) \xi(\underline{E_n}, \underline{r}) dE_0 d\Omega d^3r \tag{1.10}$$

includes all the instrumental and geometrical factors. The instrument-averaged singly DCS is now indicated as the abbreviation DCS. This is the quantity obtained from the measurement (n may now refer to a collection of experimentally indistinguishable processes). Following Trajmar and McConkey [35] we have denoted the quantity in Eq. 1.10, which is often referred to as the *effective scattering volume*, as κ (instead of $\tilde{V}_{\text{eff}}$) to highlight that it does not have dimensions of cm^3.

In general, Eq. 1.10 can be simplified by assuming that the energy distribution of electrons is independent of $\underline{r}$ and that the detector response depends only on E_n and not $\underline{r}$. Under these conditions

$$f(E_0, \underline{r}) = I_e f(E_0) f(\underline{r}) \tag{1.11}$$

and

$$\xi(E_n, \underline{r}) = \xi(E_n),$$

where I_e is the incident electron beam current. In this case the integrals over the energy distributions and coordinates can be separated to give

$$\kappa(\overline{E}_n, \theta, \phi) = I_e C(\overline{E}_n)\tilde{V}_{\text{eff}}(\theta, \phi), \tag{1.12}$$

with

$$\tilde{V}_{\text{eff}}(\theta, \phi) = \int_r \int_{\Omega(\underline{r})} \rho(\underline{r}) f(\underline{r}) d\Omega(\underline{r}) d^3r \tag{1.13}$$

and

$$C(\overline{E}_n) = \int_{E_n} \int_{E_0} f(E_0)\xi(E_n) dE_n dE_0. \tag{1.14}$$

In Eq. 1.14, $C(\overline{E}_n)$ denotes the energy dependence of the detector response function and is, in general, a quite complicated function. A futher simplification occurs in that for conventional scattering measurements the molecules are randomly oriented (or the scattering is orientation independent) so that the ϕ dependence disappears and Ω can be replaced by the scattered electron angle (with respect to the incident electron beam direction) θ. In this case Eqs. 1.9, 1.12, and 1.13 become

$$N_n(\overline{E}_0, \overline{E}_n, \overline{\theta}) = \kappa(\overline{E}_n, \overline{\theta})\text{DCS}_n(\overline{E}_0, \overline{\theta}), \tag{1.15}$$

with

$$\kappa(\overline{E}_n, \overline{\theta}) = I_e C(\overline{E}_n)\tilde{V}_{\text{eff}}(\overline{\theta}) \tag{1.16}$$

and

$$\tilde{V}_{\text{eff}}(\theta) = 2\pi \int_r \int_{\theta'(\underline{r})} \rho(\underline{r}) f(\underline{r}) \sin\theta'(\underline{r}) d\theta'(\underline{r}) d^3r. \tag{1.17}$$

$\tilde{V}_{\text{eff}}(\overline{\theta})$ can be calculated from a knowledge of the density distribution of the intersecting electron and molecular beams and the acceptance angles of the scattered electron detector, as done by Brinkmann and Trajmar [36]. Until very recently, it was always thought that the practical determination of absolute values of the overlap integral $\tilde{V}_{\text{eff}}$ would be too difficult to obtain, since they would require the determination of both the molecular beam density and the electron beam fluxes. Furthermore, even if $\tilde{V}_{\text{eff}}$ could be extracted absolutely, one would still need to characterize $C(\overline{E}_n)$ to obtain the absolute cross sections. However very recently, for the elastic channel (here the incident and scattered electron energies are equal and so the characterization of the response function is not an issue), Baek et al. [37] proposed a method to determine $\tilde{V}_{\text{eff}}$ by means of the current loss ΔI_e of the primary electron beam after passage through the molecular beam of interest and a knowledge of the TCS. Note that the TCS can be measured more accurately (see later) in an independent linear transmission apparatus and with an application of the Beer–Lambert law [38, 39]. In their study [37] very encouraging agreement, over a broad energy range, was found with previous elastic DCSs for electron scattering from the biomolecule tetrahydrofuran from other groups, who employed (see later) more traditional methods to set the absolute scale of their data [40–43]. Full details of this new approach can be found in Baek et al. [37], to whom the reader is referred for more details.

Notwithstanding the above, the usual practice in fixing the absolute scale for $\text{DCS}_n(\overline{E}_0, \overline{\theta})$ is to apply a normalization procedure, some of which are now discussed.

1.3 Normalization Techniques in Differential Cross-Section Measurements

1.3.1 Elastic Scattering: The Relative Flow Technique with a Continuous Molecular Beam Source

The most widely accepted strategy for the measurement of elastic DCSs, using a continuous flow molecular beam source, is to utilize a set of established elastic DCSs as standards for eliminating the response function $C(\overline{E}_n)$ (see Eq. 1.14) and the integral represented by $\tilde{V}_{\text{eff}}(\theta)$ (see Eq. 1.17). This is achieved by the relative flow technique, which relates the DCS of a particular target (T) gas, at a specific energy and scattering angle, to that of a reference or standard (S) gas under identical experimental conditions.

For low-energy electron scattering the standard cross section that is normally used is that for electron–helium elastic scattering. This cross section has been measured extensively (e.g., [44–46]), the absolute values often being obtained from the application of phase-shift analysis techniques [45]. This cross section has also been calculated in an ab initio fashion by several groups [47–49] and is widely regarded as being known to within a few percentage points for energies below the first inelastic threshold at 19.8 eV. The application by Fursa and Bray [50, 51] of their convergent close coupling (CCC) method, to two-electron atoms, along with the experiments of Register et al. [46] and Brunger et al. [45], has also established the case for helium to be considered as a standard cross section, at least to the 7% level. Boesten and Tanaka [52] have also derived a set of parameters for generating a preferred helium elastic DCS over a broad range of energies, which is very useful.

In light of several low-energy experimental and theoretical studies in neon [53–58], which are typically in very good agreement with one another, it has also been proposed [57] that neon may be used as a secondary standard for low-energy electron–molecule scattering. As such it was not intended that it would usurp the role of helium but rather provide a means to cross-check cross-section results for a given molecule of interest. Furthermore, the work of Buckman et al. [59] and Rugamas et al. [60], on the size and shape of pseudo-effusive beams, indicated that at elevated driving pressures, the width of a helium beam was smaller than that of most of the other gases they studied, even when the driving pressures were adjusted such that the mean free paths of the gases were identical. As the relative flow technique (see below) is based on experimental measurements carried out under identical scattering conditions for the target and standard gases, the smaller widths observed for helium beams was a matter for some concern. This indeed served to strengthen the case for establishing a secondary standard for low-energy scattering. The fact that neither Buckman et al. nor Rugamas et al. observed any differences of significance in the size of his neon beams, as a function of pressure, compared with a range of other molecular gases only added further weight to the case for its use as a secondary standard. Very recently, Zatsarinny and Bartschat [61] conducted some large-scale B-spline R-matrix calculations for electron–neon scattering. It is arguable that as a result of those calculations, neon can now be considered a benchmarked scattering system over a very wide range of energies. As a consequence, a set of elastic DCSs for electron–neon scattering, similar to that available for helium [52], is now available for use in conjunction with the relative flow normalization approach.

The relative flow procedure, with a continuous molecular source beam, requires carrying out elastic scattering measurements for the target and standard gases under identical experimental conditions at fixed values of E_0 and θ. Under those circumstances, we have from Eqs. 1.15–1.17, the relative scattering intensity being given by

$$\frac{N_0^{\mathrm{T}}}{N_0^{\mathrm{S}}} = \frac{\mathrm{DCS}_0^{\mathrm{T}}}{\mathrm{DCS}_0^{\mathrm{S}}} \left[\frac{I_e^{\mathrm{T}} C^{\mathrm{T}}(\overline{E}_n) \tilde{\mathrm{V}}_{\mathrm{eff}}^{\mathrm{T}}(\theta)}{I_e^{\mathrm{S}} C^{\mathrm{S}}(\overline{E}_n) \tilde{\mathrm{V}}_{\mathrm{eff}}^{\mathrm{S}}(\theta)} \right]. \tag{1.18}$$

The index zero on N and the DCS refers specifically to elastic scattering. If for a given E_0 and θ, the factors in the bracket on the right-hand side of Eq. 1.18 do not change when one switches from the target gas to the standard gas then the DCS is roughly obtained from a measurement of the elastic scattering intensities and using Eq. 1.18. However, establishing this condition is rather difficult and not really necessary as, unless we deal with highly reactive gases, to a very good approximation $C^{\mathrm{T}}(\overline{E}_{n=0}) = C^{\mathrm{S}}(\overline{E}_{n=0})$. To handle $\tilde{\mathrm{V}}_{\mathrm{eff}}(\theta)$, we follow the prescription of Nickel et al. [34] and factor the target density function as

$$\rho(\underline{r}) = n_{\mathrm{b}} n(\underline{r}), \tag{1.19}$$

where n_{b} is the total number of molecules/atoms in the beam and $n(\underline{r})$ refers to the spatial distribution function of the density. If the driving pressures of the two gases behind the beam-forming device are chosen such that the mean free paths of the gases are identical, then $n^{\mathrm{T}}(\underline{r}) \equiv n^{\mathrm{S}}(\underline{r})$. Under these circumstances Eq. 1.18 simplifies to

$$\frac{N_0^{\mathrm{T}}}{N_0^{\mathrm{S}}} = \frac{\mathrm{DCS}_0^{\mathrm{T}}}{\mathrm{DCS}_0^{\mathrm{S}}} \left[\frac{I_e^{\mathrm{T}} n_{\mathrm{b}}^{\mathrm{T}}}{I_e^{\mathrm{S}} n_{\mathrm{b}}^{\mathrm{S}}} \right]. \tag{1.20}$$

Nickel et al. [34] have further shown that the ratio $n_{\mathrm{b}}^{\mathrm{T}}/n_{\mathrm{b}}^{\mathrm{S}}$ can be related to the ratio of the flow rates ($\dot{\mathrm{F}}$) of the gases through the molecular beam former via

$$\frac{n_\mathrm{b}^\mathrm{T}}{n_\mathrm{b}^\mathrm{S}} = \frac{\dot{\mathrm{F}}^\mathrm{T} m_\mathrm{T}^{1/2}}{\dot{\mathrm{F}}^\mathrm{S} m_\mathrm{S}^{1/2}}, \tag{1.21}$$

where m is the molecular weight, so Eq. 1.20 reduces to its now familiar form (upon minor rearrangement)

$$\mathrm{DCS}_0^\mathrm{T} = \frac{I_e^\mathrm{S}}{I_e^\mathrm{T}} \cdot \frac{\dot{\mathrm{F}}^\mathrm{S} m_\mathrm{S}^{1/2}}{\dot{\mathrm{F}}^\mathrm{T} m_\mathrm{T}^{1/2}} \cdot \frac{N_0^\mathrm{T}}{N_0^\mathrm{S}} \cdot \mathrm{DCS}_0^\mathrm{S}, \tag{1.22}$$

where the electron beam currents are easily measured via a Faraday cup, the scattered elastic count rates are measured with, for example, channel electron multiplier (CEM) detectors and pulse amplification/counting techniques, and the relative flow rates are conveniently determined in a separate series of experiments, usually as a function of the beam-forming driving pressures (P), which are, in turn, monitored by an accurate pressure gauge, for example, a Baratron. These days both target and standard gases are often brought into the scattering chamber continuously. One is introduced to the scattering volume, while the other is introduced to the general vacuum chamber at another point. This procedure, particularly in the case of polar and reactive molecules, avoids the shifts in contact potential that can be encountered when one gas is removed from the vacuum chamber and another introduced.

The technique employed to measure the normalized flow rates, $\dot{\mathrm{F}}m^{1/2}$, that is adopted in most laboratories undertaking these types of measurements, is very similar to that outlined in Khakoo et al. [62]. Here it is not necessary to actually determine the absolute flow rates but rather the ratio of the normalized flow rate for the target gas to the standard gas. To do this the rate of change of pressure behind the beam-forming device, for example, a capillary array, as measured by an accurate pressure gauge, is recorded for both the target gas and the standard gas for various settings of their leak valves. For the purpose of accurately measureing the relevant flow rates an in-line volume is typically opened to reduce the pressure gradient. Data acquisition is achieved under computer control.

For an ideal gas, kinetic theory predicts that for collisionless flow the normalized flow rate should be independent of the gas species. Thus, if the normalized flow rates were plotted as a function of driving pressure for both the standard gas (e.g., helium) and the target gas of interest, and no species dependence was found, then the ratio $\dot{F}^{He} m_{He}^{1/2} / \dot{F}^{T} m_{T}^{1/2}$ in Eq. 1.22 could be conveniently replaced by the ratio of their driving pressures P_b^{He} / P_b^{T}. In practice, however, it was found [63] that this was not the case for most gases, with differences of over 20%, from the expected ratio of 1, being observed in some cases. We note that even though the extent of these differences will almost certainly be apparatus specific, they indicate that care should be taken in establishing the true relationship between the gas flow rates. Failure to do so could, in principle, see potentially serious errors being carried over into the determination of the absolute cross section. However, as a result of studies at the Australian National University over many years, for example [64], and Sophia University [65], it appears that quite significant deviations from the matching mean free path criterion (which determines the ideal driving pressure ratio) are required before any major deviation in the absolute DCSs are noticed, that is, deviations outside the uncertainty in the measured cross sections. This robustness in the measured DCS, to that criterion, is a significant asset in using the relative flow technique, with a continuous molecular beam source, to make absolute cross-section measurements.

Thus, in practice, in a typical experiment the ratio of the driving pressures for the two gases will be determined from their mean free paths. The absolute values of the pressures to be used are then decided by considerations on the basis of the type and size of the source used for the beam-forming device. Once the absolute gas pressures to be used are established, then the relative flow rates for each gas at these pressures must be determined. One approach is to determine the flow rates at a range of driving pressures about the ideal value and to fit a curve (polynomial) to the normalized flow rates versus driving pressures measured. The fitting parameters or their equivalent then enable the appropriate normalized flow rate ratio to be substituted directly into the relative flow equation (Eq. 1.22) for any practical value of the ratio of the capillary driving pressures. Consequently, all the terms on the right-hand side of Eq. 1.22 are known, and the required absolute elastic DCS is thus found for a given energy and angle. The entire procedure, with the exception of the calibration curve, is then repeated for all angles until the angular dependence of the DCS at a given energy is known.

Before moving on to discuss the application of the relative flow technique for pulsed and skimmed supersonic molecular gas expansion (jet) sources, we should note a variant on the traditional continuum beam technique, as pioneered by the CSU–Fullerton group [66]. In that method, which bypasses the critical mean free path condition referred to before, a thin collimating aperture is employed for the beam-forming source, instead of a conventional tube or capillary array. For such a source, Khakoo et al. [66] claim that provided the mean free path of the gases in question (helium is still regularly used as the standard gas) remains much greater than the thickness of the aperture, the angular distributions of the gases will be constant (i.e., a cosine distribution convoluted with the aperture's extent). Under these circumstances the normalized flow rate ratio in Eq. 1.22 can be replaced with the pressure ratio, and the absolute scale of the DCS is then set in the usual manner.

1.3.2 Elastic Scattering: The Relative Flow Technique with a Skimmed Supersonic Jet Beam

The scattering geometry with a skimmed supersonic expansion (SSE) is illustrated in Fig. 1.2 [67]. The absolute DCS ($\sigma \equiv$ DCS) for any scattering process is related to the count rate of the scattered particles (N^e) by

$$N^e = f^e \eta \sigma \int_{V,E} \rho \Delta\Omega dV dE, \tag{1.23}$$

where f^e is the flux of the incident electrons, η is the detector efficiency function, ρ is the target density at the scattering centre, $\Delta\Omega$ is the solid angle subtended by the detector, V is the volume of the intersection of the electron and molecular beams, and dE is the range of electron energies. An explicit characterization of all the terms in Eq. 1.23 is nontrivial. Absolute normalization is again usually achieved by comparing scattering intensities from the target to a reference species with a known DCS [67, 68], for example, helium or neon, as described before. This comparison, in practice, removes the need to explicitly determine the interacting beam fluxes

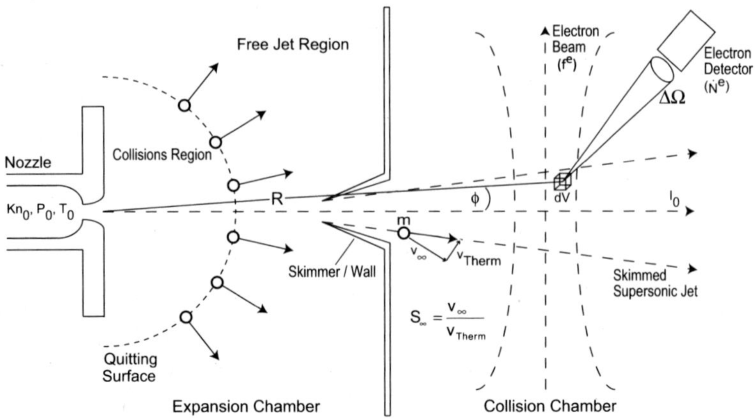

Figure 1.2 Schematic diagram for the scattering geometry with a skimmed supersonic expansion. From Ref. [67] © IOP Publishing. Reproduced with permission. All rights reserved.

and the detector efficiency function. According to Eq. 1.23 the unknown elastic DCS of the target (T), DCS_0^T, is expressed in terms of the standard (S) elastic DCS, DCS_0^S, by

$$DCS_0^T = DCS_0^S \frac{N_0^{e,T}}{N_0^{e,S}} \frac{\int \rho_S \Delta\Omega dV dE}{\int \rho_T \Delta\Omega dV dE}. \tag{1.24}$$

The beam density at the intersection of the electron beam and the target molecular beam is given by

$$\rho = \frac{I(\phi)}{\mathbf{v}_\infty R^2}, \tag{1.25}$$

where $I(\phi)$ is the target beam intensity, $\mathbf{v}_\infty$ is the target beam terminal velocity, and R is the distance to the gas source. The terminal velocity of ideal gas molecules in a SSE is given as

$$\mathbf{v}_\infty = \sqrt{\frac{2k_B T_0}{m}\left(\frac{\gamma}{\gamma - 1}\right)}, \tag{1.26}$$

where γ is the adibatic constant, T_0 is the stagnation temperature, m is the molecular mass, and k_B is Boltzmann's constant. The intensity of the gas beam is expressed as

$$I(\phi) = I_0 J(\phi), \tag{1.27}$$

where I_0 is the centreline beam intensity and $J(\phi)$ is a geometric function describing the axial intensity distribution. The geometric function for a skimmed nozzle beam [69] is written as

$$J(\phi) = \frac{1}{2\pi}\int_0^{2\pi} \left[\exp\left(-S_\infty^2 \sin^2(\phi_1)\right)\right]_{\phi_{1_{\max}}(\theta,\phi)}^{\phi_{1_{\min}}(\theta,\phi)} d\theta, \tag{1.28}$$

where θ is an angle of rotation about the jet axis and ϕ is an angle between the normal to the centre of the nozzle and a ray from that point to an off-centreline point in the scattering region (see Fig. 1.2). Evaluation of Eq. 1.28 is difficult because $\phi_{1_{\max}}$ and $\phi_{1_{\min}}$ are nontrivial functions of ϕ. However, for a molecular beam in free flow at the skimmer, $\phi_{1_{\max}}$ and $\phi_{1_{\min}}$ are determined by the terminal speed ratio (S_∞) and the relevant experimental geomtery [69]. Hence the only species-dependent term in Eq. 1.28 is the terminal speed ratio, that is, the ratio of the translational speed to the thermal speed (v_{Therm}) for those molecules expanded beyond the quitting surface [70].

The location of the quitting surface (x_q), at which free molecular flow is achieved, is given by [71]

$$x_q \approx d\left(\frac{S_\infty}{C_1}\sqrt{\frac{2}{\gamma}}\right)^{\frac{1}{\gamma-1}}, \tag{1.29}$$

with d being the nozzle diameter and C_1 a constant that is equal to 3.232 for a monatomic species, 3.606 for a diatomic molecule, and 3.971 for a polyatomic molecule [71]. Matching the geometric distribution functions for the beam of interest (target) and the standard beam removes the need for explicit evaluation of the integrals in Eq. 1.24. According to Eq. 1.28, this condition is achieved by matching the terminal speed ratio of both beams, while ensuring free molecular flow exists at the entrance orifice of the skimmer. With these requirements met, the elastic DCS of the target molecule of interest is now given as

$$\text{DCS}_0^{\text{T}} = \text{DCS}_0^{\text{S}}\left(\frac{I_0}{N_0^e \mathbf{v}_\infty}\right)_{\text{S}}\left(\frac{N_0^e \mathbf{v}_\infty}{I_0}\right)_{\text{T}}. \tag{1.30}$$

To evaluate Eq. 1.30 the terminal speed ratio and centreline intensity of both the target and standard beams at the scattering centre must be determined, with a method for evaluating those quantities [67] being outlined shortly.

The potential for the formation of clusters in a supersonic jet is an important consideration. Dimers or higher-order clusters may be formed in the beam if a sufficiently high stagnation pressure behind the nozzle (P_0) is employed. Note that significant cluster formation will perturb the expansion theory outlined here by increasing v_{Therm} due to the release of energy in forming a van der Waals bond [72]. The clusters may also be considered as an undesired contaminant species in the beam. For a monoatomic gas, dimer concentration exceeding approximately 1% is avoided if the stagnation conditions are such that the parameter D^* is less than 0.1 [71]. D^* is given as

$$D^* = \frac{P_0 \sigma^3}{\varepsilon} \left(\frac{d}{\sigma}\right)^{0.4} \left(\frac{\varepsilon}{T_0 k_B}\right)^{2.4}, \qquad (1.31)$$

where σ and ε are the standard parameters of a Lennard–Jones potential. Equation 1.31 is taken as an upper bound to the value of D^* when applied to a diatomic or a polyatomic expansion, as the internal relaxations in a molecule release heat into the expansion, which reduces clustering. A common practice for forming SSE jets, particularly in chemical physics and physical chemistry applications, is to dilute the gas under study with a lighter gas (e.g., helium or argon). By seeding the heavier target gas into a carrier, the target is accelerated to a higher velocity (and consequently cooled to a lower temperature) than would have been achieved in a pure expansion of the target under the same stagnation conditions. However, [73] notes that the increased cooling achieved in a seeded SSE also increases cluster concentration in the beam. Hence, in applications where the aim is to measure the unknown DCS of a target of interest, carrier gases should not be used.

Full details on the crossed-beam apparatus employed to measure the absolute elastic DCSs can be found in Maddern et al. [68], so only a précis is provided here. The apparatus (Figs. 1.3 and 1.4) is a crossed-beam electron scattering experiment comprising three differentially pumped chambers. The expansion chamber houses the nozzle, the pulsed solenoid valve, and the skimmer. The nozzle diameter is typically 0.8 mm, and the skimmer orifice is 1 mm in diameter. When scattering measurements with molecular radicals are to be made, the nozzle in Fig. 1.3 is replaced with the pyrolysis gas nozzle [68] of Fig. 1.4 Otherwise the other elements of the

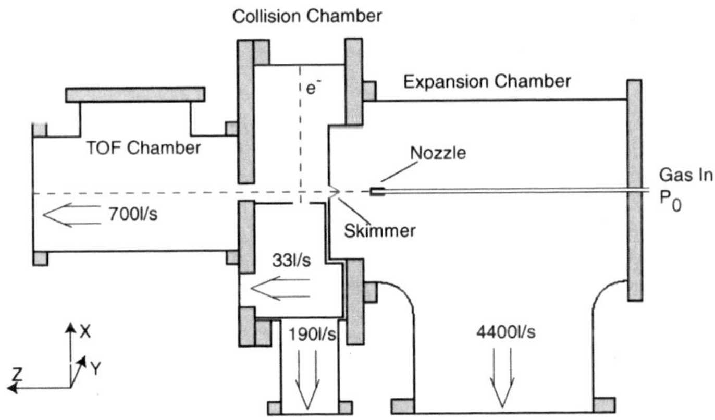

Figure 1.3 Schematic diagram of the present experimental configuration. The pumping speeds for each of the differentially pumped chambers are shown. From Ref. [67] © IOP Publishing. Reproduced with permission. All rights reserved.

spectrometer are common to both Figs. 1.3 and 1.4 The nozzle–skimmer separation is variable but is often set to 50 mm. The valve is pulsed at 10 Hz, and individual gas pulses have 3–5 ms duration. Gas is introduced to the expansion chamber at stagnation pressures in the range of 50–500 mbar.

The collision chamber houses a differentially pumped hemispherical electron Spectrometer, which produces a collimated electron beam of well-defined energy in the range of 2–50 eV and with a low-energy spread. The gas pulse enters the collision chamber through the skimmer and crosses the electron beam at 90°. The angular distribution of the scattered electrons is measured simultaneously using a fixed-position multidetector array containing up to 13 individual detectors (see Fig. 1.4). Each detector unit incorporates one CEM, with a retarding field analyzer (RFA) mounted before each CEM to prevent detection of inelastically scattered electrons. The combined energy resolution of the spectrometer and the RFA is ~1.5 eV [68].

The time-of-flight (TOF) chamber contains a linear Wiley–McLaren-type [74] orthogonal acceleration time-of-flight mass spectrometer (TOF-MS). This features 380 mm of field-free flight, two ion acceleration stages producing a second-order space focus,

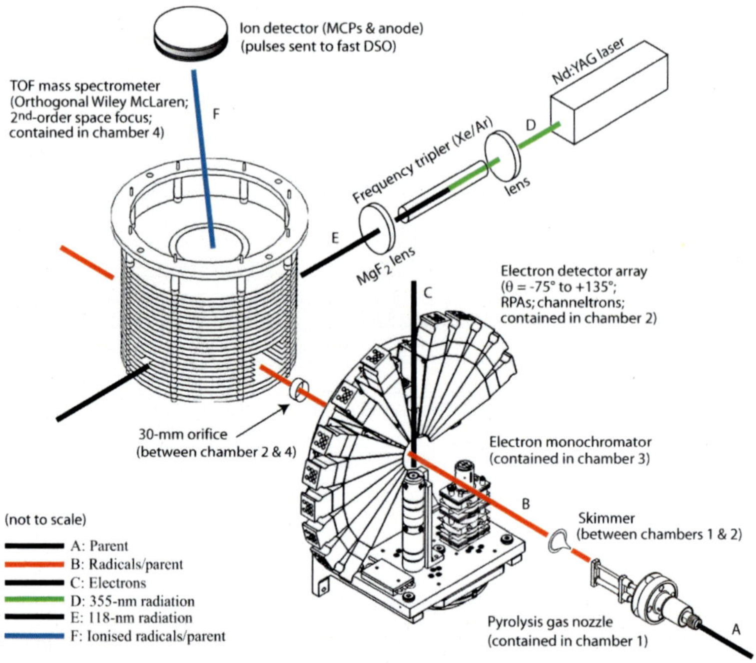

Figure 1.4 Schematic diagram of the present experimental configuration. Reprinted figure with permission from Ref. [82]. Copyright (2008) by the American Physical Society.

and a 118 nm vacuum ultraviolet (VUV) photoionization source [68]. The purpose of TOF-MS is analysis of beams containing multiple species such as molecular radicals and their parent molecules.

The method developed at Flinders University [67] for evaluating the centreline intensity and speed ratio is referred to as the pressure-rise skimmed supersonic relative density method (p-SSRDM). In p-SSRDM the centreline intensity of the gas beam downstream of the skimmer is inferred from the pressure rise in the collision chamber once gas is admitted [75]. The centreline intensity is related to the pressure rise (ΔP) by

$$I_0 = \frac{(\Delta P) S_p R^2}{\alpha k_B T_d}, \tag{1.32}$$

where S_p is the pumping speed in the collision chamber for the species under study, α is the cross-sectional area of the molecular beam at the scattering centre, and T_d is the temperature of the collision chamber.

Downstream of the skimmer, the cross-sectional area of the gas beam at the scattering centre is given by

$$\alpha = \pi\left(\frac{x_d - x_s}{S_\infty} + r_s\right)^2, \tag{1.33}$$

with x_d being the distance to the scattering centre, x_s the distance to the skimmer (both these distances are relative to the nozzle), and r_s the skimmer radius. The terminal speed ratio for diatomic and polyatomic molecules is given to within 10% [76] by

$$S_\infty \approx 5.4(P_0 d)^{0.32} \tag{1.34}$$

and with $P_0 d$ in units of Torr cm. The corresponding expression for the terminal speed ratio for monatomic gases can be found in Miller [71] and Hargreaves et al. [67]. At Flinders University, ΔP was measured with a Bayard–Alpert-type ion gauge. Therefore to account for the gauge sensitivity to different species, the total ionization cross section Q for the species of interest at 150 eV was used [76]. Hence the corrected measured pressure reading (ΔP_d^*) is given by

$$\Delta P = \Delta P_d^* \frac{Q_{N_2}}{Q}, \tag{1.35}$$

where Q_{N_2} is the total ionization cross section of molecular nitrogen (N_2) at 150 eV. N_2 was used as the reference species to correct the gauge reading as the gauge itself was originally calibrated for N_2. Ionization cross sections for a range of targets, including radicals and biomolecules, for example, [77, 78], are available in the literature. To further account for species dependence, pumping speeds are scaled as $m^{-1/2}$ [75]. The DCS of the target species is finally given as

$$\mathrm{DCS}_0^{\mathrm{T}} = \mathrm{DCS}_0^{\mathrm{S}} \left(\frac{\Delta P_d^*}{Q\mathbf{v}_\infty m^{1/2} N_0^e}\right)_S \left(\frac{Q\mathbf{v}_\infty m^{1/2} N_0^e}{\Delta P_d^*}\right)_T. \tag{1.36}$$

Equation 1.36 thus represents the working equation for the p-SSRDM approach. The group at Flinders University has tested this approach on several standard systems, with the results from the standard relative flow technique and the p-SSRDM approach being in generally excellent accord [67] at all common incident electron energies and scattered electron angles.

In general, for atomic and molecular targets, the uncertainties in the p-SSRDM data are taken as the quadrature sum of the statistical errors (1%–2%), uncertainty in the pressure rise (5% for each gas), uncertainty in the speed ratios (~10% for polyatomics), uncertainty in the ionization cross sections (2%–4%), and uncertainty on the standard gas DCS, which is ~7% for helium. The total uncertainty in those DCSs is therfore in the range of 25%–30%. The main limitations in reducing the absolute uncertainty on the p-SSRDM DCS are the often a small pressure rise (ΔP) measured in the collision chamber and uncertainty in the calculated speed ratio. In their present configuration [67, 68], pressure rises in the collision chamber can be as low as 20%, with a pressure gauge precision of about 2% of total gas pressure. However, a modified dump chamber, with a smaller volume to achieve a more significant pressure rise, could conceivably achieve smaller uncertainties than those noted before. Alternatively, a more detailed study into the speed ratios of polyatomic molecules would reduce the uncertainty introduced using the approximations embodied in Eq. 1.34. As elastic DCSs using the traditional relative flow method are often measured to an uncertainty in the range of 10%–20%, it is clear that the p-SSRDM will never supercede the traditional method when normal gaseous atoms or molecules are being studied. Nonetheless this approach has been successfully employed to study several scattering systems, including CH_3I [79] and CF_3Br [80]. However, when the pyrolysis nozzle is incorporated into the apparatus seen in Figs. 1.3 and 1.4, the p-SSRDM allows one to measure absolute elastic cross sections for molecular radicals, which the traditional relative flow method cannot. Cross sections for molecular radicals are particularly important in modeling the behavior of low-temperature plasma discharges (see later in this volume), and so the p-SSRDM is a valuable, although somewhat niche, approach. Currently, the Flinders group has used its apparatus [67, 68] to report an absolute elastic DCS for electron scattering from the molecular radicals CF_2 [81, 82] and CF_3 [83, 84], and the atomic radical I [85]. In addition, the group has also recently employed its apparatus to explicitly demonstrate the thermal stability of uracil and a range of uracil derivatives, which have potential application as a radionuclide in positron emission tomography diagnostics up to temperatures of 280°C [86]. This was an important result, as for many biomolecules some heating of the sample is required to generate a molecular beam and up until Maddern et al. [86] no studies on their thermal decomposition properties had been explicitly undertaken.

1.3.3 Inelastic Scattering Differential Cross Sections

In principle, when we refer to inelastic scattering in an electron–molecule collision context, we are referring to rotational, vibrational, and electronic-state excitation of the target molecule. In practice, however, typical crossed-beam electron spectrometers are operated with energy resolutions in the range of 10–65 meV (full width at half maximum [FWHM]), so with the possible exception of H_2 rotational excitation cannot be resolved in any energy loss spectrum for the elastic peak and any inelastic (vibrational and electronic-state) features. In this sense all current elastic and inelastic DCSs can be viewed as being *rotationally averaged.*

At this time, it appears that the best way for determining absolute inelastic cross sections, for the target gas of interest, is to measure relative inelastic-to-elastic scattering intensities and then achieve normalization through the presumably known target gas absolute elastic DCS. Normalization of the inelastic-to-elastic signal in the target gas can be achieved by two methods. In the first method, the angular distribution of the inelastic process (i.e., at a fixed energy loss) is first measured, and subsequently, the inelastic angular distribution is normalized to the elastic cross section at one or two angles [87]. In the second method, energy loss spectra, encompassing both elastic and inelastic features, are obtained at each scattering angle (for a given energy) and the inelastic-to-elastic DCS normalization is performed at each angle. The first method requires a knowledge of the overlap geometry of the gas beam, detector viewcone, and electron beam in order to correct for a possible angular dependence of the effective scattering volume. Such an effect can be greatly minimized [36] by ensuring that the gas beam is well collimated, but usually any angle-dependent effects are calibrated out by comparison to the known angular variation of a standard cross section at the energy of interest. The second method is insensitive to an effective path length correction factor, since both elastic and inelastic signals are affected equally. As a consequence, it is the second method that is typically empolyed by the scattering community in order to determine the absolute inelastic DCS.

In any event, the major problem here is the establishment of the relative detector response function for the residual energy (or, equivalently, energy loss) range corresponding to the elastic and inelastic events. From Eqs. 1.15–1.17 it follows that the expression for the measured inelastic-to-elastic scattering ratio, at fixed E_0 and θ, can be written as follows:

$$\frac{N_n^{\mathrm{T}}(E_0, \theta)}{N_0^{\mathrm{T}}(E_0, \theta)} = \frac{\mathrm{DCS}_n^{\mathrm{T}}(E_0, \theta)}{\mathrm{DCS}_0^{\mathrm{T}}(E_0, \theta)} \cdot \frac{C^{\mathrm{T}}(E_n)}{C^{\mathrm{T}}(E_{n=0})}. \tag{1.37}$$

Thus from Eq. 1.37, one *only* needs the relative response function of the detector (relative to that corresponding to elastic signal detection), $C^{\mathrm{T}}(E_n)_{\mathrm{rel}}$, in order to determine the absolute inelastic DCSs.

At high-impact energies (compared with the energy loss ΔE), the instrumental response can usually be considered to be independent of E_n [35], particularly if relevant potentials on the electron optic lens elements are continuously adjusted during the energy loss sweep to maintain the focal properties of the electron beam [88–90]. In this case $C^{\mathrm{T}}(E_n)_{\mathrm{rel}} = 1$. The situation, however, rapidly becomes more complicated as the incident electron energy is lowered as under these circumstances the response function may change drastically with residual energy. One can, in principle, correct for this effect by designing and maintaining the electron optics achromatic. However, even though there are now several powerful numerical packages commercially available, which greatly assist the design and simulation of electron transport throughout an electron spectrometer, including for space charge effects, they cannot account for the surface conditions in an operational spectrometer. As a consequence while well-designed systems can now greatly reduce this effect, an actual experimental determination for the relative response function $C^{\mathrm{T}}(E_n)_{\mathrm{rel}}$ for the system under experimental tuning and operating conditions is needed.

A previously popular approach to characterize $C^{\mathrm{T}}(E_n)_{\mathrm{rel}}$ relied on measuring the yield of electrons (ejected and scattered) from helium, following near-threshold ionization [91]. As a consequence of the Wannier theory for near-threshold ionization [92], the yield of detected electrons is expected to be independent of energy for values of excess energy ($E_{\mathrm{exc}} = E_0-$ ionization potential) up to some value, $E_{\max}$, above the ionization threshold. Furthermore, this criterion was expected to be met for any angle of observation. Thus in a measurement of the yield of scattered and ejected electrons resulting from electron impact ionization at an appropriately low energy, one may expect to observe a flat spectrum in all situations where the excess energy is less than $E_{\max}$. Hence, any deviations from flatness directly reflect the distortion introduced by the non-uniform response function of the analyzer. Unfortunately, experimental verification of the range of $E_{\max}$ remains uncertain and indeed controversial. Furthermore, as noted by Allan [93], the underlying assumption of a flat electron distribution is probably only valid at angles around 90° (or thereabouts) and so is not really applicable at small and large scattering angles. From a practical viewpoint, the helium ionization continuum is weak so that long accumulation times are required. As a consequence of these issues, this approach to determine $C^{\mathrm{T}}(E_n)_{\mathrm{rel}}$ is now largely out of favor within the experimental electron scattering community. In a seminal contribution to this issue, Allan [93] outlined a new and rigorous philosophy in order to specify the energy and angular dependence of the detector response. This approach is well described in his paper [93], and so we do not repeat those details again here. In essence, however, it relies on the availability of accurate elastic (and inelastic) theoretical DCSs, over a wide range of energies and scattering angles, such as from the convergent close coupling [47, 50] and B-spline R-matrix approaches [61], to enable the calibration of $C^{\mathrm{T}}(E_n)_{\mathrm{rel}}$ to be accurately performed. The philosophy behind this approach is now routinely used in electron-scattering experiments, using a standard crossed-beam apparatus, at Sophia University (e.g., [94]) and Flinders University (e.g., [95]), and it has been a very important development toward determining accurate and reliable DCSs in inelastic measurements.

A somewhat different approach to overcome the problems associated with determining the analyzer transmission function, for inelastic scattering measurements, is a TOF approach developed by Le Clair et al. [96]. In principle, this technique ensures $C^{\mathrm{T}}(E_n)_{\mathrm{rel}} = 1$ and so enables a direct measurement of the ratios of elastic-to-inelastic scattering intensities, even over a broad range of energy loss values. In it, a pulsed electron beam is incident on a well-defined atomic or molecular beam. Electrons of different energies that arise from the collision process are temporarily dispersed over a long ($\sim$20 cm) drift tube and detected with a large-area channel plate detector. The result is a timing spectrum consisting of a number of peaks due to either elastic or inelastic scattering. In the absence of any electric or magnetic fields, the integrated intensity under each peak is directly proportional to the scattering cross section for the particular process. With careful attention to detail, Le Clair et al. demonstrated how this approach might be used to measure near-threshold electronic excitation cross sections by measuring ratios of elastic-to-inelastic scattering and using the standard relative flow method to establish the relevant elastic DCS. In their apparatus its application was somewhat limited by the fact that they used an electron beam with a relatively broad, thermal energy distribution. However, they did successfully demonstrate the technique for a number of atoms and molecules [97]. A similar apparatus, although with far superior energy resolution, was subsequently developed by the group at the Australian National University [98]. In one of its first applications of this apparatus [98], absolute cross sections for near-threshold excitation of the $n = 2$ and $n = 3$ electronic states in helium were measured [99]. Excellent agreement with state-of-the-art calculations was found, suggesting that the presumption of $C^{\mathrm{T}}(E_n)_{\mathrm{rel}} = 1$ for this type of experimental configuration was valid. This is a very important point, as when this group subsequently studied near-threshold electron impact excitation of argon [100], it found some significant disagreements with corresponding traditional crossed-beam results from Khakoo et al. [101]. Note that the best-available theories at that time clearly supported the DCS from the ANU group [100], with the clear imputation being that these disagreements were caused by an imperfect analyzer transmission response calibration by Khakoo et al. [101]. The ANU group has also measured data on N_2 [102], which interestingly shows significant disagreement with like results from Le Clair and Trajmar [97]. As both these sets of data were taken with a TOF apparatus, an explanation for the observed discrepancies is awaited with interest. In the interim, it would appear to be premature to use the ratios of Le Clair and Trajmar in order to *renormalize* measurements made with a crossed-beam apparatus for the same excitation processes in the same target molecules (or atoms).

There is one last difficulty, associated with the measurement of DCSs for inelastic processes, that needs to be described briefly. Specifically we refer to the electron impact excitation of electronic states from the ground molecular state, where even for the most simple homonuclear diatomics, there can be appreciable overlap in the vibrational bands of two or more of these excited electronic states. This is clearly illustrated for N_2 in Fig. 1.5 The unraveling of these overlapping states has seen the development of spectral deconvolution techniques, which have been described in detail by several authors elsewhere [88, 103, 104] and to whom the reader

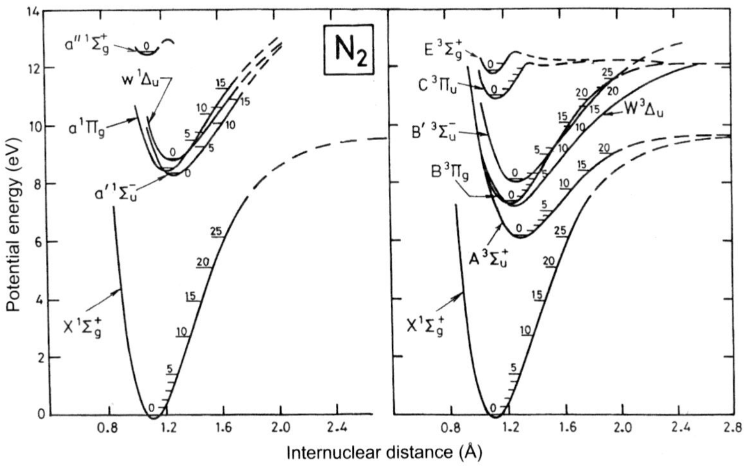

Figure 1.5 Schematic diagram of the energy levels (E_n < 12.4 eV) for the electronic states of N_2. For clarity the singlet and triplet states have been partitioned. Reprinted figure with permission from Ref. [88]. Copyright (1990) by the American Physical Society.

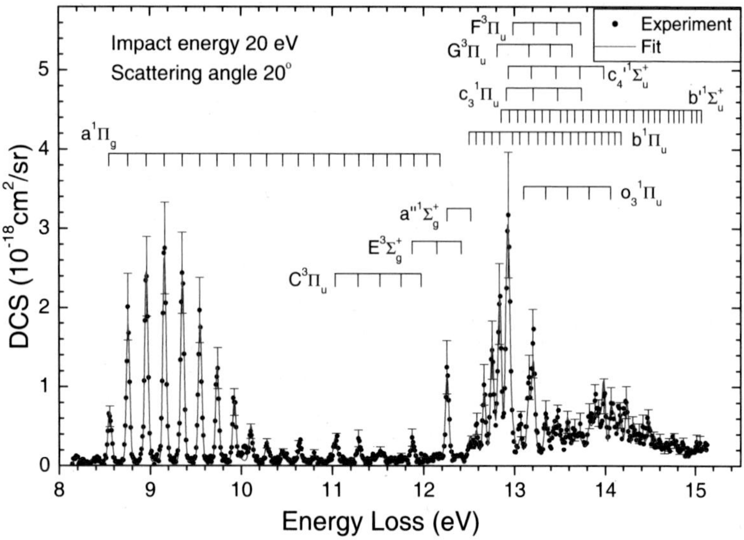

Figure 1.6 Typical energy loss spectrum for electron impact excitation of the relevant electronic states of N_2, at an impact energy of 20 eV, and at a scattering angle of 20°. Also shown in this figure is our spectral deconvolution of this energy loss spectrum and the various vibrational sublevels of each electronic state. Reprinted figure with permission from Ref. [105]. Copyright (2010) by the American Physical Society.

is referred for more details. An example of a spectral deconvolution fit to N_2, for an incident electron energy of 20 eV and a scattered electron angle of 20°, is given in Fig. 1.6. It is clear from Fig. 1.6 that a reasonable fit to the measured energy loss spectra can be achieved, with the inelastic cross sections for the various electronic states then being able to be derived. However, there is a major hurdle in respect to the uniqueness of the inelastic cross sections that are so derived. Certainly, the higher your energy resolution, the more likely you are to find isolated features to fit, which, in turn, very much aids your ability to uniquely extract the required data. Nonetheless, there in N_2 remain important differences between cross-section data originating from Flinders University and data originating from the Jet Propulsion Laboratory (JPL)/California State University (CSU)–Fullerton, which at least in part might be attributed to uniqueness issues associated with their respective deconvolutions.

Some progress in this regard was made relatively recently, where a joint Sophia University–Flinders University investigation [105] into some of the higher-lying energy loss states in N_2 was able to help clarify some existing discrepancies between data from Khakoo et al. [106] and Chutjian et al. [107]. However, the overall picture in N_2, a vital molecule if we are to quantitatively understand the role of electron-driven processes in our atmosphere (see later), remains unclear and in need of further work. Finally, we conclude this section by quickly noting that studies of polyatomic molecules often produce broad bands of states in their energy loss spectra, unlike the many unique features in Fig. 1.6 [108–111]. At this time it is preferable to report a cross section for the sum of all the states that constitute a given band, for which there may be many, rather than try and separate out the constituent electonic states. Such data remains very useful in kinetic transport modeling of discharges [112] and in charged particle track simulations in matter [113], to name just two examples.

1.4 Integral Cross Sections

As will undoubtedly become clear later in this volume, in attempting to simulate or model such diverse behavior as atmospheric phenomena, low-temperature plasmas, and discharges and radiation damage in matter, one of the primary inputs are the relevant ICSs that (in part) describe the nanoscale behavior of the electrons. With a couple of important exceptions, which I return to shortly, ICSs are generally derived from the corresponding DCS measurements via

$$\mathrm{ICS}_n(E_0) = 2\pi \int_0^{\pi} \mathrm{DCS}_n(E_0, \theta) \sin\theta d\theta, \tag{1.38}$$

where n again denotes the scattering channel of interest. Note that this equation is equivalent to that given earlier in Eq. 1.4. Implicit in Eq. 1.38 is that the limits of integration are from 0° to 180°, whereas for many crossed-beam spectrometers the angular range of the analyzers is restricted, usually by the physical dimensions of the monochromator and analyzer, to scattered electron angles between $\sim 10^\circ$ to 130°. To overome this limitation, experimental groups have largely relied on two main approaches. In the first case we characterize the approach as *extrapolation procedures*, while in the second case we have direct measurements of the cross sections via *magnetic rotation of the scattering geometry*. We now consider both of these methods, in turn, next.

In the case of extrapolation procedures, three different methods appear popular amongst the currently active experimental groups. In the first, a reliable theory might be used to extrapolate the DCS to 0° and 180°. There are today several ab initiolevel theories that, when coupled with high-performance computing resources, are able (particularly in elastic scattering) to provide such reliable data in order to perform the extrapolation and, subsequently, use Eq. 1.38 to determine the required ICS. The Flinders University [114] and Sophia University [115] groups also use what they call a *modified phase shift analysis* method in order to undertake the required extrapolation. Their approaches are both based on that for central potential scattering, as originally outlined by Andrick and Bitsch [44], and later extended by Brunger et al. [45], where a partial wave expansion is fitted to a measured cross section in order to extract the s-, p-, d-wave (and so on) scattering phase shifts. Having obtained these phase shifts, and making use of the Born formulation for the higher-order partial wave values, a DCS from 0° to 180° could then be obtained. Equation 1.38 is then again employed to derive the ICS. This method appears quite reliable and robust and should be quite accurate for nonpolar and weakly polar molecules. For molecules with high dipole moments, however, there is a suspicion that the DCS extrapolation to 0° may be somewhat underestimated, which, even after the strong $\sin\theta$ weighting at small θ, causes the ICS to be smaller than is actually the case. Care must also be taken with this approach to ensure that unphysical oscillations in the DCS, in the regions of extrapolation, are avoided. The final extrapolation procedure, dealing specifically with the excitation of optically or dipole allowed electronic states in molecules, has been used quite extensively in recent time by the Sophia University group ([94, 116] and references therein). In this approach the values of $[\theta, \mathrm{DCS}(E_0, \theta)]$, for a given Electronic state and at a given incident electron energy, are transformed to (K^2, G_{expt}) using the standard formula [117]

$$G_{\text{expt}}(K^2) = \frac{(E/R)k_i a_0}{4a_0^2 k_f a_0} K^2 \text{DCS}(E_0, \theta), \tag{1.39}$$

where k_i and k_f are the initial and final momenta of the incident and scattered electrons, E is the excitation energy for the electronic state in question, a_0 is the Bohr radius (0.529 Å), R is the Rydberg energy (13.6 eV), $G_{\text{expt}}(K^2)$ is the experimental generalized oscillator strength (GOS), and K^2 is the momentum transfer squared defined by

$$K^2 = (k_i a_0)^2 + (k_f a_0)^2 - 2(k_i a_0)(k_f a_0)\cos\theta. \tag{1.40}$$

Vriens [118] proposed the following formula to represent the GOS for a dipole-allowed excitation based on the analytic properties defined by Lassettre [119] and Rau and Fano [120]

$$G(x) = \frac{1}{(1+x)^6}\left[\sum_{m=0}^{\infty} \frac{f_m x^m}{(1+x)^m}\right], \tag{1.41}$$

where

$$x = \frac{K^2}{\alpha^2} \tag{1.42}$$

and

$$\alpha = \sqrt{B/R} + \sqrt{(B-E)/R}. \tag{1.43}$$

Note that in Eq. 1.43, B is the binding energy of the target electron. In Eq. 1.41 the f_m are the fitting constants to be determined in a least squares fit of the experimental GOSs (which are calculated from the measured DCS from Eq. 1.39). Note that typically excellent quality fits to the data are found [94, 116]. Estimates of the experimental ICS at each energy are then obtained from Eqs. 1.41–1.43 using the standard formula [121]

$$\text{ICS}(E_0) = \frac{4\pi a_0^2}{E_0/R} \int_{K_{\text{min}}^2}^{K_{\text{max}}^2} \frac{G(K^2)}{E/R}\, d\ln(K^2), \tag{1.44}$$

with

$$K_{\text{min}}^2 = 2\frac{E_0}{R}\left[1 - \frac{E}{2E_0} - \sqrt{1 - E/E_0}\right] \tag{1.45}$$

and

$$K_{\max}^2 = 2\frac{E_0}{R}\left[1 - \frac{E}{2E_0} + \sqrt{1 - E/E_0}\right]. \tag{1.46}$$

The beauty of Vriens [118] formalism is that provided the incident elecron energy is high enough, at the $x = 0$ optical limit, the value of $G(0) \equiv f_0$ is the optical oscillator strength (OOS). This provides a nice self-consistency check for this approach, as the derived OOS can be checked against corresponding independent values from dipole (e, e') and photoabsorption measurements. An excellent summary of Scaled–Born techniques, as applied to dipole-allowed discrete electronic-state excitation and ionization, was recently provided by Tanaka et al. [122].

A major experimental advance, in overcoming the finite acessible angular range of most crossed-beam apparatus, was made by the Manchester group [123, 124]. Specifically, it developed an innovative technique for extending scattering measurements to the full range of backward scattering angles (i.e., to $\theta = 180°$). In a departure from the normal mind-set of high-resolution electron spectroscopists, who shun magnetic fields as a curse, the group has imposed a localized magnetic field upon the collision region of an electron spectrometer in order to rotate the incident and scattered electron trajectories. The magnetic field is typically applied by two pairs of coaxial solenoids with opposing currents in each pair, such that the incident electron beam trajectory can be rotated in the scattering plane to an arbitrary angle with respect to the unrotated trajectory. The fields are arranged so that the incident electrons always pass through the atomic or molecular beam, which is directed along the axis of the solenoid pairs. The fields are also arranged such that they are highly localized, dropping to negligible strength in the regions beyond the interaction region where the sensitive, low-energy, eletrostatic optics are operated. This is the key to this method, with the design details being provided in Read and Channing [123]. Groups that have successfully utilized this approach not only include that at Manchester (e.g., [125, 126]) but also include those at Fribourg (e.g., [127, 128]), Gdansk (e.g., [129], and Daejeon (e.g., [130, 131]).

In addition to the extrapolation procedures and magnetic angle–changing device described before, to convert the DCS to the ICS, there are also two direct approaches to measure ICSs. The first, in which we measure the number of emitted photons that are detected as a function of the incident electron energy (the so-called optical excitation function), has a long pedigree in atomic and molecular physics [132, 133]. The challenges in applying this method include that it is only applicable to optically allowed transitions; that the measured data is only relative so that a normalization approach, such as the Bethe–Born approximation [121], has to be used to set the absolute scale; that the emitted radiation is polarized, which leads to an angular dependence in the photon intensity; and that there can be cascade contributions to the measured (using a photon multiplier tube) flux. The last of these challenges is, in fact, probably the most problematic. Nonetheless, this approach has been successfully employed at JPL over many years (see [133] and references therein). The second approach is very recent, drawing on expertise accumulated over many years in the positron scattering community [134]. Specifically, García and colleagues [135] have constructed a new experimental system for measuring electron scattering cross sections within a strong axial magnetic field. The functionality of this apparatus is based on the magnetic confinement of the electron beam from its entrance into the collision chamber, until its energy analysis and detection. In this way, scattered as well as unscattered electrons are guided in the forward direction, while retaining all information about their energy and scattering angle, being detected together after analysis of their energy. In principle, this makes it possible to undertake simultaneous absolute measurements of electrons scattered by (almost) all angles in a fashion similar to the system described in [136]. In this context, one should interpret the main magnetic field simply as a means of translating the electrons—which exhibit the exact angle and energy that resulted from a potential collision—along the central axis to the end of the collision chamber. In the retarding potential analyzer (RPA) region the beam confinement can be changed, and the independent magnetic field B_{RPA} (see [135]) can be used to select which electrons should be analyzed depending on the type of cross section (e.g., integral elastic or inelastic or total; see also the next section) to be measured. Full details of this quite novel apparatus can be found in Fuss et al. [135].

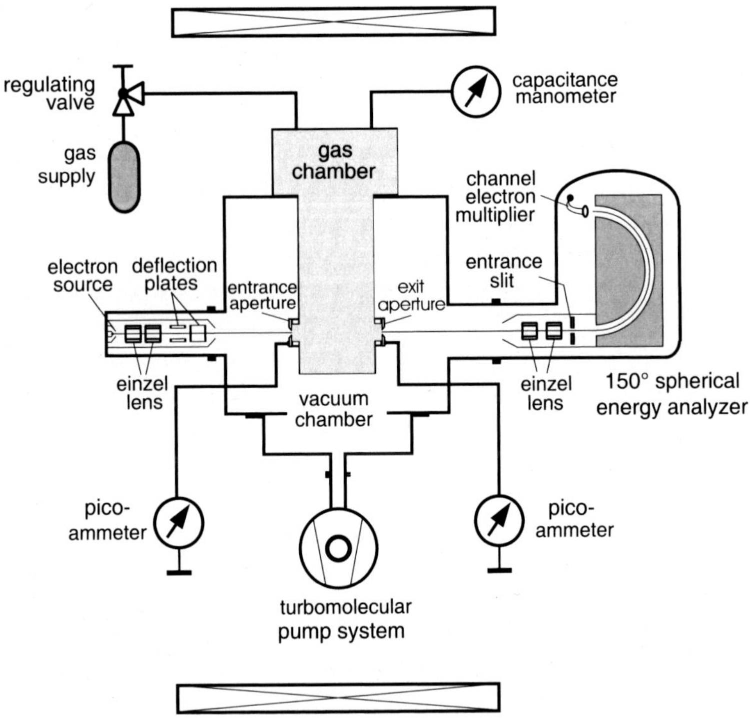

Figure 1.7 Schematic view of an experimental setup used for the measurement of the TCS. Reprinted figure with permission from Ref. [37]. Copyright (2012) by the American Physical Society.

1.5 Total Cross Sections

The principles and techniques for total electron scattering cross-section (TCS) measurements have been well documented in the literature [38, 39, 137–139]. The electron transmission method (with or without TOF discrimination) is now well established, and while it is arguable that it produces cross sections that provide the least sensitive test for the validity of the various scattering theories, the cross sections from these measurements are still the most accurate and reliable with typical error limits in the ±3%–5% range for targets that are gases at room temperature.

TCSs are deduced by measuring the attenuation of the electron beam passing through the target gas and utilizing the Beer–Lambert attenuation law:

$$I_t = I_0 \exp(-\sigma_t n L), \quad (1.47)$$

where I_t and I_0 are the electron count rates measured with and without the gas in the scattering cell, σ_t is the TCS, n is the gas number density in the scattering chamber, and L is the scattering length. A schematic diagram of an apparatus currently being employed to perform TCS measurements is shown in Fig. 1.7 and comes from [37]. The experimental problems with this technique are associated with the accurate determination of the target number density (pressure), the scattering length, and scattering to the forward direction (which contributes to the transmitted signal I_t). In the latter case, we highlight the particular problem that can occur in the study of some polar systems [37, 140]. Here the molecular dipole moment of the target significantly enhances scattering to forward angles so that the DCS is very strongly peaked in the forward direction [141]. Under these circumstances great care must be exercized in performing attenuation measurements. In the case of reactive gases it may also be necessary, for stability concerns, to have the spectrometer in a differentially pumped manifold so that the source optics, scattering cell, and detection optics are located in independently pumped chambers (e.g., [142, 143]). All these questions are addressed elsewhere [37–39, 137–139], and so we do not go into any further detail here. However, many of the above problems are now quite well understood for targets that are gases at room temperature such that with further improvements it is not inconceivable that the error limits on the TCS may fall to the $\pm1\%$–3% range or below [35]. On the other hand, for condensable (vapour) targets the pressure measurements are more difficult and subject to larger errors, which has led to some significant discrepancies amongst the cross-section data available for these targets [144, 145]. Consequently, such measurements have been limited to only a few species [35].

As noted earlier the TCS is in general defined as

$$\sigma_t = \sum_i Q_i, \quad (1.48)$$

where Q_i are the relevant ICSs for all the open reaction channels at a given incident electron energy. Namely i can represent the elastic channel (including rotational excitation due to limitations on practical spectrometer energy resolutions), vibrational excitation, electronic-state excitation, and ionization. As such, the TCS forms a very important self-consistency check for the ICS determined in one or more independent measurements. This is an important role that is sometimes overlooked. Indeed, perhaps, in part, due to that latter observation, many of the groups that made TCS measurements in the 1980s and 1990s have ceased to do so, with the most active group in recent years being based in Gdansk [146–149]. However, very recently new groups at Braunschweig [37] and Madrid [135] have begun to measure these important cross sections, which is certainly a positive development.

Acknowledgments

The author thanks the Australian Research Council for partial financial support. He also thanks Dr. L. Campbell for assistance in producing this chapter. This chapter is dedicated to Emeritus Professor Hiroshi Tanaka, who for many years led a very active group at Sophia University that has made highly accurate measurements on many very important scattering systems.

References

1. Berkowitz, J. (2002). *Atomic and Molecular Photoabsorption: Absolute Total Cross Sections*. Academic, San Diego.
2. Mota, R., Parafita, R., Giuliani, A., Hubin-Franskin, M.-J., Lourenço, J. M. C., Garcia, G., Hoffmann, S. V., Mason, N. J., Ribeiro, P. A., Raposo, M., and Limão-Vieira, P. (2005). *Chem. Phys. Lett.*, **416**, p. 152.
3. Svensson, S., Helenelund, K., and Gelius, U. (1987). *Phys. Rev. Lett.*, **58**, p. 1624.
4. Schulz, G. J. (1973). *Rev. Mod. Phys.*, **45**, p. 378.
5. Engmann, S., Stano, M., Papp, P., Brunger, M. J., Matejcík, S., and Ingólfsson, O. (2013). *J. Chem. Phys.*, **138**, p. 044305.

6. Flosadóttir, H. D., Ómarsson, B., Bald, I., and Ingólfsson, O. (2012). *Eur. Phys. J. D*, **66**, p. 13.
7. Crompton, R. W. (1994). In *Advances in Atomic, Molecular and Optical Physics*, Bederson, B. and Walther, H., eds. Academic Press, New York, p. 97.
8. Brunger, M. J., Cho, H., Tanaka, H., and Buckman, S. J. (2006). *Jpn. J. Appl. Phys.*, **45**, p. 8183.
9. Sasaki, K., Kawai, Y., Suzuki, C., and Kadota, K. (1998). *J. Appl. Phys.*, **83**, p. 7482.
10. Christophorou, L. G. and Olthoff, J. K. (2001). *J. Phys. Chem. Ref. Data*, **30**, p. 449.
11. Morris, R. A., Miller, T. M., Viggiano, A. A., and Poulson, J. F. (2001). *J. Geophys. Res.*, **100**, p. 1287.
12. Christophorou, L. G. and Olthoff, J. K. (1998). *J. Phys. Chem. Ref. Data*, **27**, p. 889.
13. Makabe, T. (2001). *Adv. At., Mol., Opt. Phys.*, **44**, p. 127.
14. Trevisan, C. S., Orel, A. E., and Rescigno, T. N. (2004). *Phys. Rev. A*, **70**, p. 012704.
15. Houghton, J. T., Meira Filho, L. G., Callander, B. A., Harris, N., Kattenberg, A., and Maskell, K., eds. (1996). *Climate Change 1995: The Science of Climate Change*. Cambridge University Press, Cambridge, UK.
16. Campbell, L. and Brunger, M. J. (2013). *Plasma Sources Sci. Technol.*, **22**, p. 013002.
17. Campbell, L. and Brunger, M. J. (2016). *Int. Rev. Phys. Chem.*, **35**, p. 297.
18. Brunger, M. J. (2017). *Int. Rev. Phys. Chem.*, **36**, p. 333.20
19. Brunger, M. J., Ratnavelu, K., Buckman, S. J., Jones, D. B., Muñoz, A., Blanco, F., and Garcia, G. (2016). *Eur. Phys. J. D*, **70**, p. 46.
20. Fuss, M. C., Ellis-Gibbings, L., Jones, D. B., Brunger, M. J., Blanco, F., Muñoz, A., Limão-Vieira, P. and García, G. (2015). *J. Appl. Phys.*, **117**, p. 214701.
21. Sanz, A. G., Fuss, M. C., Muñoz, A., Blanco, F., Limão-Vieira, P., Brunger, M. J., Buckman, S. J. and García, G. (2012). *Int. J. Rad. Biol.*, **88**, p. 71.
22. Boudaiffa, B., Cloutier, P., Hunting, D., Huels, M. A., and Sanche, L. (2000). *Science*, **287**, p. 1658.
23. Huels, M. A., Boudaiffa, B., Cloutier, P., Hunting, D., and Sanche, L. (2003). *J. Am. Chem. Soc.*, **125**, p. 4467.
24. Catana, C., Wu, Y., Judenhofer, M. S., Qi, J., Pichler, B. J., and Cherry, S. R. (2006). *J. Nucl. Med.*, **47**, p. 1968.

64. Palihawadana, P., Sullivan, J. P., Buckman, S. J., and Brunger, M. J. (2012). *J. Chem. Phys.*, **137**, p. 204307.
65. Murai, H., Ishijima, Y., Mitsumura, T., Sakamoto, Y., Kato, H., Hoshino, M., Blanco, F., García, G., Limão-Vieira, P., Brunger, M. J., Buckman, S. J., and Tanaka, H. (2013). *J. Chem. Phys.*, **138**, p. 054302.
66. Khakoo, M. A., Keane, K., Campbell, C., Guzman, N., and Hazlett, K. (2007). *J. Phys. B*, **40**, p. 3601.
67. Hargreaves, L. R., Francis-Staite, J. R., Maddern, T. M., Brunger, M. J., and Buckman, S. J. (2007). *Meas. Sci. Technol.*, **18**, p. 2783.
68. Maddern, T. M., Hargreaves, L. R., Bolorizadeh, M., Brunger, M. J., and Buckman, S. J. (2008a). *Meas. Sci. Technol.*, **19**, p. 085801.
69. Sikora, G. S. (1993). PhD Thesis, Princeton University, unpublished.
70. Powers, D. E., Hopkins, J. B., and Smalley, R. E. (1981). *J. Phys. Chem.*, **85**, p. 2711.
71. Miller, D. R. (1988). In *Atomic and Molecular Beam Methods*, Vol. 1, Scoles, G., ed. Oxford University Press, Oxford.
72. Beijerinck, H. C. W. and Verster, N. F. (1981). *Physica C*, **111**, p. 327.
73. Huang, J., Xu, D., Fink, W. H., and Jackson, W. M. (2001). *J. Chem. Phys.*, **115**, p. 6012.
74. Wiley, W. C. and McLaren, I. H. (1955). *Rev. Sci. Instrum.*, **26**, p. 1150.
75. Götte, S., Gopalan, A., Bömmels, J., Ruf, M.-W., and Hotop, H. (2000). *Rev. Sci. Instrum.*, **71**, p. 4070.
76. Holanda, R. (1973). *J. Vac. Sci. Technol.*, **10**, p. 1133.
77. Huo, W. M., Tarnovsky, V., and Becker, K. H. (2002). *Chem. Phys. Lett.*, **358**, p. 328.
78. Vinodkumar, M., Joshipura, K. N., Limbachiya, C., and Mason, N. (2006). *Phys. Rev. A*, **74**, p. 022721.
79. Hargreaves, L. R., Brunton, J. R., Prajapati, A., Hoshino, M., Blanco, F., Garcia, G., Buckman, S. J., and Brunger, M. J. (2011). *J. Phys. B*, **44**, p. 045207.
80. Hargreaves, L. R., Brunton, J. R., Maddern, T. M., and Brunger, M. J. (2015). *J. Chem. Phys.*, **142**, p. 124310.
81. Francis-Staite, J. R., Maddern, T. M., Brunger, M. J., Buckman, S. J., Winstead, C., McKoy, V., Bolorizadeh, M. A., and Cho., H. (2009). *Phys. Rev. A*, **79**, p. 052705.
82. Maddern, T. M., Hargreaves, L. R., Francis-Staite, J. R., Brunger, M. J., Buckman, S. J., Winstead, C., and McKoy, V. (2008b). *Phys. Rev. Lett.*, **100**, p. 063202.

83. Brunton, J. R., Hargreaves, L. R., Buckman, S. J., García, G., Blanco, F., Zatsarinny, O., Bartschat, K., and Brunger, M. J. (2013a). *Chem. Phys. Lett.*, **568–569**, p. 55.

84. Brunton, J. R., Hargreaves, L. R., Maddern, T. M., Buckman, S. J., García, G., Blanco, F., Zatsarinny, O., Bartschat, K., Jones, D. B., da Silva, G. B., and Brunger, M. J. (2013b). *J. Phys. B*, **46**, p. 245203.

85. Zatsarinny, O., Bartschat, K., Garcia, G., Blanco, F., Hargreaves, L. R., Jones, D. B., Murrie, R., Brunton, J. R., Brunger, M. J., Hoshino, M., and Buckman, S. J. (2011). *Phys. Rev. A*, **83**, p. 042702.

86. Maddern, T. M., Jamier, V., Brunton, J. R., Brunger, M. J., Papamicaël, C., Smith, S. V., and Buckman, S. J. (2016). *Int. J. Mass. Spectrom.*, **409**, p. 73.

87. Brunger, M. J., Buckman, S. J., Newman, D. S., and Alle, D. T. (1991). *J. Phys. B*, **24**, p. 1435.

88. Brunger, M. J. and Teubner, P. J. O. (1990). *Phys. Rev. A*, **41**, p. 1413.

89. Chutjian, A. (1974). *J. Chem. Phys.*, **61**, p. 4279.

90. Kato, H., Kawahara, H., Hoshino, M., Tanaka, H., Brunger, M. J., and Kim, Y.-K. (2007). *J. Chem. Phys.*, **126**, p. 064307.

91. Pichou, F., Huetz, A., Joyez, G., and Landau, M. (1978). *J. Phys. B*, **11**, p. 3683.

92. Wannier, G. H. (1953). *Phys. Rev.*, **90**, p. 817.

93. Allan, M. (2005). *J. Phys. B*, **38**, p. 3655.

94. Kato, H., Hoshino, M., Tanaka, H., Limão-Vieira, P., Ingólfsson, O., Campbell, L., and Brunger, M. J. (2011). *J. Chem. Phys.*, **134**, p. 134308.

95. Mašín, Z., Gorfinkiel, J. D., Jones, D. B., Bellm, S. M., and Brunger, M. J. (2012). *J. Chem. Phys.*, **136**, p. 144310.

96. Le Clair, L. R., Trajmar, S., Khakoo, M. A., and Nickel, J. C. (1996). *Rev. Sci. Instrum.*, **67**, p. 1753.

97. Le Clair, L. R. and Trajmar, S. (1996). *J. Phys. B*, **29**, p. 5543.

98. Lange, M., Matsumoto, J., Setiawan, A., Panajotović, R., Harrison, J., Lower, J. C. A., Newman, D. S., Mondal, S., and Buckman, S. J. (2008). *Rev. Sci. Instrum.*, **79**, p. 043105.

99. Lange, M., Matsumoto, J., Lower, J., Buckman, S., Zatsarinny, O., Bartschat, K., Bray, I., and Fursa, D. (2006). *J. Phys. B*, **39**, p. 4179.

100. Mondal, S., Lower, J., Buckman, S. J., McEachran, R. P., Garcia, G., Zatsarinny, O., and Bartschat, K. (2009). *PMC Phys. B*, **2**, p. 3.

101. Khakoo, M. A., Vandeventer, P., Childers, J. G., Kanik, I., Fontes, C. J., Bartschat, K., Zeman, V., Madison, D. H., Saxena, S., Srivastava, R., and Stauffer, A. D. (2004). *J. Phys. B*, **37**, p. 247.

102. Buckman, S. J. (2013). Private communication.

103. Campbell, L., Brunger, M. J., Teubner, P. J. O., Mojarrabi, B., and Cartwright, D. C. (1997). *Aust. J. Phys.*, **50**, p. 525.

104. Cartwright, D. C., Chutjian, A., Trajmar, S., and Williams, W. (1977). *Phys. Rev. A*, **16**, p. 1013.

105. Kato, H., Suzuki, D., Ohkawa, M., Hoshino, M., Tanaka, H., Campbell, L., and Brunger, M. J. (2010). *Phys. Rev. A*, **81**, p. 042717.

106. Khakoo, M. A., Malone, C. P., Johnson, P. V., Lewis, B. R., Laher, R., Wang, S., Swaminathan, V., Nuyujukian, D., and Kanik, I. (2008). *Phys. Rev. A*, **77**, p. 012704.

107. Chutjian, A., Cartwright, D. C., and Trajmar, S. (1977). *Phys. Rev. A*, **16**, p. 1052.

108. Chiari, L., Duque, H. V., Jones, D. B., Thorn, P. A., Pettifer, Z., da Silva, G. B., Limão-Vieira, P., Duflot, D., Hubin-Franskin, M.-J., Delwiche, J., Blanco, F., García, G., Lopes, M. C. A., Ratnavelu, K., White, R. D., and Brunger, M. J. (2014). *J. Chem. Phys.*, **141**, p. 024301.

109. Do, T. P. T., Leung, M., Fuss, M., Garcia, G., Blanco, F., Ratnavelu, K., and Brunger, M. J. (2011). *J. Chem. Phys.*, **134**, p. 144302.

110. Jones, D. B., Neves, R. F. C., Lopes, M. C. A., da Costa, R. F., Varella, M. T. do N., Bettega, M. H. F., Lima, M. A. P., García, G., Limão-Vieira, P., and Brunger, M. J. (2016). *J. Chem. Phys.*, **144**, p. 124309.

111. Neves, R. F. C., Jones, D. B., Lopes, M. C. A., Nixon, K. L., da Silva, G. B., Duque, H. V., de Oliveira, E. M., da Costa, R. F., Varella, M. T. do N., Bettega, M. H. F., Lima, M. A. P., Ratnavelu, K., García, G., and Brunger, M. J. (2015). *J. Chem. Phys.*, **142**, p. 104305.

112. de Urquijo, J., Basurto, E., Juárez, A. M., Ness, K. F., Robson, R. E., Brunger, M. J., and White, R. D. (2014). *J. Chem. Phys.*, **141**, p. 014308.

113. Boyle, G. J., McEachran, R. P., Cocks, D. G., Brunger, M. J., Buckman, S. J., Dujko, S., and White, R. D. (2016). *J. Phys. D*, **49**, p. 355201.

114. Campbell, L., Brunger, M. J., Nolan, A. M., Kelly, L. J., Wedding, A. B., Harrison, J., Teubner, P. J. O., Cartwright, D. C., and McLaughlin, B. (2001). *J. Phys. B*, **34**, p. 1185.

115. Hoshino, M., Horie, M., Kato, H., Blanco, F., García, G., Limão-Vieira, P., Sullivan, J. P., Brunger, M. J., and Tanaka, H. (2013). *J. Chem. Phys.*, **138**, p. 214305.

116. Kawahara, H., Suzuki, D., Kato, H., Hoshino,M., Tanaka, H., Ingólfsson, O., Campbell, L., and Brunger, M. J. (2009). *J. Chem. Phys.*, **131**, p. 114307.

117. Kim, Y.-K. (2007). *J. Chem. Phys.*, **126**, p. 064305.

118. Vriens, L. (1967). *Phys. Rev.*, **160**, p. 100.
119. Lassettre, E. N. (1965). *J. Chem. Phys.*, **43**, p. 4479.
120. Rau, A. R. P. and Fano, U. (1967). *Phys. Rev.*, **162**, p. 68.
121. Bransden, B. H. and Joachain, C. J. (1983). *Physics of Atoms and Molecules*. Longman, London.
122. Tanaka, H., Brunger, M. J., Campbell, L., Kato, H., Hoshino, M., and Rau, A. R. P. (2016). *Rev. Mod. Phys.*, **88**, p. 025004.
123. Read, F. H. and Channing, J. M. (1996). *Rev. Sci. Instrum.*, **67**, p. 2372.
124. Zubek, M., Gulley, N., King, G. C., and Read, F. H. (1996). *J. Phys. B*, **29**, p. L239.
125. Ward, R., Cubric, D., Bowring, N., King, G. C., Read, F. H., Fursa, D. V., and Bray, I. (2013). *J. Phys. B*, **46**, p. 035001.
126. Ward, R., Cubric, D., Bowring, N., King, G. C., Read, F. H., Fursa, D. V., Bray, I., Zatsarinny, O., and Bartschat, K. (2011). *J. Phys. B*, **44**, p. 045209.
127. Allan, M., Franz, K., Hotop, H., Zatsarinny, O., and Bartschat, K. (2009). *J. Phys. B*, **42**, p. 044009.
128. Allan, M., Zatsarinny, O., and Bartschat, K. (2011). *J. Phys. B*, **44**, p. 065201.
129. Linert, I., King, G. C., and Zubek, M. (2004). *J. Phys. B*, **37**, p. 4681.
130. Cho, H., Park, Y. S., Tanaka, H., and Buckman, S. J. (2004). *J. Phys. B*, **37**, p. 625.
131. Cho, H., Song, M. Y., Yoon, J. S., Hoshino, M., and Tanaka, H. (2010). *J. Phys. B*, **43**, p. 135205.
132. James, G. K., Ajello, J. M., Kanik, I., Franklin, B., and Shemansky, D. E. (1992). *J. Phys. B*, **25**, p. 1481.
133. Young, J. A., Malone, C. P., Johnson, P. V., Ajello, J. M., Liu, X., and Kanik, I. (2010). *J. Phys. B*, **43**, p. 135201.
134. Surko, C. M., Gribakin, G. F., and Buckman, S. J. (2005). *J. Phys. B*, **38**, p. R57.
135. Fuss, M. C., Sanz, A. G., Blanco, F., Oller, J. C., Limão-Vieira, P., Brunger, M. J. and García, G. (2013). *Phys. Rev. A*, **88**, p. 042702.
136. Sullivan, J. P., Gilbert, S. J., Marler, J. P., Greaves, R. G., Buckman, S. J., and Surko, C. M. (2002). *Phys. Rev. A*, **66**, p. 042708.
137. Bederson, B. and Kieffer, L. J. (1971). *Rev. Mod. Phys.*, **43**, p. 601.
138. Raith, W. (1976). *Adv. At. Mol. Phys.*, **12**, p. 281.
139. Trajmar, S. and Register, D. F. (1994). In *Electron–Molecule Collisions*, Shimamura, I. and Takayanagi, K., eds. Plenum Press, New York, Chapter 4.

140. Trajmar, S., Register, D. F., and Chutjian, A. (1983). *Phys. Rep.*, **97**, p. 219.
141. Brunger, M. J. and Buckman, S. J. (2002). *Phys. Rep.*, **357**, p. 215.
142. Alle, D. T., Brennan, M. J., and Buckman, S. J. (1996). *J. Phys. B*, **29**, p. L277.
143. Kanik, I., Trajmar, S., and Nickel, J. C. (1993). *J. Geophys. Res.*, **98**, p. 7447.
144. Sağlam, Z. and Aktekin, N. (1990). *J. Phys. B*, **23**, p. 1529.
145. Sağlam, Z. and Aktekin, N. (1991). *J. Phys. B*, **24**, p. 3491.
146. Możejko, P., Ptasińska-Denga, E., Szmytkowski, C., and Zawadzki, M. (2012). *J. Phys. B*, **45**, p. 145203.
147. Szmytkowski, C. (2010). *J. Phys. B*, **43**, p. 055201.
148. Szmytkowski, C., Możejko, P., Zawadzki, M., and Ptasińska-Denga, E. (2013). *J. Phys. B*, **46**, p. 065203.
149. Szmytkowski, C. and Ptasińska-Denga, E. (2011). *J. Phys. B*, **44**, p. 015203.

Chapter 2

Low Energy Electron–Induced Dissociation

Oddur Ingólfsson

Department of Chemistry, University of Iceland, Sæmundargata 2, 107 Reykjavík, Iceland

The Science Institute of the University of Iceland, Dunhagi 3, 107 Reykjavík, Iceland

odduring@hi.is

2.1 Introduction

In the preceding chapter, an account was given on elastic and inelastic electron scattering from molecular targets. The former of these processes is not associated with energy transfer to the respective molecules. The latter, on the other hand, describes energy transfer from the scattered electron to the electronic and/or nuclear coordinates of the molecular target. In electron scattering, the incoming projectile, the electron, may be temporarily captured by the respective molecule, forming a transient negative ion (TNI), also referred to as *resonance*. As discussed in the previous chapter, such resonances may decay by reemission of the electron, which constitutes resonant scattering. However, if the lifetime of a TNI is sufficiently long, and the thermochemical prerequisites are fulfilled,

Low-Energy Electrons: Fundamentals and Applications
Edited by Oddur Ingólfsson

ISBN 978-981-4800-00-6 (Hardback), 978-0-429-05882-0 (eBook)
www.panstanford.com

a TNI may also relax through bond rupture(s) within the molecule, thus leading to the formation of a stable negatively charged fragment along with one or more neutral fragments. Such fragmentation processes, which are generally referred to as *dissociative electron attachment* (DEA) are not restricted to single-bond ruptures and may even constitute significant molecular rearrangement reactions with multiple bond ruptures and new bond formations. Furthermore, this process can be very efficient at low incident energies (close to 0 eV) and is generally very selective with regard to which bonds are ruptured at what incident electron energies.

In the current chapter the focus is on the fragmentation of molecules upon their interaction with low-energy electrons, and special attention is given to the DEA process—a process where a temporarily captured electron effectuates nuclear displacement to an extent that leads to molecular fragmentation. Effort is made to provide a broad, basic insight into the physics of this process and use selected examples to demonstrate different aspects. Also an overview of experimental techniques used to study the DEA process is given.

For completeness, dissociative ionization (DI) of molecules upon electron impact ionization and neutral dissociation and dipolar dissociation (ND and DD, respectively) upon electronic excitation are also discussed, although to a lesser extent.

In the remainder of this chapter, first a general overview of the DEA process is given, followed by a discussion of the bond selectivity of this process, its efficiency (cross sections), the nature of the resonances formed, and the angular dependence of their formation, as well as the thermochemistry of their decay through molecular fragmentation. Following this section an overview of experimental techniques to study these aspects of the DEA process is given. Finally, in the last section a general account of other dissociative, electron-induced processes is given, that is, DI, and ND and DD upon electronic excitation.

2.2 Dissociative Electron Attachment

Different from the direct scattering processes, resonant scattering constitutes the formation of a TNI. Hence, in resonant scattering the dwell time of the incident electron in the vicinity of the target molecule is large compared to the time the electron needs to travel the distance characterizing the extent of the respective molecule. In this process, the incoming electron is incorporated in the electronic structure of a neutral molecule, temporarily forming the respective molecular anion:

$$AB + e^- \longrightarrow AB^{*\#-}. \quad (2.1)$$

Here the superscript $^{*\#}$ denotes that the TNI may be formed in an electronically (*) and/or vibronically (#) excited state, respectively.

Such resonances have finite lifetimes that may vary from a few femtoseconds, as is the case for many diatomics [1 and references therein], up to a few tens of microseconds, as is the case for SF_6 [2]. In an elastic process the molecule may return to its initial neutral (ground) state upon reemission (autodetachment) of the electron. The neural molecule may, however, also be left in a rotationally, vibronically, or electronically excited state, whereby this resonant scattering process would be rotationally, vibronically, and/or electronically inelastic (see Chapter 1).

In such resonant inelastic scattering processes the changed electronic structure of the TNI compared to the initial neutral (ground) state of the molecule is generally associated with relaxation through nuclear displacement. Thus, if the lifetime of the TNI with regard to autodetachment is sufficiently long, and the thermochemical requirements are fulfilled (see Sections 2.2.2 and 2.2.5.), such nuclear relaxation may lead to bond rupture(s) and the formation of a stable negative ion fragment and one or more (usually radical) fragment(s).

$$MX^{*\#-} \longrightarrow M + X^- \quad (2.2)$$

Figure 2.1 [3] shows this situation in terms of simplified two-dimensional potential energy curves for a quasi-diatomic molecule, MX, along its nuclear coordinate M–X. The thick vertical arrow signifies the attachment process through a vertical transition within the Franck–Condon region (shown as a shaded area). The thin vertical (downward) arrows signify the autodetachment process, the re-emission of the attached electron, and the grey corrugated arrows signify intramolecular vibrational redistribution (IVR) of the excess energy brought into the molecule through the electron

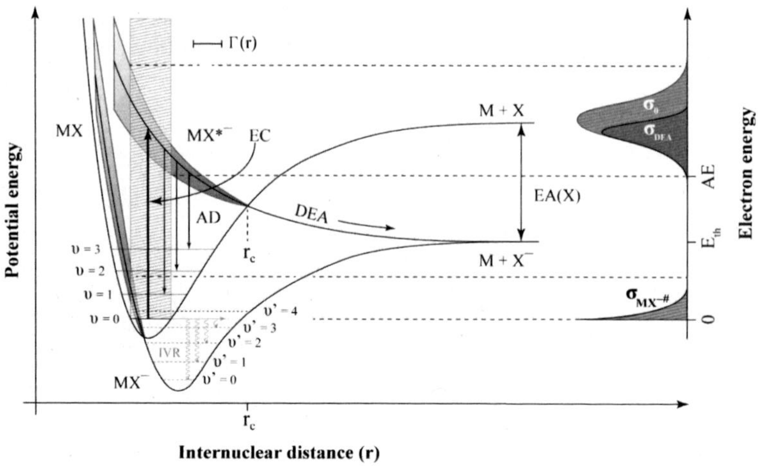

Figure 2.1 Schematic, two-dimensional, potential energy diagram depicting dissociative electron attachment to a quasi-diatomic molecule MX along its nuclear coordinate M–X. Potential energy curves are shown for the neutral ground state MX, the anionic ground state MX^-, and a purely repulsive anionic excited state MX^{*-}. The thick vertical arrow signifies the attachment process within the Franck–Condon region, which is shown as a shaded area. The thin vertical (downward) arrows signify the autodetachment process, and the grey corrugated arrows signify intermolecular vibrational redistribution (IVR) of the excess energy. The right side demonstrates the reflection principle, as is discussed in the text. Reproduced from Ref. [3] with permission.

attachment. Here Γ is the resonance width and r_c marks the crossing point of the neutral potential energy curve with the electronically excited negative ion curve. Beyond this point (r_c), autodetachment is classically not possible and the TNI initially formed is bound to dissociate. The resulting energy dependence of the cross sections for the attachment process and for negative ion formation is shown on the right-hand side axis.

The initial formation of the TNI is a vertical transition to the respective anionic state, signified by the shaded area in Fig. 2.1. A reflection of the probability for this transition on to the incident electron energy axis, shown on the right-hand side, constitutes the energy dependence of the electron attachment process; that is, the attachment cross section σ_0 (see also Section 2.2.2).

For the electronically excited anionic state depicted in Fig. 2.1, the so-formed TNI relaxes through nuclear displacement along the coordinate M–X, eventually leading to bond separation and formation of M and X^-. The energy dependence of this process, that is, the dissociation cross section σ_{DEA}, is superimposed on the attachment cross section σ_0 on the right-hand side axis, as is discussed further in Section 2.2.2. However, anywhere along the M–X coordinate, before the crossing point of the neutral and anionic curves at r_c, the TNI can reemit the attached electron and return to the neutral electronic ground state with or without vibrational excitation (thin downward-pointing arrows).

In Fig. 2.1 the DEA process is shown along a purely repulsive potential energy surface for an electronically excited state MX^{*-}, leading to the formation of a radical fragment M and a stable anionic fragment X^-. The anionic ground state, MX^-, on the other hand, is depicted as a bound state, with its vibrational ground state below that of the neutral. Hence the molecule has positive electron affinity (EA) (see Section 2.2.5). Furthermore, the dissociative asymptote of the anionic ground state, along the coordinate M–X, is above the Franck–Condon region for the formation of the TNI. Thus, dissociation of the ground state anion formed through a vertical transition from the neutral ground state is thermochemically not possible. Furthermore, the anionic ground-state potential energy curve crosses the ground-state vibrational level of the neutral, making transitions close to 0 eV most probable. If the coupling of the M–X stretch vibration ν' with other vibrational modes of MX (if polyatomic) is sufficiently efficient, IVR can take place. Such relaxation may channel vibrational energy out of the M–X coordinate in to other coordinates, as indicated by the grey, corrugated arrows in Fig. 2.1. As the M–X coordinate is the coordinate, relevant for autodetachment in the current picture, such IVR may lead to sufficient stabilization with regard to reemission of the electron for the molecular ion to be detected on the timescale of tens of microseconds. This is most commonly the case for highly symmetric molecules where strong coupling of electronic states with the nuclear coordinates is provided through the high degeneracy of the former. Good examples of such molecules are sulphur hexafluoride (SF_6) [4, 5] and hexafluorobenzene (C_6F_6) [6, 7], which both form metastable molecular anions with high cross sections close to 0 eV, as well as buckminsterfullerene, C_{60}, which forms a metastable molecular anion at electron incident energies up to about 12 eV [8–10].

Note that if M is polyatomic, the neutral fragment may be in an electronically and/or vibronically excited state.

2.2.1 Bond and Site Selectivity in DEA

The commonly cited example of bond selectivity in DEA is $CFCl_3$ [11] (Fig. 2.2), but the resonant nature of the site selectivity was firstexplicitly demonstrated by Prabhudesai et al. [12] and Ptasińska et al. [13, 14] in 2005.

In this compound, resonant electron capture (EC) close to 0 eV incident electron energy leads exclusively to the formation of Cl^- (top panel), while CCl_3^- and F^- are formed through higher-lying resonances, with the formation probability of these ions peaking close to 3.5 eV incident electron energy (see centre panel and bottom panel, respectively). Hence, while 0 eV electrons lead exclusively to cleavage of the C–Cl bond, incident electrons at around 3.5 eV lead exclusively to cleavage of the C–F bond.

Though commonly cited, the selectivity in DEA is not specific to this compound but is rather an intrinsic property of this process, largely due to its resonant nature, as is discussed here below.

A *good example* of the bond and site selectivity of DEA of more complex molecules is that of the C–H bond rupture in the pyrimidine nucleobase thymine [13, 14] (see also Chapter 6). From this fairly complicated molecule, H^- formation upon DEA is observed through four distinct resonances, giving rise to contributions in the H^- ion yield from about 4–12 eV. These are represented through Gaussian fits to the data in Fig. 2.3. The bond ruptures leading to H^- formation through these resonances may be those of a C–H bond or an N–H bond. Furthermore, these may take place from four distinct sites in this molecule, that is, from N1, N3, C6, and the methyl group CH_3 at C5. These processes are all thermochemically accessible through all four resonances. However, through selective deuteriation and methylation of different sites, selectively blocking the H^- formation, it could be shown that the underlying resonances lead to very site

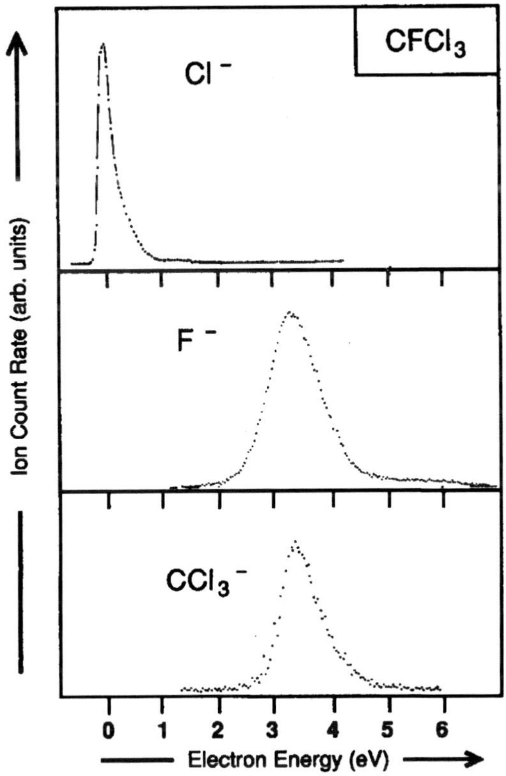

Figure 2.2 Energy dependence of negative ion formation from $CFCl_3$ upon dissociative electron attachment (DEA). This early study from the Illenberger group in Berlin shows clearly the selectivity of the DEA process; while attachment of electrons close to 0 eV leads exclusively to rupture of a C–Cl bond, attachment of electrons in the range from about 2–5 eV leads exclusively to rupture of the C–F bond. Reprinted from Ref. [11], Copyright (1989) with permission from Elsevier.

selective dissociations. That is, at 5.5 eV the H^- loss proceeds from N1, at about 6.8 eV from N3, at 8.5 eV from the C6 position, and at 10 eV from the methyl group [13, 14].

While the former of the two examples given here above demonstrates the bond selectivity in DEA, the latter shows the distinct site selectivity of this process, that is, the differentiation between N1 and N3 as well as between C6 and the methyl group.

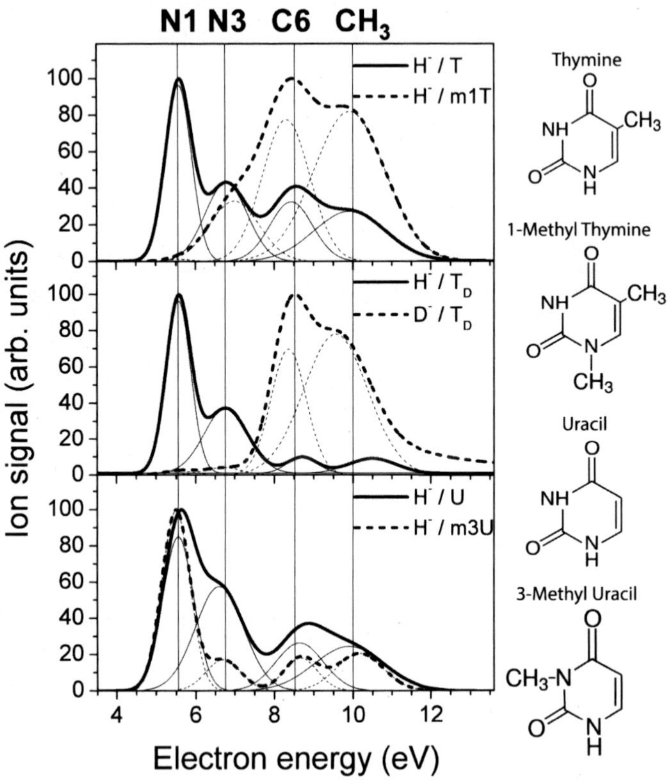

Figure 2.3 Energy dependence of H^- formation through DEA to thymine and C1-methylated thymine (top panel), H^- and D^- formation from thymine deuterated at C6 and at the methyl group at C5 (middle panel), and H^- formation from uracil and uracil methylated at the N3 position (bottom panel). For clarity the respective molecular structures are shown on the right-hand side. Reprinted (figure) with permission from Ref. [13]. Copyright (2005) by the American Physical Society.

An even more complex example of the intriguing selectivity of the DEA process is the fragmentation of hexafluoroacetone azine [15], shown in Fig. 2.4. In this molecule, resonant EC in the energy range from 0 to 1 eV and from 2 to 6 eV leads to the formation of a number of negative ion fragments that may result from a single-bond rupture, that is, $NC(CF_3)_2^-$, $C(CF_3)_2^-$, CF_3^-, and F^-. However, in the energy range from 1 to 2 eV,

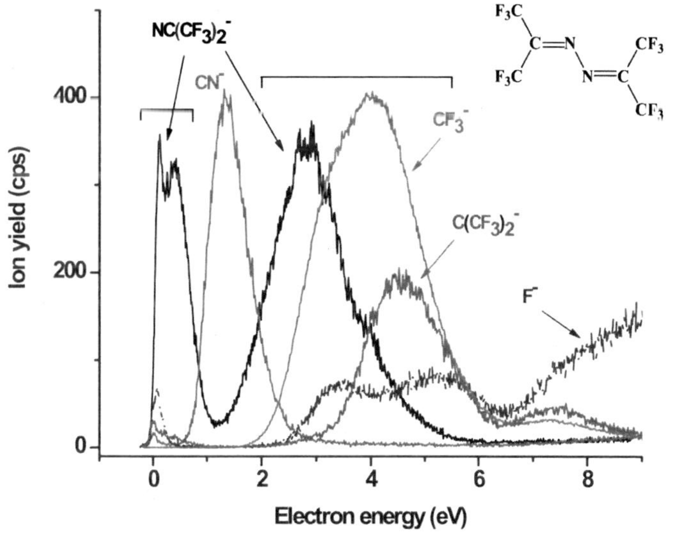

Figure 2.4 Dissociative electron attachment to hexafluoroacetone azine, $(NC(CF_3)_2)_2$. Resonant electron capture to this compound leads to the formation of a number of negative ion fragments that may result from a single bond rupture, that is, $NC(CF_3)_2^-$, $C(CF_3)_2^-$, CF_3^-, and F^-, observed in the energy range from 0 to 1 eV and from 2 to 6 eV. However, in the energy range from 1 to 2 eV, centered around 1.5 eV, the cyanide ion CN^- is formed exclusively and with considerable intensity. This channel requires multiple bond ruptures and new bond formation but is still very selective with regard to the incident electron energy [15].

centred around 1.5 eV, the cyanide ion CN^- is formed exclusively and with considerable intensity. Despite the high electron affinity (EA) of CN and the energy released through the CN triple bond formation in this process, the CN^- formation, at 1.5 eV incident electron energy, is thermochemically not possible without new bond formation. Hence, it requires a complex rearrangement process upon electron attachment that entails three bond ruptures and new bond formation (presumably $F_3C{-}CF_3$) but still remains effective and retains remarkable selectivity with regard to the incident electron energy. It is clear from this example that even in complex fragmentation processes, requiring multiple bond ruptures and new bond formation, DEA can be very extensive and energetic but intriguingly selective at the same time.

2.2.2 DEA Cross Sections

Within the Born–Oppenheimer approximation the electron attachment process can be seen as a vertical transition from the initial neutral state to the final TNI state within the Franck–Condon region [16, 17], as depicted in Fig. 2.1. In this picture, the transition probability, $P_{\text{transition}}$, reflects the overlap of the wave functions describing the populated vibrational levels of the initial neutral state ψ_i^0 and the wave function φ_i^* describing the TNI formed, hence the Franck–Condon factors defining these transitions:

$$P_{\text{transition}} \propto \left|\langle \psi_i^0 \mid \varphi_i^* \rangle\right|^2 \quad (2.3)$$

This quantity is commonly expressed in the units of area as the attachment cross section, σ_0, and can be viewed as a cross section defined by the radial distance between the molecule and the electron, within which the electron–molecule interaction leads to the formation of the respective TNI. The electron attachment cross section, on the other hand, is related to the attachment rate through the rate constant (k) for reaction (2.1):

$$k = \int_0^{\infty} \sigma_0 f(v) dv. \quad (2.4)$$

This may, in first approximation, be expresses as the product of the mean electron velocity ($\bar{v}$) and the attachment cross section averaged over the relevant incident electron energies ($\overline{\sigma_0}$):

$$k = \bar{v}\,\overline{\sigma_0}. \quad (2.5)$$

As mentioned above, the probability that a given TNI will dissociate depends on its lifetime with regard to autodetachment, τ_{AD}, and the time it takes the TNI to relax to a geometry from which autodetachment is classically not possible, τ_{diss}. In the simplified quasi-diatomic case shown in Fig. 2.1, this is the crossing point r_c. The prerequisite is obviously that the dissociation is thermochemically possible (see Section 2.2.5).

The relationship between the survival probability, $P_{\text{dissociation}}$, and the autodetachment lifetime and the relaxation time is given by

$$P_{\text{dissociation}} = e^{-\tau_{\text{diss}}/\tau_{\text{AD}}} \quad (2.6a)$$

The autodetachment lifetime τ_{AD}, on the other hand, is directly related to the width of the resonance $\Gamma(r)$ through the Heisenberg uncertainty principle

$$\tau_{AD} = \frac{\hbar}{\Gamma(r)} \tag{2.7}$$

and the relaxation time, τ_{diss}, may be expressed through the radial velocity, $v(r)$ of the departing fragments (M and X in Fig. 2.1):

$$\tau_{diss} = \int_{r}^{r_c} \frac{dr}{v(r)} \tag{2.8}$$

Accordingly Eq. 2.6a may be written as

$$P_{dissociation} = \exp\left[-\frac{1}{h}\int_{r}^{r_C} \Gamma(r)\frac{dr}{v(r)}\right] \tag{2.6b}$$

In terms of cross sections, the DEA cross section, σ_{DEA}, can be expressed as the product of the attachment cross section, σ_0, and the dissociation probability, $P_{dissociation}$:

$$\sigma_{DEA} = \sigma_0\, e^{-\tau_{diss}/\tau_{AD}}, \tag{2.9}$$

whereby the exponential term may be substituted with the right-hand side of Eq. 2.6b.

In Fig. 2.1 this is indicated in terms of the *reflection principle* [18], where the energy dependence of the attachment cross section σ_0 is shown on the right-hand side axis as a reflection of the Franck–Condon overlap of the vibrational ground state of the neutral over the potential energy curve representing the TNI state. Superimposed on the attachment cross section is the DEA cross section, which maximum is typically shifted to lower energies as compared to the attachment cross section. The reasons for this shift are that (i) the width of the resonance increases with increasing energy, causing the autodetachment lifetime to be longer at lower energies, and (ii) for transitions at lower electron incident energies, the radial distance to the crossing point is shorter, and thus the time it takes to reach the crossing point r_c. Hence, the absolute value of the exponent in Eq. 2.9 is larger.

An *elusive example* of the influence of the relaxation time on the DEA cross section is the comparison of the DEA cross section for the formation of H^- from H_2 and D^- from D_2. For these molecules, contributions assigned to the repulsive $B^2\Sigma_g^+$ state and a $2\Sigma_g^+$ Rydberg excited state are observed with maxima in the ion yields at about 10 and 14 eV, respectively. In an early study by Rapp et al. [19] the ratio of $\sigma_{D^-/D_2}/\sigma_{H^-/H_2}$ was found to be close to 1/5 for the firmer contribution and about 1/2 for the latter. In a more recent study by Krishnakumar et al. [20], with the advantage of the currently available instrumentation, this ratio was determined to be 1/18 and 1/3, respectively. Due to the heavier nuclei in D_2 as compared to H_2 the vibrational period in H_2 is about 7.5 fs as compared to 10.6 fs in D_2. Thus the relaxation time for D_2^-, needed to reach the crossing point r_c, is longer than for H_2^-. Correspondingly the exponent in Eq. 2.9 is smaller and the DEA cross section larger.

An *elusive example* of the influence of the autodetachment lifetime, on the other hand, is a comparison between DEA to chloropentafluorobenzene (C_6F_5Cl), in the gas phase under single-collision conditions and in clusters [21, 22] shown in Fig. 2.5 on the left- and right-hand side, respectively. In this case an open-channel core-excited resonance (see section 2.2.3) contributes insignificantly to the gas phase ion yield at around 6.5 eV, only visible in the $C_6F_5^-$ ion yield with low intensity. In clusters the polarization interaction stabilizes the TNI to a larger extent than the neutral parent state. This, in turn, lowers the TNIs total energy to be below that of the neutral parent state and thus extends its lifetime with respect to autodetachment and the exponent in Eq. 2.9 increases. Correspondingly this resonance is apparent in the ion yield of all fragments observed from clusters, making the Cl^- formation in this energy range significant.

For exothermic DEA processes, the cross section is generally largest at very low energies, that is, close to 0 eV. This is very clear in the case of the Cl^- formation from CCl_4 [4, 23], which is close to the s-wave attachment limit $\pi\lambda\!\!\!^-$ [24] at very low incident energies (here $\lambda\!\!\!^-$ is the reduced de Broglie wavelength). Other examples of very high cross sections at 0 eV threshold include that of SF_6 [4, 5], C_6F_6 [6, 7], and C_{60} [8–10], as discussed here above.

At such very low incident electron energies, close to the 0 eV threshold, the Wigner threshold law [24] predicts the cross section for a two-particle capture process to be

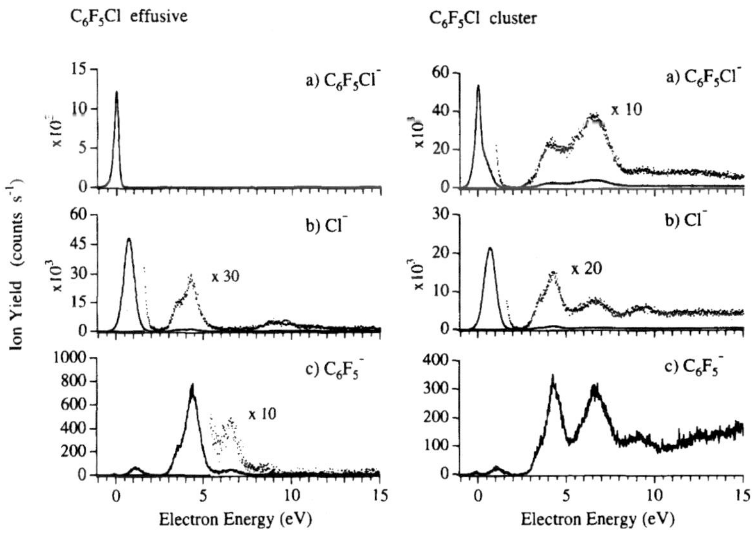

Figure 2.5 Energy dependence of negative ion formation from C_6F_5Cl upon electron attachment (DEA). The left-hand side middle and bottom panels show fragment formation in the gas phase under single-collision conditions, while the right-hand side middle and bottom panels show fragment formation from clusters. Additionally the energy dependence of the ion yield for the molecular anion under the same conditions is shown in the top panels. Reprinted from Ref. [21], Copyright (1996), with permission from Elsevier.

$$\sigma_0 (E, l) \propto E^{l-1/2} (E \rightarrow 0), \tag{2.10}$$

where l is the angular momentum quantum number describing the electron partial wave in the process and E the incident electron energy. For s-wave capture ($l = 0$), which constitutes the dominating component at 0 eV. The threshold cross section is thus predicted to be proportional to $E^{-1/2}$. For p-wave capture, on the other hand, the threshold cross section is predicted to be proportional to $E^{1/2}$. The threshold behaviour is thus highly dependent on the symmetry of the anionic state, which, in turn, is due to constructive interference required for EC (see Section 2.2.4). In most cases the s-wave component governs the threshold (0 eV) behaviour in electron attachment, but p-wave attachment has also been demonstrated at low energies for the halogens Cl_2 [25, 26] and F_2 [27].

In the Vogt–Wannier model [28] for s-wave attachment at very low energies, the attachment cross section is given as

$$\sigma_{0,VW}(E) = 4\pi\,[\alpha/2E]^{1/2}\ (E \to 0)\,. \qquad (2.11)$$

Later Klots [29] proposed a semi-empirical formula that converges to the $E^{-1/2}$ limit at very low incident energies and Klar et al. [23] used a variation of that formula to fit experimental data on the Cl^- formation from CCl_4. An excellent agreement with the experimental data was achieved in the energy range from 0.8 to about 50 meV, however, convergence to the $E^{-1/2}$ limit was only approached below 0.3 meV, which was outside the low energy limit of their study.

In the Vogt–Wannier model, the threshold behaviour for s-wave capture depends only on the electron incident energy and the polarizability and does not consider the presence of a permanent dipole moment. Hotop, Fabrikant and coworkers [30–32] derived an extension of the Vogt–Wannier model for s-wave attachment to molecules with a dipole moment lower than the critical 1.625 D and for energies where molecular rotation may be ignored:

$$\sigma_{EVW}(E) = \left(\pi k^{-2}\right)\left[4\cos^2\pi\lambda\left(b^2 + \frac{1}{b^2} + 2\cos 2\pi\lambda\right)^{-1}\right] \qquad (2.12)$$

In this extended Vogt–Wannier model, b^2 contains the polarizability (and the dipole moment) and for dipole moments μ that are larger than 0 but smaller than 1.625 D, λ takes values between 0 and –1/2.

The next chapter, Chapter 3, deals in detail with current theoretical approaches to low-energy electron–molecule interactions, but for further discussion on the Vogt–Wannier model and its extension the reader is referred to the original publication by Vogt and Wannier [28] and Refs. [30–32] and references therein.

2.2.3 Categorization of Resonances

In terms of the mechanisms behind the incorporation of the extra electron into the molecular structure, that is, what leads to the initial formation of the TNI, resonances are generally categorized as *shape resonances* or as *Feshbach resonances* [33, 34], the later also being termed *closed channel resonances*. These are then further

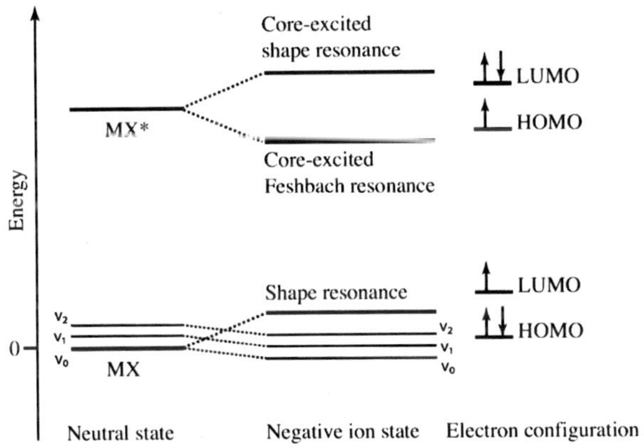

Figure 2.6 Schematic energy level diagram for different types of resonances in electron attachment. On the left-hand side energy levels of the neutral parent molecule are shown for the electronic ground state, MX, and an electronically excited state MX*. In the middle, the energy levels of the different resonances are shown relative to those of the respective parent states shown on the left. Nuclear excited Feshbach resonances are shown for the first two excited vibrational states of the anion, v_1 and v_2, while the ground-state vibrational level of the anion lies below that of the neutral (thin lines). The relevant electron configurations are given on the right-hand side using the lowest-lying orbitals, HOMO and LUMO, as examples. Adapted from Ref. [3] with permission.

divided into *single-particle resonances* and *one-particle one-hole resonances* (also referred to as *core-excited resonances*). This is shown in Fig. 2.6 in terms of a simple molecular orbital/energy level schema. On the left-hand side, the relative energies of the parent states are compared to the energies of the corresponding open-channel (*shape*) resonances and closed-channel (*Feshbach*) resonances (shown in the middle). This is shown both for a single-particle resonance and a one-particle one-hole resonance. The electron configuration for these resonances is shown in Fig. 2.6 on the right-hand side in terms of a single-electron occupation of the lowest unoccupied molecular orbital (LUMO) for the single-particle resonance and a concomitant highest occupied molecular orbital (HOMO)/LUMO transition for the one-particle one-hole (core-excited) resonance. The molecular orbitals involved may, however, be other higher-lying vitrual states, and core-excited resonances are commonly associated with Rydberg states.

2.2.3.1 Shape (open-channel) resonances

Shape resonances are resonances where the electron is temporarily captured by a potential barrier induced by the combination of the long-range, attractive polarization potential

$$V_\alpha = -\frac{\alpha e^2}{2r^4}, \tag{2.13}$$

and the centrifugal term describing the angular momentum dependence of the repulsive potential experienced by the electron

$$V_l = \frac{\hbar^2 l\,(l+1)}{2\mu r^2}. \tag{2.14}$$

In Eqs. 2.13 and 2.14, α is the polarizability of the molecule, e the elementary charge of the electron, r the electron–molecule distance, μ the reduced mass of the electron–molecule system ($\mu \approx m_e$), and $\hbar$ the reduced Planck's constant.

For l values larger than 0 the sum of these potentials

$$V_{\text{eff}} = \frac{\hbar^2 l\,(l+1)}{2\mu r^2} - \frac{\alpha e^2}{2r^4} \tag{2.15}$$

leads to a centrifugal barrier. This is shown in Fig. 2.7a for the interaction of N_2 with electrons with l values of 0, 1, and 2, that is, for s-, p-, and d-waves. For completeness Fig. 2.7b shows a hypothetical interaction potential, including the electron–electron repulsive interaction at very short distances. In this picture an incoming electron with $l > 0$ experiences a potential barrier from which the electron may be reflected. It may, however, also tunnel through the barrier and thus be trapped within the vicinity of the molecule for the time it takes it to tunnel out again. As shown in Fig. 2.6, shape resonances can be associated with a single-electron occupation of low-lying virtual molecular orbitals but may also be associated with concomitant electronic excitation within the molecule, in which case these would be core-excited shape resonances. In this context, Fig. 2.8a shows semi-schematic

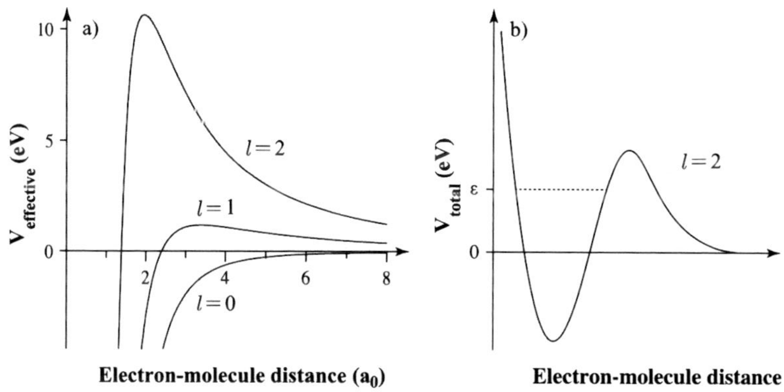

Figure 2.7 (a) Effective potentials for the interaction of N_2 with electrons with l values of 0, 1, and 2, that is, for *s*-, *p*-, and *d*-waves. (b) Hypothetical interaction potential, including the electron-electron repulsive interaction at very short distances. Reproduced from Ref. [3] with permission.

potential energy diagrams for the formation of the $^2\Pi_g$ (N_2^-) single-particle shape resonance observed in electron scattering from N_2 at around 2.3 eV, and Fig. 2.8b shows the formation of the $^2\Pi_u^+$ (N_2^-) one-particle one-hole shape resonance (core-excited) observed in electron scattering from N_2 at around 10 eV. Figure 2.8 is reprinted from Ref. [3] and is based on data from Refs. [35–38]. In the former, the $^1\Sigma_g^+$ (N_2) ground state is the parent state and autodetachment is thus an open-channel, that is, a single-particle transition to the ground state. The later is a core-excited shape resonance, that is, also an open-channel resonance; however, in this case the electronically excited $^3\Sigma_u^+$ (N_2) state is the parent state. Relaxation of these resonances to their respective parent states is thus a direct process. Single-particle and core-excited shape resonances are accordingly generally short lived and appear as broad features in scattering experiments and in DEA (see Eq. 2.7). Note that in the description of shape resonances here, only the long-range charge-induced dipole interaction is considered. The bulk of all molecules, on the other hand, have a permanent dipole (or higher-order terms). This is of fundamental importance in the consideration of low-energy electron–molecule interaction.

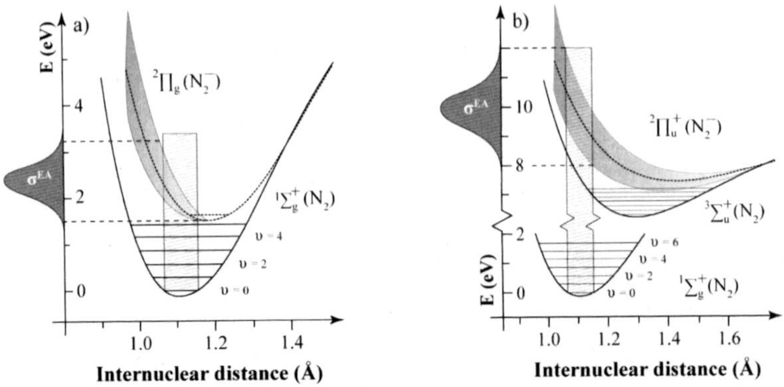

Figure 2.8 Semischematic representation of the formation of (a) the single-particle. $^2\Pi_g$ (N_2^-) shape resonance formed from N_2 at about 2.3 eV and (b) the two-particle one-hole. $^2\Pi_u^+$ (N_2^-) resonance formed at around 10 eV. In (a), the neutral. $^1\Sigma_g^+$ (N_2) ground state is the parent state, and in (b), this is the triplet excited. $^3\Sigma_u^+$ (N_2) state. Thus both resonances are open channel and can decay through a single-particle transition. Reproduced from Ref. [3] with permission, where the figures are based on data from Refs. [35–38].

2.2.3.2 Feshbach (closed-channel) resonances

Feshbach resonances are resonances where the TNI state lies energetically below the respective parent state and thus direct autodetachment through a single-particle process is energetically not accessible. Such resonances may be associated with electronic excitations, that is, core-excited Feshbach resonances, or with the electronic ground state of the molecules. In the latter case the trapping mechanism constitutes a weakly bound state enabled through strong long-range interaction with the vibrational level of the anionic state lying below the corresponding vibrational parent state of the neutral. In such cases these resonances are referred to as vibrational Feshbach resonances (also nuclear-excited Feshbach resonances) [33, 34]. Vibrational Feshbach resonances may be observed in molecules where the long-range interaction with the electron is strong, that is, molecules with high polarizability and/or large dipole moments. Such long-range interactions are discussed explicitly by Fabrikant et al. [32, 39] and are also discussed in Chapter 3 of this book. In these resonances, the long-range interaction leads to a quasi-discrete anion states in the continuum for which

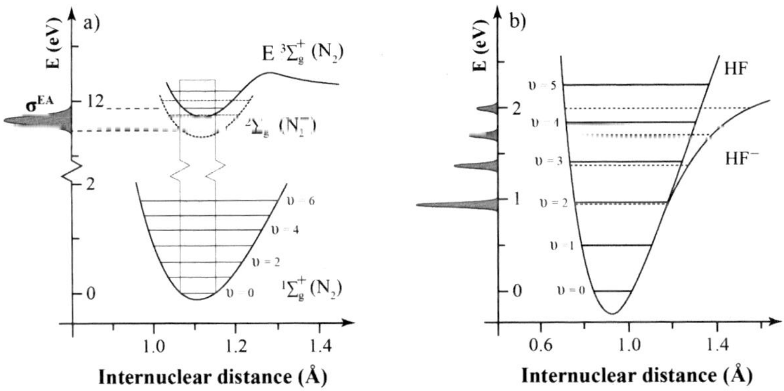

Figure 2.9 Semischematic representation of the formation of (a) two-particle one-hole. $^2\Sigma_g^-$ (N_2^-) Feshbach resonance formed from N_2 at around 11.5 eV and (b) vibrational Feshbach resonances observed in electron attachment to HF. Reproduced from Ref. [3] with permission, where figure (a) is based on data from Refs. [35–38] and figure (b) on data from Refs. [46, 47].

the low-lying vibrational levels are energetically below the corresponding vibrational levels of the neutral molecule. Such resonances were first observed in vibrational scattering from the highly polar molecules HF and HCl [40] (see also Ref. [31]). Vibrational Feshbach resonances are, however, not limited to diatomics with high dipole moments but have also been observed in molecular clusters (see Ref. [31] and references therein) and in DEA to a number of different molecules, including methyl halides such as CH_3I [41] and CH_3Br [42], as well as the nuclear bases uracil [43, 44] and thymine [45]. For further detail see, for example, Refs. [31, 32, 39].

Figure 2.9a shows a semi-schematic potential energy diagram for the formation of the $^2\Sigma_g^-$ (N_2^-) one-particle one-hole Feshbach resonance, observed in electron scattering from N_2 at around 11.5 eV. Figure 2.9b shows the formation of a vibrational Feshbach resonance observed in electron scattering from HF at around 1–2 eV. Both figures are adapted from Ref. [3], whereas Fig. 2.9a is based on the same references as Fig. 2.8, and Fig. 2.9b is based on Refs. [46, 47]. In the former case the parent state is the $^3\Sigma_g^+$ state of N_2, which lies about 0.5 eV below the respective TNI state. In the latter case the resonance constitutes a vibrational progression through a series of vibrational levels in the anionic ground state that are shifted slightly below the parent vibrational levels of the neutral. In both cases, direct autodetachment through a one-particle process is not possible without extra energy being channelled into the coordinate relevant for autodetachment. Thus Feshbach resonances are relatively long lived and appear as narrow contributions in electron scattering and DEA experiments (see Eq. 2.7).

Though the concept of single-particle and one-particle one-hole resonances serves well to describe the bulk of the electron attachment processes observed, there are cases where this concept is insufficient. Most noticeably in a recent study on the focused electron beam–induced deposition (FEBID) precursor; $HFeCo_3(CO)_{12}$, dissociative electron attachment leading to sequential CO loss is observed from the 0 eV threshold up to above 20 eV [48]. This in turn, is about 12 eV above the ionization energy of this molecule. The combined ion yield curves for the loss of 3 up to 12 CO upon electron attachment to this compound are shown on the left-hand side of Fig. 2.10. This is remarkable, not only as the electron attachment process takes place about 12 eV above the ionization limit of this molecule, but also, and maybe more so, as the TNI formed at these very high energies has sufficient lifetime to dissociate all CO units (see the lifetime discussion in Section 2.2.2). The authors tentatively explain this observation through a large density of fairly long-lived excited states associated with multiple electron excitations in the attachment process. Hence multi-particle multi-hole resonances, enabled through a high density (bands) of occupied and unoccupied molecular orbitals close to the HOMO–LUMO gap. As shown on the right-hand side of Fig. 2.10 these constitute the metal based σ-bonding d-orbitals, with some ligand orbital mixing and the π^* CO based orbitals with some metal d-orbital mixing, respectively. The sequential, "resonance-like" structure of the resulting ion yield curves is attributed to metastable decay where the preceding CO loss channel $[M - nCO]^-$ diminishes rapidly as the next channel, $[M - (n + 1)\ CO]^-$ becomes energetically accessible. For further reading on the performance of this precursor in FEBID and its interaction with low energy electrons see Refs. [48–52].

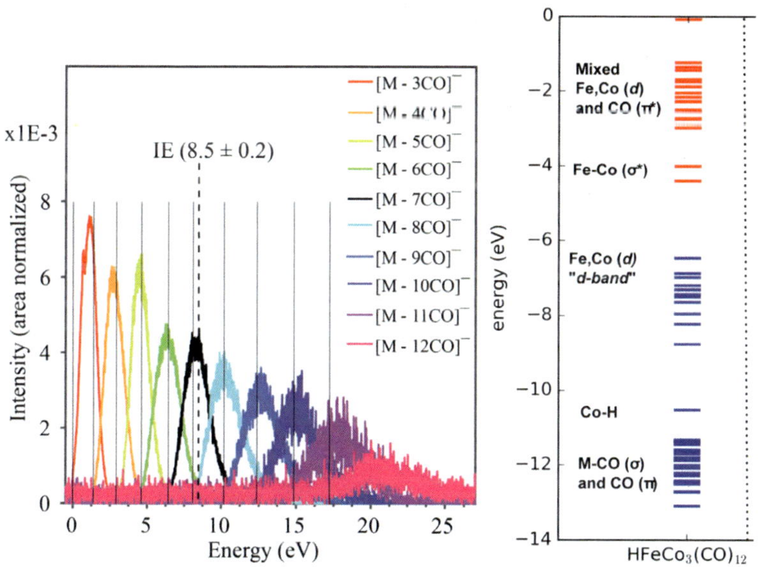

Figure 2.10 (Left) Combined negative ion yield curves for the loss of 3 up to all 12 CO ligands from $HFeCo_3(CO)_{12}$ upon electron attachment. The onset of the respective channels is signified with solid vertical lines, and the dotted vertical line marks the IE of this compound. (Right) Molecular (valence) orbital diagram of $HFeCo_3(CO)_{12}$. Occupied orbitals are depicted in blue and unoccupied in red. Orbital assignment is shown on the left-hand side of the MO levels. Reproduced from Ref. [48] under the terms of the Creative Commons Attribution-Non Commercial 3.0 Unported Licence (https://creativecommons.org/licenses/by-nc/3.0/).

2.2.4 Angular Dependence in DEA

Equal to the differential electron-scattering cross sections discussed in Chapter 1, the angular dependence of the cross section for fragment formation, through a given resonance in DEA, is generally anisotropic [53]. However, for this anisotropy to be reflected in the angular distribution of the fragments formed, there must be sufficient kinetic energy release (KER) in the process and the axial recoil approximation must hold, that is, the dissociation must proceed much faster than rotation of the respective molecule. This is a natural consequence of the fact that the angular dependence of the fragments reflects that of the TNI formation in the energy range where its formation leads to dissociation. Hence, this can only be the case if the orientation of the respective molecule, with regard to the incoming electrons momentum vector, is unchanged from attachment to dissociation. On the basis of this, and assuming that the coupling between the initial neutral state and the TNI state is due to a purely electronic matrix element, O'Malley and Taylor [54] derived an expression for the angular dependence of the DEA cross section resulting from a transition from the initial vibrational states of the neutral target χ_v to the final TNI state χ_{J_r}:

$$\frac{d\sigma}{d\Omega}(\theta,\phi) = \frac{4\pi^3}{k_i^2} e^{-\tau_{\text{diss}}/\tau_{\text{AD}}} g \sum_{\wedge_r} \left| \sum_{l=\mu}^{\infty} \langle \chi_{J_r} | V_{l|\mu|} | \chi_v \rangle Y_{l\mu}^{*}(\theta,\phi) \right|^2 \tag{2.16}$$

Here the J dependence is due to the centrifugal potential and is limited to the average of a set of contributing terms, $V_{l|\mu|}$ and $Y_{l\mu}^{*}(\Omega)$, which are the electronic transition element and spherical harmonics, respectively. The incident electron momentum is given by k_i and g is a spin-weighing factor.

For a given angular momentum of the incident electron the result is that the symmetry of the angular distribution of the DEA products, relative to the incoming electron beam, reflects that of the partial wave describing the incident electron leading to the TNI formation.

Within the pure resonant scattering (PRS) approximation, including direct scattering from a spherical potential (*potential scattering* (PS)), that is, within the PRS–PS approximation, the matrix element of Eq. 2.16 can be written as [55]

$$\langle \chi_{J_r} \mid V_{l|\mu|} \mid \chi_v \rangle = i^l e^{i\delta_l} a_{l|\mu|}(k) \tag{2.17}$$

In this form $a_{l|\mu|}$ is a weighing factor describing the contributions of individual incident electrons partial waves (l values), and δ takes (additionally) into account the phase shift introduced through the direct PS. Within the PRS–PS approximation, for a single resonance contribution, the incident electron momentum k and the angular coordinates (θ, ϕ), the angular distribution can thus be expressed in terms of the intensity as

$$I(\theta,\phi,k) \propto \left| \sum_{l=\mu}^{\infty} i^l e^{i\delta_l} a_{l|\mu|}(k) Y_{l|\mu|}(\theta,\phi) \right|^2. \tag{2.18}$$

Here the angular distribution is given by a weighted ($a_{l|\mu|}$) linear combination of the allowed l values for the TNI formation within the point group of the initial state, and the selection rule

$$\mu = \Lambda_{\mathrm{f}} - |\Lambda_{\mathrm{i}}| \tag{2.19}$$

results from conservation of angular momentum. Here Λ_{f} and Λ_{i} are the axial orbital momentum quantum numbers for the resonant state and neutral state, respectively.

Equation 2.18 is derived for the case of a diatomic dissociation, where the principal molecular axis is per definition the dissociation coordinate, but has also been shown to hold for dissociation along the principal axis of polyatomic molecules when the bulk of the excess energy is channelled into kinetic energy of the fragments (quasi-diatomic dissociation). As the partial waves representing the electron beam are defined with respect to the momentum vector of the incoming electrons, they must be rotated into the dissociation frame.

An example of the application of this diatomic model to DEA to a polyatomic molecule is the Cl^- formation from trifluorochloromethane (CF_3Cl) through a single-particle shape resonance centred around 1.3 eV [56]. CF_3Cl belongs to the C_{3v} point group, and its ground-state electron configuration can be written as $(1a_2)^2\,(7a_1)^4\,(6e)^4 \rightarrow X^1A_1$. Electron attachment in the energy range around 1.3 eV leads exclusively to Cl^- formation through a shape resonance that is associated with single-electron occupation of the σ^*(C$-$Cl) LUMO and has previously been shown to be of A_1 symmetry [57, 58]. For the C_{3v} point group, there are three irreducible representation: A_1, for which the allowed l values for the incident incoming electron are 0, 1, 2, and 3; A_2, where only f-wave ($l = 3$) contribution is allowed; and E, for which the l values 1, 2, and 3 are allowed [59]. Figure 2.11 shows the angular distribution of Cl^- formed through this resonance at 0.6, 1.2, and 1.8 eV (filled circles), adapted from Ref. [60]. From Fig. 2.11 it is clear that there is a finite contribution over the whole angular range, as is to be expected for an s-wave contribution. Further, there are two minima, which, in turn, indicates a d-wave contribution. The fits to these distributions are obtained by using Eq. 2.18 and by assuming A_1 symmetry, where s-, p-, d-, and f-waves are allowed. These are shown as solid lines

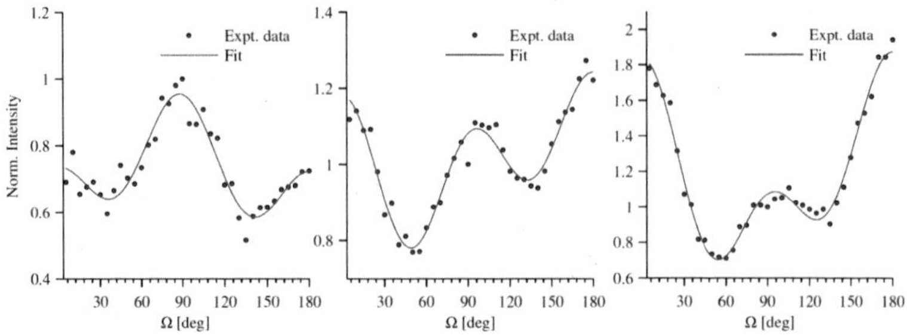

Figure 2.11 Angular dependence of Cl^- formation from CF_3Cl at (a) 0.6 eV, (b) 1.2 eV, and (c) 1.8 eV. Experimental data is shown as filled circles, while the blue line shows fits to the angular distributions using the functional form shown in Eq. 2.18 for diatomic dissociation and the l values 0, 1, and 2 allowed for the A_1 irreducible representation of the C_{3v} point group. The f-wave component is not included in these fits. Adapted from Ref. [60] with permission.

for weighted contributions from the s, p, and d partial waves. As is evident from Fig. 2.11, a good fit to the angular distribution is achieved with this approach, confirming the A_1 symmetry of the underlying resonance.

Though the diatomic model may be successfully applied for specific cases of DEA to polyatomic molecules, the cases where this approach is not justifiable are more numerous. In this context Azria et al. [61] extended the expression derived for the angular dependence of fragment formation from diatomic molecules to a description for polyatomic molecules. Similar to the diatomic model, rotation is assumed to be slow compared to the dissociation process, and the angular dependence is given by the matrix element $\langle \varphi_r \mid H_e \mid \varphi_i E_i \rangle$ (composed of the wave function describing the TNI state, the electronic Hamiltonian, and the wave function of the initial state, respectively). Here the wave function of the initial state, $\varphi_i E_i$, contains the incident plane wave, $e^{ik_i r_e}$, describing the incident electron. To introduce selection rules for polyatomic molecule, $e^{ik_i r_e}$ is expanded in a linear combination of spherical harmonics forming a basis for the irreducible representations of the respective point groups. These functions are chosen real and represented here as φ_{lm}^{ϵ} with m $>$1 and $\varepsilon = \pm 1$. The differential cross section in dependence on the orientation of the electron momentum vector k_i in the molecular frame is then described by

$$\frac{d\sigma_{\text{DEA}}}{d\Omega} \propto \left| \sum_{l,m,e} i^l e^{i\delta_l} a^{\epsilon}_{lm} \varphi^{\epsilon *}_{lm} (k_i) \right|^2, \tag{2.20}$$

with the weighing factor a^{ϵ}_{lm} being given by

$$a^{\epsilon}_{lm} = \int \varphi^*_r H_e \phi_i j_l(k_i r_e) \varphi^{\epsilon}_{lm} (r_e) dr \tag{2.21}$$

Here $j_l(k_i r_e)$ are spherical Bessel functions. In the PRS approximation the phase shift, δ_l, is 0, and thus a^{ϵ}_{lm} is real and is equal to 0 if φ^{ϵ}_{lm} is not the irreducible representations of the resonant and target states.

In the lab frame, the final expression of the fragments' angular dependence in DEA to polyatomic molecules takes the form

$$I(\theta) \propto \int_0^{2\pi} \left| \sum_{l,m,e} i^l e^{i\delta_l} a^{\epsilon}_{lm} \chi^{\epsilon}_{lm} (\theta, \phi) \right|^2 d\phi, \tag{2.22}$$

where $\varphi^{\epsilon *}_{lm} (\mathbf{k}_i)$ has been transformed into χ^{ϵ}_{lm}, with θ and ϕ being the polar angles in the dissociation frame, with ϕ describing the angle between the electron momentum vector and the direction of dissociation. For further reading on the derivation of the models outlined here, the reader is referred to the original literature [53–55, 57–59, 61, 62].

Measurements of the angular distribution, and at the same time the KER, in DEA processes have been revitalized in recent years through the introduction of position-sensitive detectors in DEA measurements [63]. In the meantime such angular-dependent measurements have been realized by a number of research groups [64–67], enabling a deeper understanding of the electronic and nuclear dynamics of the DEA process (see also Section 2.2.6.3). Such studies include angular dependence and KER measurements on CF_4, revealing quantum superposition of the target and the final state (the TNI) in the electron attachment process [68], detailed investigations of the role of conical intersections and the Renner–Teller effect in dissociation of the H_2O TNI [64, 69], and recently velocity slice imaging (VSI) was used to show symmetry breaking by quantum coherence in single-electron attachment to H_2 [70].

These are only few of the recent angular dependence and KER studies, but a good overview is given by Slaughter et al. in a 2016 topical review [71]. Also, in a 2017 review on DEA, by Fabrikant et al. [39], a good account on a number of VSI experiments is given.

A *school book example* of the value of such angular dependence measurements in combination with measurements of the KER in DEA is a recent study on DEA to CF_3I through electron attachment around 4 eV [72]. This molecule has been extensively studied with regard to photodissociation, showing that the photolysis of CF_3I at around 350–200 nm is dominated by the A_1 symmetry $^3Q_0$ state (Mulliken notation [73]) and a diabatic dissociation of the CF_3—I bond along the respective $^3Q_0$ potential surface, leading to the formation of iodine in its I ($P_{1/2}$) excited state (see, for example, Refs. [74–76]). This path crosses the $^1Q_1$ state, which dissociation asymptote leads to the formation of iodine in its I ($^2P_{3/2}$) ground state, as is shown with solid lines in Fig. 2.12d. The angular dependence of the formation of CF_3^- in DEA to CF_3I, shown in Fig. 2.12a and b, reveals that DEA at around 4 eV leads predominantly to the formation the $^2A_1$ TNI state. This TNI state corresponds to the neutral $^3Q_0$. However, analysis of the KER in this process, which is shown in Fig. 2.12c, reveals that the bulk of these, $^2A_1$ TNIs, dissociate along the ^2E state corresponding to the neutral $^1Q_1$ asymptote, hence leading to the formation of CF_3^- and iodine in its I ($P_{3/2}$) ground state. This situation is depicted in Fig. 2.12d with dotted lines. These measurements of the angular dependence of DEA to CF_3I and of the KER in the process reveal that though the initial state formed in the DEA process corresponds to the dominating state formed in photoexcitation to the A-band of this compound, the dissociation path is completely different. In fact, this can readily be rationalized, as in the case of photolysis a spin flip is necessary at the crossing of the $^3Q_0$ and $^1Q_1$ states for dissociation to lead to the I ($^2P_{3/2}$) ground state. This is not the case for the $^2A_1$ TNI as the asymptote leading to the ground-state iodine I ($^2P_{3/2}$) is also that of a doublet state, that is, the ^2E state corresponding to the $^1Q_1$ neutral state.

The energy difference between the I ($^2P_{3/2}$) ground state and the I ($^2P_{1/2}$) excited state of iodine is about 0.94 eV [77].

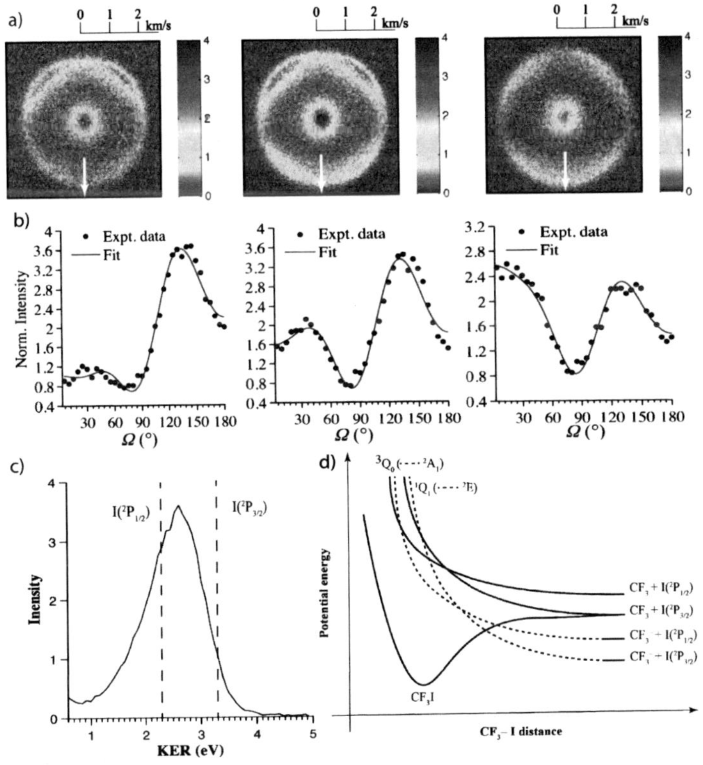

Figure 2.12 (a) Velocity slice images of CF_3^- formation from CF_3I through DEA. (b) Experimentally determined angular distributions and corresponding fits to those using Eq. 2.18. (c) Total kinetic energy release in the process recorded at 3.8 eV incident electron energy, whereby the dashed lines signify the maximum KER associated with the formation of iodine in the excited I ($P_{1/2}$) and ground I ($P_{3/2}$) states, respectively. (d) Schematic potential energy curves of the neutral $^3Q_0$ state dominating the A-band transitions in photodissociation of CF_3I at 350–200 nm and the $^1Q_1$ state leading to the ground state I ($P_{3/2}$) (solid lines) along with the corresponding $^2A_1$ and ^2E TNI states, respectively. Adapted from Ref. [72] with permission. Copyright (2014) by John Wiley and Sons.

Correspondingly the information revealed here through the angular dependence and kinetic energy release measurements is important for modeling of CF_3I in plasma processing, where the partitioning of the excess energy as internal and kinetic energy is important.

2.2.5 Thermochemisty in DEA

Figure 2.13 shows a simplified two-dimensional potential energy diagram for dissociation of the hypothetical quasi-diatomic molecule, MX, through electron attachment (DEA). In Fig. 2.13 the anionic ground state, marked MX^-, is shown to cross the vibrational ground state of the neutral. Further, the dissociative asymptote leading to M^- and X from the anionic ground state lies below the ground-state vibrational level of MX. The total energy of MX^- at its geometrical equilibrium lies below that of the neutral ground state, showing that MX possesses positive EA. In addition to the anionic ground state, an electronically excited anionic state MX^{*-} is shown as a purely repulsive state, which dissociative asymptote leads to the formation of $M + X^-$ and crosses the neutral ground state at the

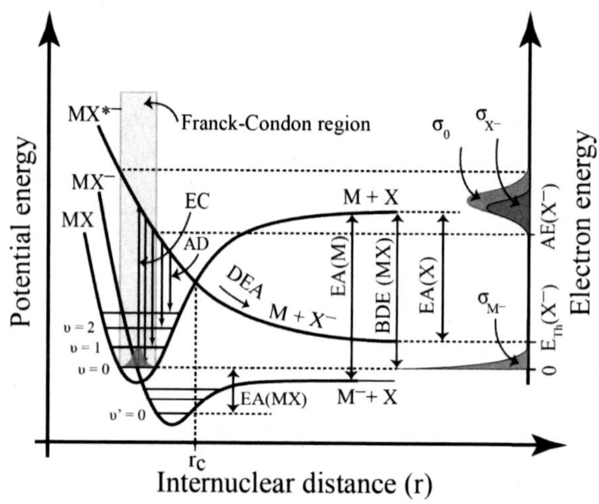

Figure 2.13 Schematic, two-dimensional potential energy curves depicting dissociative electron attachment to a quasi-diatomic molecule MX along its nuclear coordinate M–X. The Franck–Condon region and the attachment processes as well as the reflection principle are symbolized in the same way as in Fig. 2.1. Here the dissociative asymptote leading to the formation of M^- lies below the neutral ground state, leading to the formation of M^- at 0 eV threshold. The electron affinities (EAs) of MX, M, and X are depicted in relation to the respective anionic potential energy curves, and the bond dissociation energy of the neutral is shown in relation to the neutral potential energy curve. Adapted from Ref. [3] with permission.

nuclear distance r_c. In this picture EC at 0 eV incident energy can lead to the formation of the TNI MX^- in its electronic ground state, while a vertical transition within the Franck–Condon region (shaded area) at higher energies leads to the formation of the TNA MX^{*-} in an electronically excited state. These TNIs can then either relax through autodetachment, as discussed here above, or through nuclear relaxation along the respective dissociative asymptotes.

The thermochemical threshold for such a DEA process that involves only a single-bond rupture is given by the respective bond dissociation energy (BDE) less the EA of the precursor of the negative ion fragment formed, in this case M or X. The thermochemical threshold for, for example, the formation of $M^- +$ X from MX is thus given by

$$E_{\text{th}} = \text{BDE}\,(\text{M}-\text{X}) - \text{EA}\,(\text{M}). \tag{2.23}$$

In this case (Fig. 2.13) the EA of M is larger than BDE(M−X), rendering this process exothermic, and it can thus readily take place at 0 eV incident electron energy. The EA of X, on the other hand, is lower than BDE(M−X), making the X^- formation at 0 eV thermochemically inaccessible.

An *illustrative example* of the thermochemistry of such single-bond rupture processes in a quasi-diatomic system is that of X^- and $C_6F_5^-$ formation upon electron attachment to C_6F_5X (X = Cl, Br, or I) at around threshold [21, 22]. Figure 2.14 compares the ion yields for the formation of $C_6F_5^-$ and the halogen anion from these compounds. From C_6F_5Cl, the Cl^- formation peaks at about 0.7 eV and that of the $C_6F_5^-$ slightly above 1 eV. Furthermore, the intensity ratio of these fragments is about 150:1 in favour of the Cl^- formation. For C_6F_5Br this ratio is 4:1, with both contributions peaking at about 0 eV, and for C_6F_5I the ratio is 1:13 in favour of $C_6F_5^-$ formation.

The EAs of the halogens are well established as 3.615 ± 0.004, 3.364 ± 0.004, and 3.059 ± 0.001 eV for Cl, Br, and I, respectively [78]. The C_6F_5−X BDE is also fairly well known and is reported to be 4.0 ± 0.1, ~3.4, and 2.87 ± 0.05 eV for X = Cl, Br, and I, respectively. According to Eq. 2.23 this puts the threshold for the X^- formation from C_6F_5Cl, C_6F_5Br, and C_6F_5I at 0.4 ± 0.1, at ~0.0, and at − 0.19 ± 0.05 eV, respectively. The EA of C_6F_5 is not as well

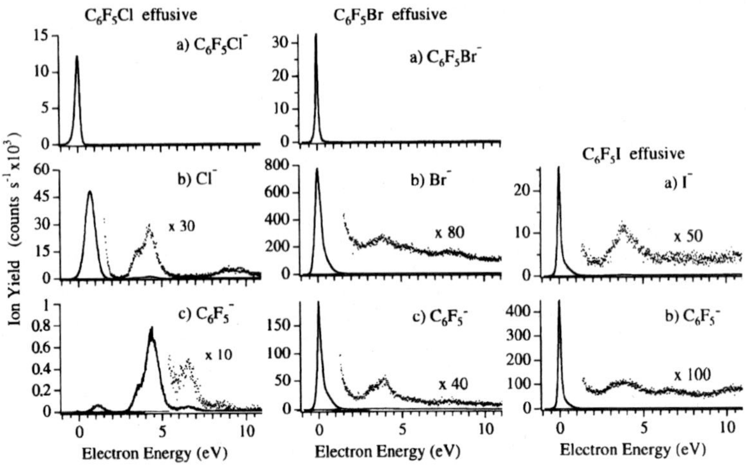

Figure 2.14 Energy dependence of negative ion formation from C_6F_5Cl (left panels), C_6F_5Br (center panels), and C_6F_5I (right panels) upon electron attachment (DEA) under single-collision conditions. For C_6F_5Cl and C_6F_5Br, panels (a) show the energy dependence of the formation of the respective molecular anions, panels (b) that of the Cl^- and Br^- formation, respectively, and panels (c) that of $C_6F_5^-$ formation. For C_6F_5I, panel (a) shows the energy dependence of I^- formation and panel (b) that of $C_6F_5^-$ formation. Reproduced from Ref. [21], Copyright (1996), with permission from Elsevier.

established, but from electron attachment to fluorinated benzenes it was estimated to be $\geq 3.4 \pm 0.3$ eV [6]. The observations in Ref. [21] indicate an EA of about 3.2. eV. The shifts of the onset and peak positions of the X^- and the $C_6F_5^-$ contributions are a clear demonstration of Eq. 2.23, when proceeding from C_6F_5Cl to C_6F_5Br and C_6F_5I. Furthermore, the branching ratios for X^- formation as compared to $C_6F_5^-$ formation also change dramatically. For C_6F_5Cl the inverse energy dependence of the *s*-wave attachment process also plays a significant role in determining these branching ratios. The $C_6F_5^-$ formation is only accessible at about 0.8 eV, while the Cl^- channel is already accessible at about 0.4 eV. For C_6F_5Br and C_6F_5I, the situation is more convoluted as both channels are accessible close to 0 eV and the branching ratios are rather determined by the energetics and dynamics of the dissociation process. It is noted, that the electron energy resolution in the experiments discussed here is about 150 meV, which in turn is reflected in the apparent cross sections as these emerge in the ion yield curves.

It is also interesting to compare the branching ratio of the observation of the intact molecular ion as compared to rupture of the C_6F_5-X bond shown in Fig. 2.14. For C_6F_5Cl the ratio of the intensity of the molecular anion as compared to the C_6F_5-Cl bond rupture fragments is about 1:4, for C_6F_5Br this ratio is about 1:33, and from C_6F_5I the molecular anion is not observed any more. The ratio of the intensity of the molecular anion as compared to that of the C_6F_5-X bond rupture channels reflects the survival probability of the TNI, whereby IVR plays a substantial role.

As the initial TNI formation is a vertical transition within the Franck–Condon region, the onset of the observed ion yield (the appearance energy [AE]) generally reflects the low-energy onset of this overlap rather than the threshold energy. Thus there may be considerable excess energy in the process. In the case of a diatomic parent molecule, this excess energy may appear in electronic excitation or as translational energy of the departing fragments. Equally, in the case of polyatomic molecules MX, this excess energy may also be partitioned into rotational and vibrational degrees of freedom of the fragments formed. For a simple, single-bond rupture process the AE may thus, more correctly, be written as

$$\mathrm{AE} = \mathrm{BDE}\,(\mathrm{M}-\mathrm{X}) - \mathrm{EA}\,(\mathrm{X}) + \mathrm{E}^*, \tag{2.24}$$

whereby the excess energy, E^*, is the difference between the threshold energy and the incident electron energy and is partitioned into the available degrees of freedom of the fragments formed.

In complex rearrangement reactions, where DEA constitutes multiple bond ruptures and new bond formations, the threshold energy is approximately equal to the reaction enthalpy for the process. For the formation of X^- through complex rearrangement reactions in DEA to a hypothetic, polyatomic molecule, Eq. 2.23 can be written as

$$E_{\mathrm{th}}\left(\mathrm{X}^-\right) \approx \Delta H_{\mathrm{react}} = \sum \mathrm{BDE}\,(\mathrm{educt}) - \sum \mathrm{BDE}\,(\mathrm{product}) - \mathrm{EA}\,(\mathrm{X})\,. \tag{2.25}$$

Here, the first term on the right-hand side refers to the sum of all BDEs in the target molecule, termed "educt," as customary in chemistry. The difference, as compared to Eq. 2.24, is that the energy needed for multiple bond ruptures and the energy gain through new bond formation(s) must also be considered.

Equal to the diatomic dissociation, multiple fragmentation processes may also take place above their threshold, in which case the excess energy is distributed into the translational energy and the internal degrees of freedom of the fragments formed. The AE for a rearrangement process is thus given by

$$\mathrm{AE} = \sum \mathrm{BDE}\,(\mathrm{educt}) - \sum \mathrm{BDE}\,(\mathrm{product}) - \mathrm{EA}\,(\mathrm{X}) + E^{*}. \quad (2.26)$$

For such rearrangement reactions, and in general in DEA to polyatomics, the observed m/z ratios of the negative ion fragments do not often allow conclusive identification of the neutral fragments. However, several commercially available computational packages allow for calculations of the thermochemical thresholds for potential reaction paths.

A *remarkable example* of extensive rearrangement reactions in DEA, which have been interpreted with the aid of thermochemical threshold calculations, is that of the fragmentation of tetrafluoroquinon shown in Fig. 2.15 [79]. There are two progressions observed in DEA to this compound. The first is the excision of one CO and the loss of zero up to three fluorine atoms, that is, the formation of $[\mathrm{M} - \mathrm{CO} - n\mathrm{F}]^{-}$ with $n = 0 - 3$, as shown on the left-hand side of Fig. 2.15. The second is the excision of two CO units and the loss of one to four hexafluoride atoms, that is, the formation of $[\mathrm{M} - 2\mathrm{CO} - n\mathrm{F}]^{-}$ with $n = 0 - 4$, as shown on the right-hand side of Fig. 2.15. For both progressions the initial CO loss is the most intense fragmentation channel with the lowest AE. The additional loss of the first fluorine appears with significantly lower intensity and the AE is shifted to higher energies. This is not surprising, as additional energy is needed to cleave the extra C–F bond. However, when a second fluorine is lost in addition to the CO units, the intensity rises significantly and the AE is again shifted to lower energy values. The loss of the third fluorine then again causes considerable drop in the ion yield and significant shift in the AEs, up to about 4 eV. Remarkably the AE for the loss of the fourth fluorine along with two CO units (right side, bottom panel) is also markedly shifted to lower energy as compared to the proceeding channel, the loss of two CO units and three fluorine atoms. Moreover, this channel constitutes

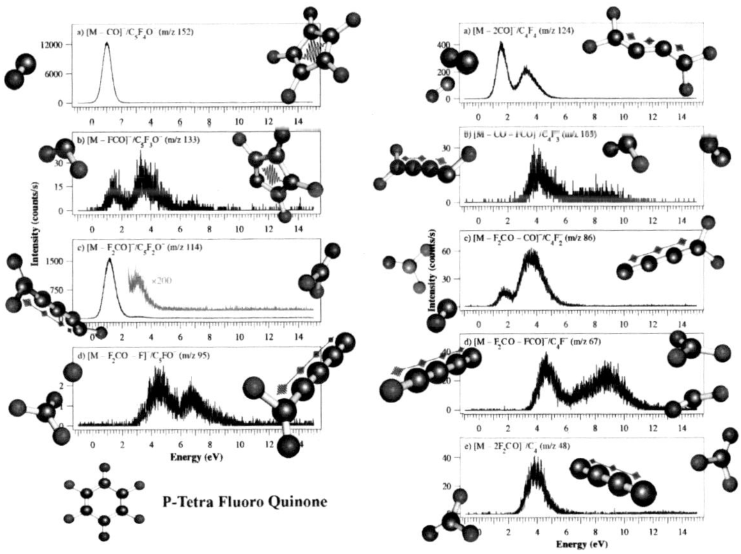

Figure 2.15 Energy dependence of negative ion formation from tetrafluoroquinone ($C_6F_4O_2$), upon dissociative electron attachment (DEA). The sticks and balls represent the most stable fragments, as determined by the Orca 2.9 computational chemistry software using density functional theory (DFT) calculations based on the B3LYP hybrid functional approximation and the ma-TZVP40 basis set. The molecular structure is shown as an insert on the bottom left. Ion yield curves adapted from Ref. [79] with permission from the PCCP Owner Societies.

the formation of C_4^-, with an AE below 3 eV—a channel that requires the cleavage of four C—C bonds and four C—F bonds. Clearly, from the observation of C_4^- a new C—C bond is formed, but the energy gained from this bond formation alone is far from being sufficient to enable this extensive fragmentation process at 3 eV incident electron energy without further bond formation.

In this case, calculations at the density functional theory (DFT) level [79] revealed that the threshold energies for these processes are in best agreement with the observed fragment formation under the assumption of the formation of difluoroformaldehyde for the loss of one or two CO and two or more hexafluoride atoms. Similarly, the channels where an uneven number of fluorine is lost are associated with the formation of FCO. The formation of C_4^- may thus be described through the reaction

$$C_6F_4O_2 + e^- \longrightarrow C_4^{\,-} + 2F_2CO; \quad \Delta H = 2.71 \text{ eV}. \qquad (2.27)$$

Hence, a single electron with less than 3 eV incident energy enables the rupture of four C—C bonds and four C—F bonds and the formation of one new C—C bond and four new C—F bonds.

These considerations show clearly that quantum mechanical calculations provide a valuable tool for the interpretation of DEA processes, where only the negative ion fragments are observed. However, the calculations discussed here are purely thermochemical and consider only the initial and final states in the process, ignoring eventual reaction barriers on the path. Such reaction barriers, however, are always present in rearrangement reactions and may be higher than the actual thermochemical thresholds and thus determining for the branching ratios observed.

A recent combination of experiment and theoretical calculations on the coupled cluster level of theory addresses this in a comparative study of DEA to *para*- and *ortho*-tetrafluorohydroquinone [80] (*p*-TFHQ and o-TFHQ, respectively). Figure 2.16 shows the ion yields observed in this study for *p*-TFHQ (left) and *o*-TFHQ (right) in the energy range from about 0 to 15 eV. Both compounds show significant contributions peaking at around 0 eV and about 1 to 1.5 eV. These are attributed to single-electron occupation of the lowest-lying $a_2(\pi^*)$ and $b_21(\pi^*)$ orbitals, respectively.

The dominant contribution from both compounds is the loss of one HF at about 0 eV incident energy. This channel is in both cases close to thermo-neutral with insignificant activation barriers. However, the loss of the second HF unit is significant at around 0 eV for *p*-TFHQ but only accessible through the higher-energy resonance in *o*-TFHQ, at about 1.5 eV. On the other hand, the loss of one HF and one CO unit is significant for *o*-TFHQ but is not observed for *p*-TFHQ. Simple thermochemical calculations show that the thermochemical threshold for the loss of the second HF unit is similar for *o*-TFHQ and *p*-TFHQ, and the loss of one HF and one CO unit is exothermic in both cases. The difference in the fragmentation patterns of these very similar compounds can thus not be explained from the thermochemical thresholds alone.

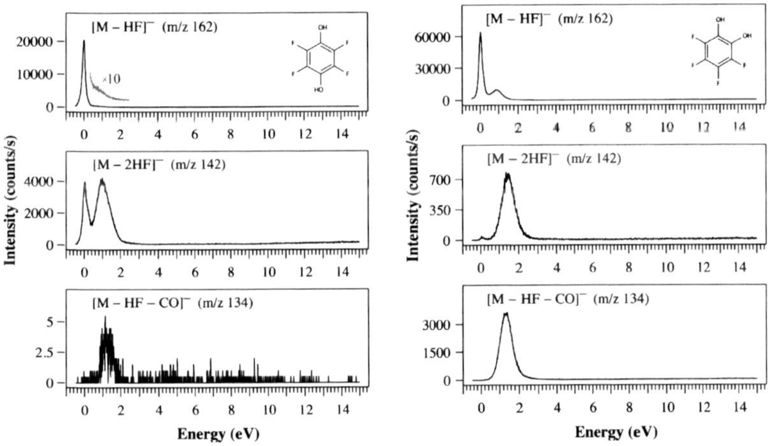

Figure 2.16 Ion yield curves for the principal fragmentation channels observed in DEA to *p*-TFHQ (left) and *o*-TFHQ (right). Reprinted with permission from Ref. [80], Copyright (2017) American Chemical Society.

Calculations of the respective reaction paths, however, give a consistent picture of these processes, in good agreement with the experimental observations. These are shown in Fig. 2.17, starting from the intermediate 1e and 2e formed after the first HF loss from *p*-TFHQ and *o*-THFQ, respectively. While the barrier for direct loss of the second HF unit from *p*-TFHQ is only about 0.6 eV (1f on the orange path in Fig. 2.17a), the barrier for the second HF loss from *o*-THFQ is about 2.5 eV (2h″ on the red path in Fig. 2.17b). Direct formation of $[M - 2HF]^-$ from *p*-TFHQ through the higher-lying resonance at about 1 eV is thus energetically accessible, while direct second HF loss from the intermediate 2e is not possible from *o*-TFHQ through this resonance. Conversely, for *o*-TFHQ, a proton-shuttling pathway from $[M - HF]^-$ with a barrier of only 0.6 eV is open through the higher-energy resonance (2j on the blue path in Fig. 2.17b). This pathway, in turn, leads to an intermediate from which both $[M - 2HF]^-$ and $[M - HF - CO]^-$ may be form (not shown here). These are competing channels and the barrier for the former is calculated to be about 1.4 eV and the latter about 1.1 eV. These are thus accessible through the higher-lying resonance as is reflected in the ion yield curves in Fig. 2.16. From *p*-TFHQ, on the

study the angular dependence of O^- formation from O_2 was reported. Several studies followed, for example, on O^- formation from N_2O [133] and CO [134], on H^- and D^- formation from H_2O and D_2O [135], and H^- formation from HBr [136]. The setups for these measurements were constructed in a manner similar to those described in Chapter 1 for measurement of angular differential electron-scattering cross sections. For measurement of the angular dependence of ion formation, however, the relative angle of the electron beam is rotatable with respect to the ion detection system (see, for example, Ref. [137]). The KER in the respective processes could be explored by, for example, momentum analysis using a magnetic sector field [137] or by using electrostatic energy filters [136]. Similar to the electron-scattering "turn table" experiments these experiments all had a limited angle of acceptance due to the finite physical dimensions of the electron guns and ion detection systems.

Though ion imaging in photo-fragmentation experiments was already introduced in the 1980s [138, 139], it was not until 2005 that this approach was applied to study the angular dependence in DEA [63]. This was realized by the group of E. Krishnakumar at the Tata Institute of Fundamental Research (TIFR), Mumbai, India, by adapting the VSI technique. In this setup, a pulsed (25 ns to 1 μs, 20 kHz) magnetically collimated electron beam with a resolution of about 0.5 eV crosses an effusive beam of the target gas between the repeller and the extractor of a ToF-MS. A time delay after the electron beam allows the ionic fragments formed to expand to a Newton sphere, reflecting the angular distribution and the kinetic energy of the fragments. After a typical delay of about 200 ns the Newton sphere of ions is extracted toward a ToF-MS to be detected on a pulsed, position-sensitive detector. The Newton sphere expands further during the flight in the field-free region, and by pulsing the detector appropriately a centre slice of the Newton sphere may be selected. This centre slice directly reflects the angular distribution of the fragment ions formed over the full 360° with respect to the electron beam. Further, the expansion, that is, the radius of the Newton sphere, contains direct information on the KER in the process. A variation of this setup is shown schematically in Fig. 2.21. Here the expansion of the Newton's sphere is shown along the flight

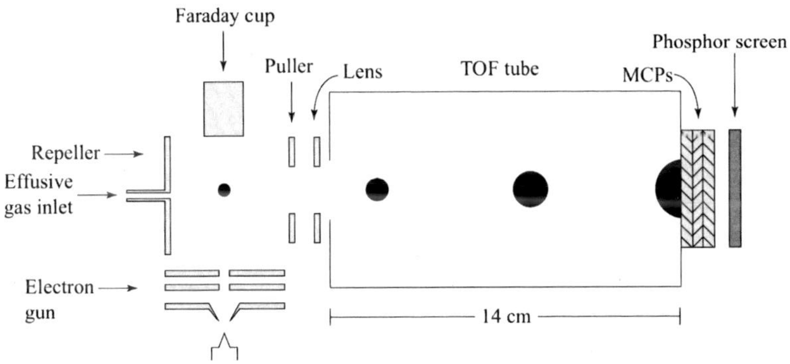

Figure 2.21 Schematic representation of the velocity slice imaging (VSI) setup installed at the Open University in collaboration with E. Krishnakumar's group at TIFR. The apparatus is equipped with a three-lens element electron gun, an effusive gas inlet, a ToF-MS, and a triple microchannel plate (MCP) detector with a phosphor screen and a charge-coupled device (CCD) camera (not shown). The black-filled circles represent the expansion of the Newton sphere along the field-free region of the ToF. Reproduced from Ref. [60] with permission.

path. In the original setup [63] the detector consisted of a *Z* stack of three 48 mm microchannel plates (MCPs) and a wedge-and-strip anode was used in combination with the appropriate electronics for the two-dimensional position information. In a later setup, constructed at the group of Nigel Mason at the Open University in Milton Keynes in collaboration with the TIFR group, the wedge-and-strip anode was replaced by a phosphorescence screen [68]. Further VSI instruments are currently in operation at the Hefei National Laboratory in China [67] and at the Indian Institutes of Science Education and Research (IISER) in Kolkata, India [65].

A different approach was taken at the Lawrence Berkeley National Laboratory in Berkeley, California [140], by using four-π momentum imaging rather than velocity slicing. In this setup a pulsed electron beam, collimated with a coaxial magnetic field, is also used and the electron–molecule interaction region, the delay time, and the initial ion extraction optics are comparable to those used in the VSI setups. However, before the field-free region the ions pass a stack of 27 resistively coupled electrical lens elements with 5 mm spacing and a typical voltage drop from about 120 V to 0 V along the stack. With this position focusing element, ions with the same momentum but different positions of origin are focused on the same spot on a 80 mm time- and position-sensitive detector composed of two MCPs in chevron configuration and a position-sensitive delay line anode. Currently such momentum imaging setups are also in operation at the Auburn University in Alabama [141] and at the Max Planck Institute for nuclear physics in Heidelberg, Germany [66, 142].

For further insight into ion mapping in DEA and a more extensive account of the studies conducted, the reader is referred to a recent review on momentum imaging by Slaughter et al. [71] and a general review on DEA by Fabrikant et al. [39] that discuss many of the recent angular dependence DEA studies.

2.2.6.4 Photoelectron attachment and high-Rydberg-state electron transfer for high-resolution studies

For high-energy resolution in the low-energy range, electron transfer from atoms prepared in high Rydberg states or electron attachment of free electrons generated through photoionization, mostly in a resonant multiphoton ionization process, has been applied.

The *high-Rydberg-state electron transfer* experiments are based on the assumption that high-Rydberg-state electrons can be viewed as free electrons with regard to transfer to an electron-attaching target gas [143]. This approach is in many senses similar to the swarm experiments discussed here above, that is, high Rydberg states are prepared in a selective multi-photon excitation of atoms such as Xe and Rb in the presence of a target gas (see, for example, Refs. [144, 145] and references therein). The electron energy is determined by the Rydberg state prepared ($\approx 13.6/n^2$ eV) and for $n = 10$ the electron energy is ≈ 136 meV. However, through excitation to very high Rydberg states, even as high as $n = 1000$ [146, 147], the electron energy range has been extended down to the micro electron volt range.

Similar to swarm experiments, the rate constants are acquired by varying the interaction time (t) of the target molecules with the high-Rydberg-state atoms in a field-free region using selective field ionization to follow their depletion. The depletion of high-Rydberg-state atoms with time is related to the rate constant for electron attachment by

$$N(t) = N(0)e^{-t/\tau}, \quad (2.37)$$

whereby

$$\frac{1}{\tau} = \rho k_p + \frac{1}{\tau_{eff}}. \quad (2.38)$$

Here k_p is the rate constant for the depletion of the Rydberg atoms, ρ is the target gas density, and τ_{eff} is the lifetime of the Rydberg atoms without the presence of the target gas [145]. The rate constant for Rydberg atoms' depletion is directly related to the rate constant for electron attachment, which, in turn, may be translated to the respective cross sections by using Eq. 2.4.

In *photoionization electron attachment studies*, selective threshold ionization of rear gas or alkaline atoms is achieved with a narrow-band irradiation and the kinetic energy of the photoelectrons may be varied within a narrow energy range by scanning the wavelength of the ionizing light. This approach was initially introduced by Ajello and Chutjian in 1979 [148] using monochromatized vacuum ultraviolet (VUV) light from a continuous source to ionize Kr or Xe to their $^2P_{1/2}$ first excited cationic states. Similar to the Rydberg electron attachment experiments described here above, the photoionization of the rear gas or alkaline atoms took place in the presence of an electron-attaching target gas. Here, however, the negative ions formed in the attachment process were extracted with a weak electric field and mass-analyzed using a QMS. With this approach Chutjian et al. studied electron attachment to a significant number of molecules in the electron energy range from a few meV up to about 150 meV with a resolution of 6–8 meV. Much of this work is reviewed in Ref. [82], where theoretical and experimental approaches to study electron attachment to molecules at low electron energies are reviewed.

Meanwhile, the approach introduced by Chutjian et al. has undergone considerable refinements, most noticeably by the group of Hartmund Hotop in Kaiserslautern. A comprehensive account of the development of this technique, termed "laser photoelectron attachment" (LPA) is given by Hotop, Ruf, Allan, and Fabrikant in a 2003 review [31], and thus only a brief account is given here.

In the approach by Hotop et al., in Kaiserslautern, photoelectrons were generated from laser-excited $^{40}Ar^*(4p^3D_3)$ or $^{39}K^*(4p_{3/2})$ atoms with a narrow-bandwidth dye laser, allowing electron attachment experiments with sub-meV resolution down to energies as low as 20 μeV [149]. To avoid degradation of resolution through the extraction voltage this setup was run in pulsed mode, separating the attachment period and the anion extraction. In a later approach, Hotop et al. used two-step photoionization of K atoms in a collimated beam in the wavelength range from 455.2 to 424 nm. In this approach a continuous electron energy variation from a few hundred μeV to about 200 meV with electron currents of 20–50 pA and a resolution of about 30 μeV/pA was achieved [150]. A later combination of this setup with supersonic expansion also allowed for the first time studies of electron attachment to van der Waals cluster with sub-meV electron energy resolution [151]. Finally, a considerable extension of the energy range accessible with LPA was achieved by post-acceleration of the photoelectrons prior to attachment. In this approach, termed "extended laser photoelectron attachment" (EXLPA), a typical energy resolution of 14 meV was retained up to electron energies of about 2 eV [4].

2.2.7 Dissociative Ionization and Neutral and Dipolar Dissociation upon Electron Excitation

As discussed in the preceding chapter, direct, nonresonant, inelastic scattering may lead to electronic excitation within the respective molecules or to ionization. Such inelastic scattering processes may, in turn, lead to dissociation and production of ionic or neutral fragments, often radical species, which may also be in vibrationally or electronically excited states. These fragmentation processes can be very efficient and play a significant, and even determining, role in many low energy electron–induced processes.

In *dissociative ionization* (DI), electron impact initially leads to ionization of the target molecule, which then dissociates to form a positively charged fragment and one or more neutral counterpart(s). At a given energy, such DI generally leads to characteristic fragmentation and has in the past decades served as a fingerprint for the identification of organic compounds, with more than 30.000 DI spectra being available in different databases today [35].

For single-bond rupture upon ionization of the quasi-diatomic molecule MX, the DI process is represented by the formation of a positively charged fragment and one neutral counterpart:

$$MX + e^- \longrightarrow M^{*\#} + X^+ + 2e^- \qquad (2.39)$$

Here the $^{*\#}$ superscript indicates that the polyatomic rest, M, may be in an electronic or nuclear excited state. Figure 2.22 shows this situation with schematic potential energy curves along the nuclear coordinate M–X [152]. The interaction with the incoming electron in DI initially initiates an electronic transition from the ground-state neutral molecule to the ground-state molecular cation or any other cationic state accessible within the Franck–Condon region (shown as grey-shaded area in Fig. 2.22). The probability of the transition is defined by the Franck–Condon factors, as discussed for DEA here above (Eq. 2.3). Different to the DEA process, however, DI is a nonresonant process in which direct energy transfer from the incident electron causes removal of a bound electron from the respective molecule. This constitutes a transition of the originally bound electron to the ionization continuum, leading to the formation of a discrete cationic state. Figure 2.22 depicts this situation for the ground-state cation and two excited states, MX^{*+} and MX'^{*+}. Here the ground state is depicted such that its equilibrium M–X bond length is significantly larger than that of the neutral molecule, representing ionization from a bonding orbital. The consequence is that the transition probability from the neutral ground state to the higher-lying vibrational states of the cationic ground state is significant.

In this quasi-diatomic model the threshold for the DI process is given by the sum of the bond dissociation energy (BDE) of the bond broken and the ionization energy (IE) of the precursor to the cationic fragment formed. For the formation of X^+:

$$E_{th} = BDE\,(M - X) + IE(X) \qquad (2.40)$$

For more complex rearrangement reactions, BDE(M–X) is substituted by the sum of the BDEs of all bonds broken, less the BDEs for all bonds formed, as shown for DEA in Eq. 2.25.

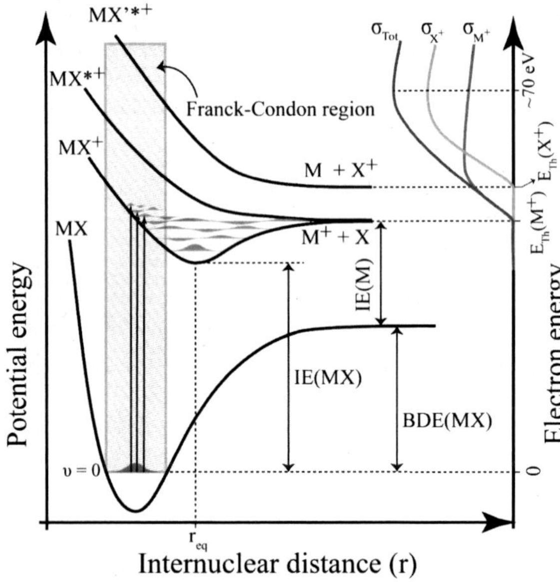

Figure 2.22 Schematic two-dimensional potential energy curves depicting dissociative ionization of a quasi-diatomic molecule, MX, along its nuclear coordinate M–X. Potential energy curves are shown for the neutral ground state MX, the cationic ground state MX^+, and two purely repulsive cationic excited states MX^{*+} and MX'^{*+}. The thin vertical arrows signify transitions to vibrational exited states of the cationic ground state within the Franck–Condon region. On the right-hand side axis the energy dependence of the individual and total cross sections are demonstrated as a reflection of the transition probability to the respective states. Reproduced from Ref. [152] under the terms of the Creative Commons Attribution 2.0 (https://creativecommons.org/licenses/by/2.0/).

Consequently, in Fig. 2.22, transitions to the three lowest-lying vibrational states lead to the observation of the molecular cation, while transitions to higher-lying vibrational levels of the cationic ground state may lead to dissociation. In addition to the ground state, two purely dissociative, excited cationic states are shown. From these, the lower-lying state leads to the formation of M^+, while the higher-lying state leads to the formation of X^+. Correspondingly, the threshold for bond dissociation from the lower-lying state is equal to that of the ground state, while that for the higher-lying excited state is the sum of the BDE (M−X) and the IE (X), as shown in Eq. 2.40.

The energy dependence of the cross section for the formation of the fragments M^+ and X^+ is shown on the right-hand side axis, along with the total DI cross section σ_{tot}, which represents the sum of the cross sections for all accessible channels (here $\sigma_M{}^+$ and $\sigma_X{}^+$).

The onset of DI is generally slightly above the IE of the respective molecules, and with increasing incident energy more fragmentation pathways become available, the fragmentation becomes more extensive, and the total cross section increases. Generally the total cross section reaches a maximum around 50 eV and then gradually decreases as energy transfer becomes less efficient.

Dissociation upon electron excitation is also a nonresonant process. Here, however, the initial electron molecule interaction constitutes an electronic excitation:

$$\mathrm{MX} + \mathrm{e}^-(\varepsilon_1) \quad \longrightarrow \quad \mathrm{MX}^* + \mathrm{e}^-(\varepsilon_2{<}\varepsilon_1), \tag{2.41}$$

where * signifies an electronic excitation, ε_1 is the incident electron energy, and ε_2 is the electron energy after the inelastic scattering process.

This electronically excited state may then dissociate through homolytic bond rupture(s), resulting in two neutral, generally radical fragments:

$$\mathrm{MX} \longrightarrow \mathrm{M}^{*\#} + \mathrm{X}, \tag{2.42}$$

that is, neutral dissociation (ND). These may be in their ground state or vibrationally or electronically excited states, as signified by the $^{*\#}$ superscript.

The dissociation can also proceed through heterolytic bond rupture leading to the formation of a negative ion fragment and a positively charged fragment:

$$\mathrm{MX} + \mathrm{e}^-(\varepsilon_1) \longrightarrow \mathrm{M}^{*\#+} + \mathrm{X}^- + e - (\varepsilon_2 < \varepsilon_1), \tag{2.43}$$

that is, dipolar dissociation (DD).

Figure 2.23 shows a schematic representation of ND upon electron excitation [152]. Again, the process is depicted for the quasi-diatomic molecule MX along the nuclear coordinate M–X. The process is shown for three electronically excited states

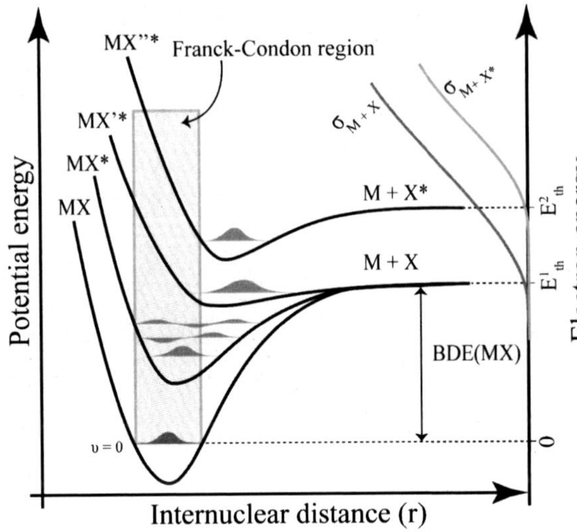

Figure 2.23 Schematic two-dimensional potential energy curves depicting neutral dissociation of a quasi-diatomic molecule MX upon electronic excitation. Dissociation is shown along the nuclear coordinate M–X for the ground state and three excided states: MX*, MX′*, and MX″*. While dissociation of MX* and MX′* is shown to lead to the ground state neutral fragments M and X, dissociation of MX″* is shown to lead to an electronically excited fragment X*. The resulting partial cross sections are shown on the right-hand side axis of the representation. Reproduced from Ref. [152] under the terms of the Creative Commons Attribution 2.0 (https://creativecommons.org/licenses/by/2.0/).

MX*, MX′*, and MX″*, which all represent bound states with an equilibrium bond length larger than that of the neutral ground state. These excitations thus all represent transitions from bonding orbitals and/or transitions to antibonding orbitals. For ND the thermochemical threshold is equal to the BDE (M–X) and the initial electronic excitation may lead to dissociation if it is energetically above the dissociation limit. The threshold for ND is thus lower than that for DI, the difference being equal to the IE of the neutral precursor to the charged fragment formed in the DI process.

In both cases the resulting fragments may be in an electronically or vibrationally excited states, as signified by the superscripts *# at the polyatomic fragments M.

Generally the onset for ND is close to the first HOMO–LUMO transitions. Above the threshold, the cross section for ND then gradually increases with increasing incident electron energy, hence, as more higher lying states become accessible. Similarly more dissociation channels open up, contributing to the total cross section. In many senses ND is comparable to DEA, as the underlying electronic excitations constitute vacancy in a previously occupied molecular orbital and occupation of a previously vacant orbital. Further, the energy transfer is at large confined to the respective excitation energies and thus single-bond ruptures along repulsive potential energy surfaces are rather expected than extensive fragmentation. However, depending on the molecule in question the excitation cross sections can be very high and the initial thresholds may be as low as 3–4 eV. Furthermore, ND, like DI, is active over a vide energy range with an onset close to the threshold of the lowest electronic transitions and a maxima around 50–100 eV before falling gradually (see, for example, Refs. [153, 154]). The energy integral cross sections may thus be significant for processes where the relevant electron energy spread is large.

Unfortunately, experimental data on ND is very limited due to the difficulties associated with detection of neutral fragments. Furthermore, as the energy range where ND is active overlaps with that of both DEA and DI, distinction between neutral fragments originating from these processes and from ND is not trivial and often not possible. Nonetheless, some experimental approaches to determine cross sections for ND processes may be found in the literature, including chemical gettering [155], formation and detection of organic tellurides through reactions with the neutral radicals formed in the ND process [154], postionization of neutral fragments [156], and laser-induced fluorescence [153]. Fast neutral beams have also been used to study ND from O_2 [157], CO [158], and N_2 [159], with the clear advantage that these address the partial cross sections and are selective toward ND. The number of targets studied is, however, very limited and mainly confined to diatomics and small organic molecules.

It is, nonetheless, clear that ND may play a significant role in a number of relevant processes and the development of experimental methods to access energy-dependent cross for this process is an important task.

A good example of the potential significance of ND in applications is that of the focused electron beam–induced deposition precursor $Pt(PF_3)_4$, discussed in Chapter 5. For this precursor, both the DEA [127] and the electronic excitation cross sections [160] were found to be very high, but while the partial cross sections for the DEA process could be determined directly, this is not the case for ND, leaving the fate of the electronic excited states unclear and thus the potential role of ND in this deposition process. In a recent theoretical study, however, Zlatar et al. [161] showed that the excited states observed in the experiments were either directly dissociative or dissociative via conical intersections, an indication that the quantum yield for dissociation from these states may be close to 100% under single collision conditions. This in turn demonstrates the potential importance of ND both in natural phenomena and in applications as well as the need to develop reliable experimental approaches to be able to study this process.

Finally, DD upon electron excitation is expected to proceed at large from the same electronic excited states as ND. Here, however, the bond rupture is heterolytic, leading to one positively charged and one negatively charged fragment, as shown in Eq. 2.43. Correspondingly the threshold for this process is given by

$$E_{\text{th}} = \text{BDE}\,(\text{M}{-}\text{X}) + \text{IE}(\text{M}) - \text{EA}(\text{X}), \qquad (2.44)$$

if M^+ and X^- are formed.

The relevant IE in these processes can generally be expected to be at least 4–6 eV higher than the relevant electron affinities, making the threshold for this process considerably higher than that for the respective ND process. This process is, however, generally not found to be very efficient and can be monitored fairly easily through detection of the negative ion fragment formed. Hence, as DD usually sets in above the energy range where DEA is observed and is characterized by a gradual increase in negative ion formation above the respective threshold, these processes can easily be studied in conventional crossed-beam setups.

Acknowledgments

I thank Reza Tafrishi for all the help with the cover and for getting the figures in printable format. Benedikt Ómarsson and Frímann Hauk Ómarsson have my thanks for making original figures from their PhD dissertations available. Support from the Icelandic Center of Research (RANNIS) through grant nos. 13049305 and 141218051 and previous grants, as well as support from the University of Iceland Research Fund, are duly acknowledged, in addition to the COST Actions: P9; RADAM, CM0601; ECCL and CM1301; CELINA and the Horizon 2020 MSCA ITN; ELENA (grant agreement no. 722149). These collaboration platforms have been instrumental in promoting collaborative studies on low-energy electron–molecule interactions, both from the fundamental point of view and in relation to the role such interactions play in different fields.

References

1. Christophorou, L. G. (1978). The lifetimes of metastable negative ions. *Adv. Electron. El. Phys.*, **46**, pp. 55–129.
2. Compton, R., Christophorou, L., Hurst, G., and Reinhardt, P. (1966). Nondissociative electron capture in complex molecules and negative-ion lifetimes. *J. Chem. Phys.*, **45**(12), pp. 4634–4639.
3. Ómarsson, B. (2013). Promoting reaction channels in dissociative electron attachment through bond formation and rearrangement. PhD Thesis, Faculty of Physical Sciences, Department of Chemistry, University of Iceland, Reykjavík.
4. Braun, M., Marienfeld, S., Ruf, M., and Hotop, H. (2009). High-resolution electron attachment to the molecules CCl_4 and SF_6 over extended energy ranges with the (EX) LPA method. *J. Phys. B: At. Mol. Opt. Phys.*, **42**(12), p. 125202.
5. Christophorou, L. G. and Olthoff, J. K. (2001). Electron attachment cross sections and negative ion states of SF_6. *Int. J. Mass Spectrom.*, **205**(1), pp. 27–41.
6. Fenzlaff, H.-P. and Illenberger, E. (1984). Low energy electron impact on benzene and the fluorobenzenes. Formation and dissociation of negative ions. *Int. J. Mass Spectrom. Ion Processes*, **59**(2), pp. 185–202.

7. Gant, K. and Christophorou, L. (1976). Attachment of slow electrons to hexafluorobenzene. *J. Chem. Phys.*, **65**(8), pp. 2977–2981.
8. Jaffke, T., Illenberger, E., Lezius, M., Matejcik, S., Smith, D., and Märk, T. D. (1994). Formation of C_{60}^- and C_{70}^- by free electron capture. Activation energy and effect of the internal energy on lifetime. *Chem. Phys. Lett.*, **226**(1–2), pp. 213–218.
9. Lezius, M., Scheier, P., and Märk, T. (1993). Free electron attachment to C_{60} and C_{70}. *Chem. Phys. Lett.*, **203**(2–3), pp. 232–236.
10. Matejčik, Š., Märk, T. D., Španěl, P., Smith, D., Jaffke, T., and Illenberger, E. (1995). Formation and decay of C_{60}^- following free electron capture by C_{60}. *J. Chem. Phys.*, **102**(6), pp. 2516–2521.
11. Oster, T., Kühn, A., and Illenberger, E. (1989). Gas phase negative ion chemistry. *Int. J. Mass Spectrom. Ion Processes*, **89**(1), pp. 1–72.
12. Prabhudesai, V. S., Kelkar, A. H., Nandi, D., and Krishnakumar, E. (2005). Functional group dependent site specific fragmentation of molecules by low energy electrons. *Phys. Rev. Lett.*, **95**(14), p. 143202.
13. Ptasińska, S., Denifl, S., Grill, V., Märk, T. D., Illenberger, E., and Scheier, P. (2005). Bond-and site-selective loss of H^- from pyrimidine bases. *Phys. Rev. Lett.*, **95**(9), p. 093201.
14. Ptasińska, S., Denifl, S., Mróz, B., Probst, M., Grill, V., Illenberger, E., Scheier, P., and Märk, T. (2005). Bond selective dissociative electron attachment to thymine. *J. Chem. Phys.*, **123**(12), p. 124302.
15. Bald, I., Dąbkowska, I., Illenberger, E., and Ingólfsson, O. (2007). Energy selective excision of CN^- following electron attachment to hexafluoroacetone azine ($(CF_3)_2C{=}N{-}N{=}C(CF_3)_2$). *Phys. Chem. Chem. Phys.*, **9**(23), pp. 2983–2990.
16. Condon, E. (1926). A theory of intensity distribution in band systems. *Phys. Rev.*, **28**(6), p. 1182.
17. Franck, J. and Dymond, E. (1926). Elementary processes of photochemical reactions. *Trans. Faraday Soc.*, **21**, pp. 536–542.
18. Child, M. S. (1978). Condon reflection principle in collision dynamics. *Mol. Phys.*, **35**(3), pp. 759–770.
19. Rapp, D., Sharp, T., and Briglia, D. (1965). Large isotope effect in the formation of H^- or D^- by electron impact on H_2, HD, and D_2. *Phys. Rev. Lett.*, **14**(14), p. 533.
20. Krishnakumar, E., Denifl, S., Čadež, I., Markelj, S., and Mason, N. (2011). Dissociative electron attachment cross sections for H_2 and D_2. *Phys. Rev. Lett.*, **106**(24), p. 243201.

21. Ingólfsson, O., Weik, F., and Illenberger, E. (1996). The reactivity of slow electrons with molecules at different degrees of aggregation: gas phase, clusters and condensed phase. *Int. J. Mass Spectrom. Ion Processes*, **155**(1–2), pp. 1–68.

22. Tegeder, P., Lehmann, L., Ingólfsson, O., and Illenberger, E. (1996). Strong solvation effects in the reactivity of $C_6F_5X^-$ anions (X = Cl, Br, I). Competition between dissociative and associative processes following electron capture. *Z. Phys. Chem.*, **195**(1–2), pp. 217–236.

23. Klar, D., Ruf, M.-W., and Hotop, H. (2001). Dissociative electron attachment to CCl_4 molecules at low electron energies with meV resolution. *Int. J. Mass Spectrom.*, **205**(1), pp. 93–110.

24. Wigner, E. P. (1948). On the behavior of cross sections near thresholds. *Phys. Rev.*, **73**(9), p. 1002.

25. Barsotti, S., Ruf, M.-W., and Hotop, H. (2002). Clear experimental evidence for p-wave attachment-threshold behavior in electron attachment to chlorine molecules. *Phys. Rev. Lett.*, **89**(8), p. 083201.

26. Ruf, M., Barsotti, S., Braun, M., Hotop, H., and Fabrikant, I. (2003). Dissociative attachment and vibrational excitation in low-energy electron collisions with chlorine molecules. *J. Phys. B: At. Mol. Opt. Phys.*, **37**(1), p. 41.

27. Braun, M., Ruf, M.-W., Fabrikant, I. I., and Hotop, H. (2007). Observation of p-wave threshold behavior in electron attachment to F_2 molecules. *Phys. Rev. Lett.*, **99**(25), p. 253202.

28. Vogt, E. and Wannier, G. H. (1954). Scattering of ions by polarization forces. *Phys. Rev.*, **95**(5), p. 1190.

29. Klots, C. E. (1976). Rate constants for unimolecular decomposition at threshold. *Chem. Phys. Lett.*, **38**, pp. 61–64.

30. Fabrikant, I. I. and Hotop, H. (2001). Low-energy behavior of exothermic dissociative electron attachment. *Phys. Rev. A*, **63**(2), p. 022706.

31. Hotop, H., Ruf, M.-W., Allan, M., and Fabrikant, I. (2003). Resonance and threshold phenomena in low-energy electron collisions with molecules and clusters. *Adv. At. Mol. Opt. Phys.*, **49**, pp. 85–216.

32. Fabrikant, I. I. (2016). Long-range effects in electron scattering by polar molecules. *J. Phys. B: At. Mol. Opt. Phys.*, **49**(22), p. 222005.

33. Schulz, G. J. (1973). Resonances in electron impact on diatomic molecules. *Rev. Mod. Phys.*, **45**(3), p. 423.

34. Schulz, G. J. (1973). Resonances in electron impact on atoms. *Rev. Mod. Phys.*, **45**(3), p. 378.

35. Christophorou, L. G. (1984). *Electron—Molecule Interactions and Their Applications*. Vol. 1 and 2. Academic Press.

36. Comer, J. and Read, F. (1971). Electron impact studies of a resonant state of N_2. *J. Phys. B: At. Mol. Phys.*, **4**(8), p. 1055.

37. Gilmore, F. R. (1965). Potential energy curves for N_2, NO, O_2 and corresponding ions. *J. Quant. Spectrosc. Radiat. Transfer*, **5**(2), pp. 369IN1–389IN3.

38. Huetz, A., Čadež, I., Gresteau, F., Hall, R., Vichon, D., and Mazeau, J. (1980). Formation and decay of the first $^2\Pi u$ state of N_2^-. *Phys. Rev. A*, **21**(2), p. 622.

39. Fabrikant, I. I., Eden, S., Mason, N. J., and Fedor, J. (2017). Recent progress in dissociative electron attachment: from diatomics to biomolecules. *Adv. At. Mol. Opt. Phys.*, **66**, pp. 545–657.

40. Rohr, K. and Linder, F. (1976). Vibrational excitation of polar molecules by electron impact. I. Threshold resonance in HF and HCl. *J. Phys. B: At. Mol. Phys.*, **9**(14), p. 2521.

41. Schramm, A., Fabrikant, I., Weber, J., Leber, E., Ruf, M., and Hotop, H. (1999). Vibrational resonance and threshold effects in inelastic electron collisions with methyl iodide molecules. *J. Phys. B: At. Mol. Opt. Phys.*, **32**(9), p. 2153.

42. Braun, M., Fabrikant, I. I., Ruf, M., and Hotop, H. (2007). Low-energy electron collisions with CH_3Br: the dependence of elastic scattering, vibrational excitation, and dissociative attachment on the initial vibrational energy. *J. Phys. B: At. Mol. Opt. Phys.*, **40**(4), p. 659.

43. Hanel, G., Gstir, B., Denifl, S., Scheier, P., Probst, M., Farizon, B., Farizon, M., Illenberger, E., and Märk, T. (2003). Electron attachment to uracil: Effective destruction at subexcitation energies. *Phys. Rev. Lett.*, **90**(18), p. 188104.

44. Scheer, A. M., Aflatooni, K., Gallup, G. A., and Burrow, P. (2004). Bond breaking and temporary anion states in uracil and halouracils: Implications for the DNA bases. *Phys. Rev. Lett.*, **92**(6), p. 068102.

45. Denifl, S., Ptasińska, S., Probst, M., Hrušák, J., Scheier, P., and Märk, T. (2004). Electron attachment to the gas-phase DNA bases cytosine and thymine. *J. Phys. Chem. A*, **108**(31), pp. 6562–6569.

46. Cizek, M., Horacek, J., Allan, M., Fabrikant, I., and Domcke, W. (2003). Vibrational excitation of hydrogen fluoride by low-energy electrons: theory and experiment. *J. Phys. B: At. Mol. Opt. Phys.*, **36**(13), p. 2837.

47. Knoth, G., Gote, M., Rädle, M., Jung, K., and Ehrhardt, H. (1989). Nuclear-excited Feshbach resonances in the electron scattering from hydrogen halides. *Phys. Rev. Lett.*, **62**(15), p. 1735.

48. Ragesh Kumar, T. P., Bjornsson, R., Barth, S., and Ingólfsson, O. (2017). Formation and decay of negative ion states up to 11 eV above the ionization energy of the nanofabrication precursor $HFeCo_3(CO)_{12}$. *Chem. Sci.*, **9**(9), pp. 5949–5952.
49. Ragesh Kumar, T. P., Barth, S., Bjornsson, R., and Ingólfsson, O. (2016). Structure and energetics in dissociative electron attachment to $HFeCo_3(CO)_{12}$. *Eur. Phys. J. D*, **70**(8), p. 163.
50. Ragesh Kumar, T. P., Unlu, I., Barth, S., Ingólfsson, O., and Fairbrother, D. H. (2018). Electron induced surface reactions of $HFeCo_3(CO)_{12}$, a bimetallic precursor for focused electron beam induced deposition (FEBID). *J. Phys. Chem. C*, **122**(5), pp. 2648–2660.
51. Ragesh Kumar, T. P., Weirich, P., Hrachowina, L., Hanefeld, M., Bjornsson, R., Hródmarsson, H. R., Barth, S., Fairbrother, D. H., Huth, M., and Ingólfsson, O. (2018). Electron interactions with the heteronuclear carbonyl precursor $H_2FeRu_3(CO)_{13}$ and comparison with $HFeCo_3(CO)_{12}$: from fundamental gas phase and surface science studies to focused electron beam induced deposition. *Beilstein J. Nanotechnol.*, **9**, pp. 555–579.
52. Porrati, F., Pohlit, M., Muller, J., Barth, S., Biegger, F., Gspan, C., Plank, H., and Huth, M. (2015). Direct writing of CoFe alloy nanostructures by focused electron beam induced deposition from a heteronuclear precursor. *Nanotechnology*, **26**(47).
53. Dunn, G. H. (1962). Anisotropies in angular distributions of molecular dissociation products. *Phys. Rev. Lett.*, **8**(2), p. 62.
54. O'Malley, T. F. and Taylor, H. S. (1968). Angular dependence of scattering products in electron-molecule resonant excitation and in dissociative attachment. *Phys. Rev.*, **176**(1), p. 207.
55. Tronc, M., Schermann, C., Hall, R., and Fiquet-Fayard, F. (1977). Differential cross sections and angular distributions of H^- from dissociative electron attachment to H_2 between 3.75 eV and 13 eV. *J. Phys. B: At. Mol. Phys.*, **10**(2), p. 305.
56. Ómarsson, F., Ingólfsson, O., Mason, N., and Krishnakumar, E. (2012). Dissociative electron attachment to CF_3Cl. *Eur. Phys. J. D*, **66**(2), pp. 1–7.
57. Underwood-Lemons, T., Gergel, T. J., and Moore, J. H. (1995). Dissociative electron attachment cross sections for halofluoromethanes. *J. Chem. Phys.*, **102**(1), pp. 119–123.
58. Underwood-Lemons, T., Winkler, D. C., Tossell, J. A., and Moore, J. H. (1994). Low-energy electron scattering cross sections of halofluorocarbons. *J. Chem. Phys.*, **100**(12), pp. 9117–9122.

59. Read, F. (1968). Angular distributions for resonant scattering of electrons by molecules. *J. Phys. B: At. Mol. Phys.*, **1**(5), p. 893.
60. Ómarsson, F. H. (2014). Symmetries, dynamics and energetics in dissociative electron attachment to selected group IV halides-Velocity Slice Imaging and crossed beam studies. PhD Thesis, Faculty of Physical Sciences, Department of Chemistry, University of Iceland, Reykjavík.
61. Azria, R., Le Coat, Y., Lefevre, G., and Simon, D. (1979). Dissociative electron attachment on H_2S: energy and angular distributions of H^- ions. *J. Phys. B: At. Mol. Phys.*, **12**(4), p. 679.
62. O'Malley, T. (1966). Theory of dissociative attachment. *Phys. Rev.*, **150**(1), p. 14.
63. Nandi, D., Prabhudesai, V. S., Krishnakumar, E., and Chatterjee, A. (2005). Velocity slice imaging for dissociative electron attachment. *Rev. Sci. Instrum.*, **76**(5), p. 053107.
64. Adaniya, H., Rudek, B., Osipov, T., Haxton, D., Weber, T., Rescigno, T. N., McCurdy, C., and Belkacem, A. (2009). Imaging the molecular dynamics of dissociative electron attachment to water. *Phys. Rev. Lett.*, **103**(23), p. 233201.
65. Nag, P. and Nandi, D. (2015). Dissociation dynamics in the dissociative electron attachment to carbon dioxide. *Phys. Rev. A*, **91**(5), p. 052705.
66. Weyland, M., Ren, X., Pflüger, T., Dornt, A., Baek, W. Y., and Rabus, H. (2015). Momentum imaging of dissociative electron attachment in biologically relevant molecules. *J. Phys. Conf. Ser.*, **635**, p. 072013.
67. Wu, B., Xia, L., Wang, Y.-F., Li, H.-K., Zeng, X.-J., and Tian, S. X. (2012). Renner-teller effect on dissociative electron attachment to carbon dioxide. *Phys. Rev. A*, **85**(5), p. 052709.
68. Ómarsson, F. H., Szymanska, E., Mason, N. J., Krishnakumar, E., and Ingólfsson, O. (2013). Quantum superposition of target and product states in reactive electron scattering from CF_4 revealed through velocity slice imaging. *Phys. Rev. Lett.*, **111**(6), p. 063201.
69. Haxton, D., Adaniya, H., Slaughter, D., Rudek, B., Osipov, T., Weber, T., Rescigno, T., McCurdy, C., and Belkacem, A. (2011). Observation of the dynamics leading to a conical intersection in dissociative electron attachment to water. *Phys. Rev. A*, **84**(3), p. 030701.
70. Krishnakumar, E., Prabhudesai, V. S., and Mason, N. J. (2017). Symmetry breaking by quantum coherence in single electron attachment. *Nat. Phys.*, **14**, pp. 149–153.

71. Slaughter, D., Belkacem, A., McCurdy, C., Rescigno, T., and Haxton, D. (2016). Ion-momentum imaging of dissociative attachment of electrons to molecules. *J. Phys. B: At. Mol. Opt. Phys.*, **49**(22), p. 222001.

72. Ómarsson, F. H., Mason, N. J., Krishnakumar, E., and Ingólfsson, O. (2014). State selectivity and dynamics in dissociative electron attachment to CF_3I revealed through velocity slice imaging. *Angew. Chem. Int. Ed.*, **53**(45), pp. 12051–12054.

73. Mulliken, R. S. (1940). Intensities in molecular electronic spectra X. Calculations on mixed-halogen, hydrogen halide, alkyl halide, and hydroxyl spectra. *J. Chem. Phys.*, **8**(5), pp. 382–395.

74. Gedanken, A. (1987). The magnetic circular dichroism of the A band in CF_3I, C_2H_5I and t-BuI. *Chem. Phys. Lett.*, **137**(5), pp. 462–466.

75. Lokhman, V., Ogurok, D., and Ryabov, E. (2008). Multiphoton ionization of iodine atoms and CF_3I molecules by XeCl laser radiation. *Eur. Phys. J. D*, **46**(1), pp. 59–67.

76. Mason, N., Vieira, P. L., Eden, S., Kendall, P., Pathak, S., Dawes, A., Tennyson, J., Tegeder, P., Kitajima, M., and Okamoto, M. (2003). VUV and low energy electron impact study of electronic state spectroscopy of CF_3I. *Int. J. Mass Spectrom.*, **223**, pp. 647–660.

77. Steinfeld, J., Zare, R., Jones, L., Lesk, M., and Klemperer, W. (1965). Spectroscopic constants and vibrational assignment for the $B_3\Pi_{0_u}+$ state of iodine. *J. Chem. Phys.*, **42**(1), pp. 25–33.

78. Mead, R. D., Stevens, A. E., and Lineberger, W. C. (1984). In *Gas Phase Ion Chemistry*. Bowers, M. T., ed., Academic Press, Orlando.

79. Ómarsson, B. and Ingólfsson, O. (2013). Stabilization, fragmentation and rearrangement reactions in low-energy electron interaction with tetrafluoro-para-benzoquinone: a combined theoretical and experimental study. *Phys. Chem. Chem. Phys.*, **15**(39), pp. 16758–16767.

80. Ómarsson, B., Bjornsson, R., and Ingólfsson, O. (2017). Proton shuttling and reaction paths in dissociative electron attachment to o- and p-tetrafluorohydroquinone, an experimental and theoretical study. *J. Phys. Chem. A*, **121**(30), pp. 5580–5585.

81. Lindinger, W., Märk, T. D., and Howorka, F. (2012). Swarms of ions and electrons in gases. Springer Science & Business Media.

82. Chutjian, A., Garscadden, A., and Wadehra, J. M. (1996). Electron attachment to molecules at low electron energies. *Phys. Rep.*, **264**(6), pp. 393–470.

83. Hunter, S. and Christophorou, L. (1984). Electron attachment to the perfluoroalkanes ${}_nCNF_{N+2}$ (N = 1–6) using high pressure swarm techniques. *J. Chem. Phys.*, **80**(12), pp. 6150–6164.

84. Datskos, P., Christophorou, L., and Carter, J. (1993). Temperature-enhanced electron detachment from $C_6F_6^-$ negative ions. *J. Chem. Phys.*, **98**(10), pp. 7875–7882.

85. Christophorou, L., McCorkle, D., and Anderson, V. (1971). Swarm-determined electron attachment cross sections as a function of electron energy. *J. Phys. B: At. Mol. Phys.*, **4**(9), pp. 1163.

86. Spyrou, S. and Christophorou, L. (1985). Effect of temperature on the dissociative electron attachment to $CClF_3$ and C_2F_6. *J. Chem. Phys.*, **82**(6), pp. 2620–2629.

87. Datskos, P., Christophorou, L., and Carter, J. (1993). Temperature dependence of electron attachment and detachment in SF_6 and c-C_4F_6. *J. Chem. Phys.*, **99**(11), pp. 8607–8616.

88. Davis, F., Compton, R., and Nelson, D. (1973). Thermal energy electron attachment rate constants for some polyatomic molecules. *J. Chem. Phys.*, **59**(5), pp. 2324–2329.

89. Brouillard, F. and McGowan, J. W. (1983). *Physics of Ion-Ion and Electron-Ion Collisions*. Vol. 83. Springer Science & Business Media.

90. Smith, D. and Adams, N. (1984). Elementary interactions between charged and neutral species in plasmas. *Pure Appl. Chem.*, **56**(2), pp. 175–188.

91. Smith, D., Adams, N., Dean, A., and Church, M. (1975). The application of Langmuir probes to the study of flowing afterglow plasmas. *J. Phys. D: Appl. Phys.*, 1975. **8**(2), p. 141.

92. Smith, D. and Plumb, I. (1972). An appraisal of the single Langmuir probe technique in the study of afterglow plasmas. *J. Phys. D: Appl. Phys.*, **5**(7), p. 1226.

93. Shimamori, H. and Hatano, Y. (1976). Thermal electron attachment to O_2 in H_2 and D_2. *Chem. Phys. Lett.*, **38**(2), pp. 242–247.

94. Shimamori, H. and Hotta, H. (1986). Examination of the effects of van der Waals molecules on the thermal electron attachment to NO_2 at relatively high pressures. *J. Chem. Phys.*, **85**(2), pp. 887–889.

95. Toriumi, M. and Hatano, Y. (1985). Thermal electron attachment to van der Waals molecules ($O_2 \cdot N_2$). *J. Chem. Phys.*, **82**(1), pp. 254–259.

96. Adams, N., Church, M., and Smith, D. (1975). An experimental and theoretical investigation of the dynamics of a flowing afterglow plasma. *J. Phys. D: Appl. Phys.*, **8**(12), p. 1409.

97. Smith, D., Adams, N., and Alge, E. (1984). Attachment coefficients for the reactions of electrons with CCl_4, CCl_3F, CCl_2F_2, $CHCl_3$, Cl_2 and SF_6

determined between 200 and 600 K using the FALP technique. *J. Phys. B: At. Mol. Phys.*, **17**(3), p. 461.

98. Smith, D. and Ŝpanêl, P. (1994). Studies of electron attachment at thermal energies using the flowing Afterglow–Langmuir probe technique. *Adv. At. Mol. Opt. Phys.*, **32**, pp. 307–343.

99. Shimamori, H. and Fessenden, R. W. (1979). Electron attachment to N_2O at pressures near one atmosphere. *J. Chem. Phys.*, **71**(7), pp. 3009–3015.

100. Shimamori, H., Tatsumi, Y., Ogawa, Y., and Sunagawa, T. (1992). Low-energy electron attachment to molecules studied by pulse-radiolysis microwave-cavity technique combined with microwave heating. *J. Chem. Phys.*, **97**(9), pp. 6335–6347.

101. Shimamori, H., Tatsumi, Y., Ogawa, Y., and Sunagawa, T. (1992). Electron-energy dependence of electron attachment to c-C_7F_{14}, CH_3I and CH_2Br_2 studied by the pulse-radiolysis microwave-cavity technique combined with microwave heating. *Chem. Phys. Lett.*, **194**(3), pp. 223–229.

102. Sunagawa, T. and Shimamori, H. (1995). Low energy electron attachment to brominated ethanes and ethylenes. *Int. J. Mass Spectrom. Ion Processes*, **149**, pp. 123–129.

103. Sunagawa, T. and Shimamori, H. (1997). Low-energy electron attachment to brominated methanes. *J. Chem. Phys.*, **107**(19), pp. 7876–7883.

104. Shimamori, H., Tatsumi, Y., and Sunagawa, T. (1993). Thermal electron attachment to C_6F_5X and C_6H_5X (X=I, Br, Cl, and F). *J. Chem. Phys.*, **99**(10), pp. 7787–7792.

105. Bjarnason, E. H., Ómarsson, B., Engmann, S., Ómarsson, F. H., and Ingólfsson, O. (2014). Dissociative electron attachment to titatinum tetrachloride and titanium tetraisopropoxide. *Eur. Phys. J. D*, **68**(5), p. 121.

106. Matejčık, Š., Foltin, V., Stano, M., and Skalný, J. (2003). Temperature dependencies in dissociative electron attachment to CCl_4, CCl_2F_2, $CHCl_3$ and $CHBr_3$. *Int. J. Mass Spectrom.*, **223**, pp. 9–19.

107. Kopyra, J., König-Lehmann, C., Illenberger, E., Warneke, J., and Swiderek, P. (2016). Low energy electron induced reactions in fluorinated acetamide–probing negative ions and neutral stable counterparts. *Eur. Phys. J. D*, **70**(6), pp. 1–7.

108. Kopyra, J., Maciejewska, P., and Maljković, J. (2017). Dissociative electron attachment to coordination complexes of chromium: chromium

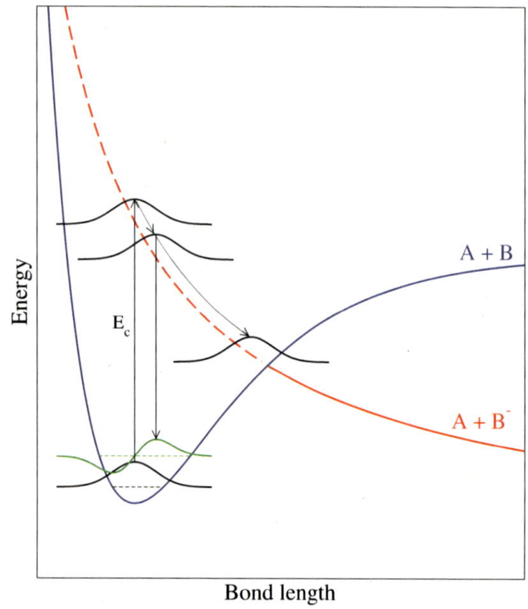

Figure 3.1 Schematics of resonant vibrational excitation and DEA processes. See text for details.

or its dissociation, for example, via dissociative electron attachment (DEA). Resonances can be characterized as one-electron resonances, that is, shape Resonances, or two-particle one-hole resonances (core-excited shape and Feshbach resonances).

One way to understand the role of resonances in molecular nuclear dynamics initiated by low-energy electron impact is to look at shape resonances and the diagram shown in Fig. 3.1. The blue line denotes the potential curve of the neutral target molecule, while the red line is for the negative ion. For the DEA process to occur it is necessary that the additional electron in the system stabilizes the molecule at larger internuclear distances, that is, one of the molecular fragments (fragment B in Fig. 3.1) has a positive electron affinity. Hence the full red line describes the energy of the stable negative ion, while the dashed red line shows the energy of the resonance immersed in the continuum. Figure 3.1 depicts a process in which a nuclear wave packet corresponding to the (nuclear) ground state of the target molecule gets vertically excited on to the dissociative resonant curve of the transient negative ion. Then, depending on the lifetime of the resonant state, the nuclear wave function may, during its evolution, reach distances beyond the crossing of the neutral and anionic curves, leading to a bound negative ion system and dissociation. Another possible path is for the scattered electron to autodetach shortly after the initial vertical excitation; in this case, the short evolution on the resonant surface favors final vibrationally excited states (according to the Franck–Condon principle [1, 2]). Thus, the resonant mechanism provides also a very efficient way to transfer the scattered electron energy (cooling of free electrons in a plasma environment) to the vibrational degrees of freedom of the target molecule (heating of the molecules).

This qualitative description of the nuclear dynamics driven by a resonant scattering process (proposed half a century ago [3]) indicates that the theoretical study of electron impact–induced nuclear motion can be performed in two steps. The first step involves the calculation of electronic properties (energies) for a fixed molecular geometry (fixed-nuclei approximation). The resulting energy surfaces are then utilized in the second step to describe the evolution of the nuclear wave function. In contrast to quantum chemistry applications, the electronic potential energy surfaces are often complex (i.e., the resonances have both an energy and a lifetime, or width, associated to them), allowing for the autodetachment process. Moreover, complex energies alone are not generally sufficient as some of the more elaborated models (e.g., nonlocal resonance model [4]) require quantities calculated from scattering wave functions.

The calculation of the real (i.e., bound) segment of the potential energy surfaces is, at present, a reasonably straightforward task. These calculations describe a stable negative ion or a neutral molecule and at least a dozen commercially available quantum chemistry packages are up to the task. On the other hand, the situation is less straightforward for the complex (resonant) parts of the surfaces. The electronic resonant states are not directly linked to bound states that can be obtained by variational or nonvariational methods used in quantum chemistry computations. The molecular Hamiltonian often requires a modification in the form of a complex absorbing potential [5], rotation of electronic coordinates into a complex plane (complex scaling methods [6, 7], or an additional external field that binds the resonant state (analytical continuation in coupling constant methods [8]). A recent review summarizes some of these techniques [9].

Another option for determination of the resonance parameters (i.e., energy and width), avoiding altering the Hamiltonian, is to carry out scattering calculations. In scattering calculations the resonant state is searched directly among the continuum states. Solving the scattering problem can be done within a time-dependent approach (in the context of electron–molecule scattering, this has been done for ionization of H_2^+) or within a time-independent approach. The latter is more usual in low-energy scattering.

In practice, as we will see in this chapter, this entails the calculation of a quantity (scattering amplitude, K-matrix, S-matrix, T-matrix) that contains all the information regarding the collision. These quantities are all related, and from them, one can calculate cross sections (integral and differential), determine the parameters (energy and width) of a resonance, etc. For example, the scattering amplitudes $f_{\mathbf{k}_i,\mathbf{k}_f}$ (see Section 3.2.1) allow us to directly calculate the angular differential cross section for scattering from channel i to channel j:

$$\frac{\mathrm{d}\sigma_{ij}}{\mathrm{d}\Omega} = |f_{\mathbf{k}_i,\mathbf{k}_f}|^2 \tag{3.1}$$

Similarly, the T-matrices (see Section 3.3.2) allow us to calculate integral cross sections:

$$\sigma_{ij} = \frac{\pi}{k_i^2}\sum \frac{2S+1}{2(2s_i+1)}|T_{ij}(E)|^2, \tag{3.2}$$

where T_{ij} are elements of the T-matrix, S is the total spin of the target + electron system, and s_i is the spin of the target state associated to channel i. (Note that to obtain a cross section that can be compared to experimental results, one needs to sum over all the channels associated to a specific initial and final target state).

Finally, T-, S-, and K-matrices are trivially related:

$$\mathbf{T} = \mathbf{S} + \mathbf{1} = \frac{2i\mathbf{K}}{\mathbf{1} - i\mathbf{K}}, \tag{3.3}$$

so once the K-matrix is determined, differential and integral cross sections can be trivially calculated.

The resonance parameters can be obtained in a number of ways. One possibility is to calculate the eigenphase sum from the K-matrices,

$$\delta_{\text{sum}}(E) = \sum_i \arctan(K_{ii}(E)), \tag{3.4}$$

and then fit it using the Breit–Wigner formula [10] that uses the fact that an isolated resonance manifests itself as a jump by π in the eigenphase sum. Another possibility is to use the S-matrix to determine the time delay [11]. The presence of a resonance will show as a Lorentzian peak in the eigenvalues of the time delay matrix. Again, fitting these peaks one can obtain the energy and width of the resonance. Finally, it is also possible to extract resonance parameters directly from the S-matrix, but this in principle requires the determination of the S-matrix for complex energies, so this approach is rarely used.

In this chapter, we present three low-energy methods to solve the scattering problem. Other methodologies are available: for example, the complex Kohn variational method [12, 13] used by the UC Davis/LBNL group, who have had significant success studying dissociative electron attachment to small molecules, or the single centre expansion approach [14] of the Rome group that has been applied to study a number of collisions with biologically relevant molecules.

In addition, electron scattering from molecules at higher energies (at least a few tens of electron volts above the ionization threshold) can be studied using simpler techniques. These methods (e.g., the independent-atom model with screening-corrected additivity rule [IAM-SCAR] method [15] developed by the Madrid group) are geared toward the calculation cross sections (elastic, electronically inelastic, ionization), since they tend to fail for those scattering energies for which resonance formation is significant.

3.1.1 Notation

Tables 3.1 and 3.2 summarize the abbreviations and symbols regularly used in this chapter.

Table 3.1 Abbreviations used in this chapter

HF	Hartree–Fock
CSF	configuration state function
SE	static-exchange
SEP	static-exchange plus polarization
ICS	integral cross section
DCS	differential cross section

Table 3.2 Symbols used frequently in this chapter and their meaning

Φ_m	electronic wave function for state m of the N-electron target
ε_m	energy of state m of the N-electron target
H_N	Hamiltonian for the N-electron target
$H = H(\mathbf{x}_i, \ldots .\mathbf{x}_{N+1})$	Hamiltonian for the $N+1$-electron (target + projectile) system
$V = V(\mathbf{x}_i, \ldots .\mathbf{x}_{N+1})$	electron–molecule scattering potential
$E, \mathbf{k}$	total electronic energy and wave vector of the $N+1$-electron system
$\mathbf{k}_i$	wave vector of the scattering electron in channel i
$f_{\mathbf{k}_i,\mathbf{k}_f}$	scattering amplitude for initial channel i and final channel f
$\mathbf{r}_i$	coordinate of the ith electron
$\mathbf{r}_{N+1}$	coordinate of the scattering electron
g_α	Gaussian-type orbital with exponent α

3.2 Schwinger Multichannel Method

The Schwinger multichannel method (SMC) is a variational approach to the scattering amplitude based on the Lippmann–Schwinger integral equation. The method was originally designed to account for the electronic excitation channels in electron–molecule collisions and can be viewed as an alternative derivation of the Schwinger variational principle (SVP) suitable for numerical implementation. The key differences between the usual SVP and the SMC arise from the definition of a projection operator onto the energetically open electronic states of the target molecule,

$$P = \sum_{m=0}^{\text{open}} |\Phi_m\rangle\langle\Phi_m| \,, \tag{3.5}$$

where $m = 0$ denotes the target ground state. The P operator, introduced by Takatsuka and McKoy [16, 22], plays a central role in the SMC method as it defines the electronic channels that will be included in the calculation. In the following section, we neglect the vibrational motion and ionization channels, such that the asymptotic behavior of the scattering wave function $\Psi^{(+)}_{\mathbf{k}_i}$ is given by

$$\Psi^{(+)}_{\mathbf{k}_i}(\mathbf{r}_1, \cdots, \mathbf{r}_{N+1}) \overset{r_{N+1}\to\infty}{\longrightarrow} S_{\mathbf{k}_i}(\mathbf{r}_1, \cdots, \mathbf{r}_{N+1})$$

$$+ \sum_{f}^{\text{open}} f_{\mathbf{k}_i,\mathbf{k}_f} \Phi_f(\mathbf{r}_1, \cdots, \mathbf{r}_N) \frac{\exp(ik_f r_{N+1})}{r_{N+1}}, \tag{3.6}$$

where $f_{\mathbf{k}_i,\mathbf{k}_f}$ is the scattering amplitude, $\mathbf{k}_{i(j)}$ is the incident (final) wave vector of the scattering particle, and Φ_f is an electronic target state, such that energy conservation imposes the condition $k_i^2 + 2\varepsilon_i = k_f^2 + 2\varepsilon_f$ for the f-th collision channel, where $\varepsilon_{i(f)}$ is the energy of the initial (final) target state. $\mathbf{r}_i$ denotes the electronic coordinates (spin degrees of freedom are omitted for simplicity), N is the number of target electrons, such that $i = 1, \cdots, (N+1)$, and $S_{\mathbf{k}_i} = \Phi_i(\mathbf{r}_1, \cdots, \mathbf{r}_N)\exp(i\mathbf{k}_i \cdot \mathbf{r}_{N+1})$ is an asymptotic state, that is, a solution of the non-interacting Hamiltonian,

$$H_0 = T_{N+1} + H_N, \tag{3.7}$$

where T_{N+1} is the kinetic energy of the incident electron and H_N is the Hamiltonian of the N-electron target.

3.2.1 Schwinger Variational Principle

The scattering amplitude can be written in two equivalent forms employing scattering states with the outgoing (+) or the incoming (−) boundary conditions (see, for instance, [17]),

$$f_{\mathbf{k}_i,\mathbf{k}_f} = -\frac{1}{2\pi}\langle S_{\mathbf{k}_f}|V|\Psi^{(+)}_{\mathbf{k}_i}\rangle = -\frac{1}{2\pi}\langle \Psi^{(-)}_{\mathbf{k}_f}|V|S_{\mathbf{k}_i}\rangle. \tag{3.8}$$

A third expression for the scattering amplitude can be obtained from the Lippmann–Schwinger equation,

$$|\Psi^{(\pm)}_{\mathbf{k}_i}\rangle = |S_{\mathbf{k}_i}\rangle + G_0^{(\pm)} V |\Psi^{(\pm)}_{\mathbf{k}_i}\rangle, \tag{3.9}$$

where V is the electron–molecule scattering potential, given by the sum over all electron–electron and electron–nuclei Coulomb interactions, and $G_0^{(\pm)}$ is the free-particle Green's operator,

$$G_0^{(\pm)} = \lim_{\eta \to 0} \frac{1}{E - H_0 \pm i\eta} \,. \tag{3.10}$$

By multiplying both sides of Eq. 3.9 by V, choosing the outgoing condition (+) for convenience and substituting $V|S_{\mathbf{k}_i}\rangle$ into the last expression in Eq. 3.8, we obtain

$$f_{\mathbf{k}_i,\mathbf{k}_f} = -\frac{1}{2\pi} \langle \Psi_{\mathbf{k}_f}^{(-)} | A^{(+)} | \Psi_{\mathbf{k}_i}^{(+)} \rangle \,, \tag{3.11}$$

with

$$A^{(+)} = V - V G_0^{(+)} V \,. \tag{3.12}$$

From the three expressions given in Eqs. 3.8 and 3.11, the scattering amplitude can be written in the following exact form:

$$f_{\mathbf{k}_f,\mathbf{k}_i} = -\frac{1}{2\pi} \left[\langle S_{\mathbf{k}_f} | V | \Psi_{\mathbf{k}_i}^{(+)} \rangle + \langle \Psi_{\mathbf{k}_f}^{(-)} | V | S_{\mathbf{k}_i} \rangle - \langle \Psi_{\mathbf{k}_f}^{(-)} | A^{(+)} | \Psi_{\mathbf{k}_i}^{(+)} \rangle \right] \,, \tag{3.13}$$

which can be viewed as a functional of the scattering states. By taking an arbitrary variation of the bra, $\langle \Psi_{\mathbf{k}_f}^{(-)} | \rightarrow \langle \Psi_{\mathbf{k}_f}^{(-)} | + \langle \delta \Psi_{\mathbf{k}_f}^{(-)} |$, we readily conclude that the scattering amplitude has extrema ($\delta f_{\mathbf{k}_f,\mathbf{k}_i} = 0$) at the scattering eigenstates,

$$\delta f_{\mathbf{k}_f,\mathbf{k}_i} = 0 \quad \text{if} \quad A^{(+)} | \Psi_{\mathbf{k}_i}^{(+)} \rangle = V | S_{\mathbf{k}_i} \rangle \,, \tag{3.14}$$

where the Lippmann–Schwinger equation multiplied by V should be recognized with the help of Eqs. 3.9 and 3.12. Let us now consider the variation of the ket, $|\Psi_{\mathbf{k}_i}^{(+)}\rangle \rightarrow |\Psi_{\mathbf{k}_i}^{(+)}\rangle + |\delta\Psi_{\mathbf{k}_i}^{(+)}\rangle$. The extremum condition is now obtained as

$$\delta f_{\mathbf{k}_f,\mathbf{k}_i} = 0 \quad \text{if} \quad \langle \Psi_{\mathbf{k}_f}^{(-)} | A^{(+)} = \langle S_{\mathbf{k}_f} | V \implies A^{(+)\dagger} | \Psi_{\mathbf{k}_f}^{(-)} \rangle = V | S_{\mathbf{k}_f} \rangle \,. \tag{3.15}$$

The incoming scattering states $\Psi_{\mathbf{k}_f}^{(-)}$ will be extrema of the functional if the last expression is equivalent to the Lippmann–Schwinger equation with the (−) boundary condition, that is, provided that

$$A^{(+)\dagger} = A^{(-)} \,. \tag{3.16}$$

Let us now assume that the scattering state can be expanded in some arbitrary trial basis $\{|\chi_\mu\rangle\}$, such that

$$|\Psi_{\mathbf{k}}^{(\pm)}\rangle = \sum_{\mu} c_{\mu}^{(\pm)}(\mathbf{k}) \, |\chi_\mu\rangle \,. \tag{3.17}$$

By substituting the above expansion in Eq. 3.13 and imposing the stationary condition $\delta f_{\mathbf{k}_i,\mathbf{k}_f} = 0$, analytical expressions for the variational coefficients $c_\mu^{(+)}(\mathbf{k}_i)$ and $c_\mu^{(-)}(\mathbf{k}_f)$ can be readily obtained. Substitution of these expressions back into Eq. 3.13 leads to a representation of the scattering amplitude in the trial basis,

$$f_{\mathbf{k}_i,\mathbf{k}_f} = -\frac{1}{2\pi}\sum_{\mu\nu}\langle S_{\mathbf{k}_f}|V|\chi_\mu\rangle\,(d^{-1})_{\mu\nu}\,\langle\chi_\nu|V|S_{\mathbf{k}_i}\rangle\,, \tag{3.18}$$

with

$$d_{\mu\nu} = \langle\chi_\mu|A^{(+)}|\chi_\nu\rangle\,. \tag{3.19}$$

3.2.2 Schwinger Multichannel Method

Equations 3.13 and 3.18 raise a question on the possibility of obtaining a solution for the electron–molecule collision problem from the SVP. In any practical sense, the answer would be negative in view of the following. To represent the Green's operator in Eq. 3.10, we should employ the identity operator in the $(N+1)$-electron space from the complete sets of plane waves and of target eigenstates,

$$G_0^{(\pm)} = \lim_{\eta\to 0}\sum_m\!\!\!\!\!\!\int\int d^3k\frac{|\Phi_m\mathbf{k}\rangle\langle\mathbf{k}\Phi_m|}{\frac{1}{2}k_m^2 - \frac{1}{2}k^2 \pm i\eta}\,, \tag{3.20}$$

where $k_m = [k_i^2 - 2(\varepsilon_m - \varepsilon_i)]^{1/2}$ arises from the energy conservation.[a] The symbol $\sum\!\!\!\!\int$ stands for the sum over the discrete target spectrum (electronic bound states) and integration over the continuum spectrum (ionization states). The inclusion of the whole spectrum is essential from the formal standpoint, as the continuum states are necessary to ensure the antisymmetry on the right-hand side of Eq. 3.9, as pointed out by Geltman [18].[b] However, the numerical description of the complete spectrum would obviously be impractical, and this is the reason why the projector operator

[a] In Eq. 3.10 and the collision energy is given by $E = k_i^2/2 + \varepsilon_i$. From the Hamiltonian matrix elements in the $\{|\Phi_m\mathbf{k}\rangle\}$ basis, the denominator becomes $k_i^2/2 + \varepsilon_i - (k^2/2 + \varepsilon_m)$. Introducing the final wave vector in the m-th collision channel, $k_m = [k_i^2 - 2(\varepsilon_m - \varepsilon_i)]^{1/2}$, we obtain Eq. 3.20.

[b] Note that the $(N+1)$-th electron is not antysimmetrized to the target electrons in the $|S_{\mathbf{k}_i}\rangle$ asymptotic state.

in Eq. 3.5 comes into play. If we project the Lippmann–Schwinger equation on the P space,

$$P|\Psi_{\mathbf{k}_i}^{(\pm)}\rangle = |S_{\mathbf{k}_i}\rangle + G_P^{(\pm)}V|\Psi_{\mathbf{k}_i}^{(\pm)}\rangle\,, \qquad (3.21)$$

where $P|S_{\mathbf{k}_i}\rangle = |S_{\mathbf{k}_i}\rangle$ and $G_P^{(\pm)} = PG_0^{(\pm)}$, and follow the same steps that lead to Eq. 3.13, the SVP will be obtained with the operator $A^{(\pm)} = PV - VG_P^{(\pm)}V$. However, since $PV \neq VP$, the latter will not satisfy the condition in Eq. 3.16.

The central point in the development of the SMC approach is obtaining the $A^{(\pm)}$ operator in a form that incorporates the projected Green's operator without violating the stability condition. To this end, the scattering Schrödinger equation can be written in the following form [19],

$$(E - H)\,[(1 - aP) + aP]\,|\Psi_{\mathbf{k}_i}^{(\pm)}\rangle = 0\,, \qquad (3.22)$$

which is exact for any value of the real arbitrary parameter a, as the $(1 - aP)$ term recovers the information on the closed channels projected out by aP. By noting that $[H_0, P] = 0$ and $G_0^{(\pm)}H_0 = 1$, we obtain, after some manipulation of Eq. 3.22, the alternative form

$$A^{(\pm)}\,|\Psi_{\mathbf{k}_i}^{(\pm)}\rangle = V|S_{\mathbf{k}_i}\rangle\,, \qquad (3.23)$$

with

$$A^{(\pm)} = \frac{1}{2}(PV + VP) - VG_P^{(\pm)}V + \frac{1}{a}\left[\hat{H} - \frac{a}{2}\left(\hat{H}P + P\hat{H}\right)\right]\,, \qquad (3.24)$$

and $\hat{H} = (E - H)$. This expression is equivalent to the Lippmann–Schwinger equation multiplied by V, according to Eq. 3.14, although now the $A^{(\pm)}$ operator conveniently incorporates the symmetrized potential $(PV + VP)$ and the projected Green's operator. To make sure that the condition in Eq. 3.16 will be fulfilled, an advantage may be taken from the parameter a. We assume that the trial basis employed in the computations will have the form

$$|\chi_\mu\rangle \equiv |\chi_{ij}\rangle = \mathcal{A}|\Phi_i\rangle|\varphi_j\rangle\,, \qquad (3.25)$$

where Φ_i is an N-electron target state, φ_j is a single-electron state, and the operator $\mathcal{A}$ imposes the antisymmetrization among the $(N + 1)$ electrons.[c] We further assume that the $\{\varphi_j\}$ basis comprises

---

[c]We avoid the double-index notation, χ_{ij}, to keep the matrix notation as simple as possible. However, in Eq. 3.25 we point out that the subscript μ actually accounts for the target state (i) and the scattering orbital (j) that build up the $(N + 1)$-electron configuration χ_μ

both square integrable orbitals and continuum orbitals with the proper scattering boundary condition. In principle, the first would be convenient to describe short-range effects, such as exchange and correlation, while the latter would account for long-range interactions and the boundary condition. Let us now consider the representation of the $A^{(\pm)}$ operator in this trial basis. The variational stability condition, Eq. 3.16, will be immediately satisfied for the $\langle \chi_\mu|(PV + VP)|\chi_\nu\rangle$ matrix elements, since the interaction operator has been properly symmetrized, and also in view of the asymptotic behavior $V \to 0$, as the $r \to \infty$ limit is taken for any electron.[d] Likewise, since the Green's operator satisfies $G_0^{(+)\dagger} = G_0^{(-)}$ [17], it may be easily realized that the $\langle \chi_\mu|VG_P^{(\pm)}V|\chi_\nu\rangle$ matrix elements are also compatible with the stability condition. The third term on the right-hand side of Eq. 3.24 would be a little more complicated because $\hat{H}$ contains the T_{N+1} kinetic energy operator that would give rise to surface terms in the continuum–continuum matrix elements[e] that is, in case $\varphi_\mu(\mathbf{r}_{N+1})$ and $\varphi_\nu(\mathbf{r}_{N+1})$ are both continuum orbitals. Recalling that Eqs. 3.22 to 3.24 are exact for any value of the arbitrary parameter, it is straightforward to verify that $a = (N+1)$ zeroes out the contribution from the continuum–continuum matrix elements, such that a working expression for the SMC scattering amplitude is given by Eqs. 3.18 and 3.24, with $a = (N+1)$.

3.2.3 Implementation

As the SMC method is based on the integral formulation of the collision problem, the Green's function accounts for the scattering boundary condition. In addition, the matrix elements in Eq. 3.18, with $A^{(+)}$ given in Eq. 3.24, involve operators that vanish asymptotically, as $r_{N+1} \to \infty$. The only exception would be the T_{N+1} kinetic energy operator, but the choice for the arbitrary parameter a has zeroed out the contribution from the continuum–continuum matrix elements. As a result, the trial basis $\{\chi_\mu\}$ does not *necessarily*

---

[d]Since we assume the interaction of an electron with a neutral molecule, V asymptotically tends to zero as r^{-k} with $k \geq 2$.

[e]In case $\varphi_{\mu,\nu}(\mathbf{r})$ are continuum orbitals, the integrand of the matrix element $\int d\mathbf{r}\, \varphi_\mu^*(\mathbf{r})\nabla^2\varphi_\nu(\mathbf{r})$ does not vanish as $r \to \infty$, thus giving rise to the surface terms. Since $\langle \varphi_\mu|T_{N+1}|\varphi_\nu\rangle \neq \langle \varphi_\nu|T_{N+1}|\varphi_\mu\rangle$, the stability condition would be violated.

need to include continuum states. Even though the scattering orbitals would be more diffuse than those representing the bound electrons, a square-integrable trial basis can be employed in the SMC approach, that is, the $\{\varphi_\mu\}$ set in Eq. 3.25 may only comprise L^2 orbitals.

In practice, the bound and scattering orbitals are expanded in Gaussian-type orbitals (GTOs), since these are often employed in electronic structure calculations. In principle, we would be inclined to use highly sophisticated methods to describe the isolated target molecule. However, care must be taken to balance accuracy with numerical effort, as the scattering states are more computationally demanding than bound states. In electron scattering from biomolecules, we are usually concerned with the description of temporary negative ion states, namely shape and core-excited resonances that may lead to dissociative attachment and resonant vibrational excitation. An accurate calculation thus requires a balanced description of electronic correlation in the $(N+1)$- and N-electron states. When a high-level method is employed to describe the isolated target, a similar description might not be possible for the scattering state, due to computational limitations. As a result, the energy of the anion states would be overestimated, as well as the widths. These resonance parameters would likewise be underestimated in the case when the anion state was overcorrelated with respect to the target state. In this sense, it is convenient to describe the target molecule at the Hartree–Fock (HF) level, in view of the low cost. The simplest approximation to the scattering solution would be

$$|\Psi_{\mathbf{k}}^{(\pm)}\rangle = \sum_{\mu} c_{0\,\mu}^{(\pm)}(\mathbf{k})\,\mathcal{A}|\Phi_0\rangle|\varphi_\mu\rangle\,, \tag{3.26}$$

where Φ_0 is the HF target ground state, $c_{0\,\mu}^{(\pm)}(\mathbf{k})$ is a variational coefficient, and φ_μ is a scattering orbital, as in Eq. 3.25. The spin adapted $(N+1)$-electron Slater determinants, given by products of target states and scattering orbitals, are called configuration state functions (CSFs). We refer to Eq. 3.26 as the static-exchange (SE) approximation, since the target is frozen in the ground state, implying no dynamical response to the incident electron. To improve the description of the anion state, though assuming elastic collisions for the time being ($P = |\Phi_0\rangle\langle\Phi_0|$), we take into account virtual excitations of the target by promoting a single electron from a hole orbital to a particle orbital. Denoting this singly excited virtual state by Φ_i, we obtain the static-exchange plus polarization (SEP) approximation,

$$|\Psi_{\mathbf{k}}^{(\pm)}\rangle = \sum_{\mu} c_{0\mu}^{(\pm)}(\mathbf{k})\, \mathcal{A}|\Phi_0\rangle|\varphi_\mu\rangle + \sum_{i\nu} c_{i\nu}^{(\pm)}(\mathbf{k})\, \mathcal{A}|\Phi_i\rangle|\varphi_\nu\rangle\,. \quad (3.27)$$

The number of excited states and scattering orbitals in Eq. 3.27 is important, as it may result in under- or overcorrelated scattering states with respect to the HF target. A comprehensive discussion on the description of polarization effects in the SMC framework is presently out of scope, although it is available elsewhere [20, 21]. We finally mention that an HF target might not always be convenient, for instance, in situations where a single Slater determinant does not provide a reasonable description of the initial state.

While an L^2 approximation to the scattering state is very convenient, the matrix elements of the projected Green's operator are computationally demanding. In the early implementations of the method [16, 19, 22], these matrix elements were computed with the insertion of numerically complete Gaussian basis sets on both sides of the $G_P^{(+)}$ operator, but the slow convergence required very large sets. To overcome this problem, an alternative algorithm for the off-the-energy-shell integration in momentum space was proposed [23],

$$\langle\chi_\mu|VG_P^{(+)}V|\chi_\nu\rangle = \sum_{m}^{\text{open}} \left[\mathrm{P}\int_0^\infty k^2 dk\, \frac{g_{\mu\nu}^m(k)}{\frac{1}{2}k_m^2 - \frac{1}{2}k^2} + i\pi g_{\mu\nu}^m(k) \right], \quad (3.28)$$

where P denotes the Cauchy's principal value, and

$$g_{\mu\nu}^m(k) = \int d\hat{\mathbf{k}}\, \langle\chi_\mu|\Phi_m\mathbf{k}\rangle\langle\mathbf{k}\Phi_m|\chi_\nu\rangle\,. \quad (3.29)$$

The radial k-space is discretized on the quadrature points $\{k_j\}$, where the $g_{\mu\nu}^m(k_j)$ matrix elements are calculated, and $\{\hat{\mathbf{k}}_l\}$ angular quadratures are also employed to obtain the $g_{\mu\nu}^m$ elements, according to Eq. 3.29.

Apart from the Green's operator matrix elements, all integrals in the denominator of Eq. 3.18, with $A^{(+)}$ given by Eq. 3.24, can be computed with standard quantum chemistry techniques. The numerator involves the matrix elements $\langle \mathbf{k}\Phi_m|V|\chi_\mu\rangle$ that can be reduced to one- and two-electron integrals involving Gaussian functions and plane waves. The two-electron integrals are given by

$$\int d^3r_1 \int d^3r_2\, e^{-i\mathbf{k}\cdot\mathbf{r}_1} g_\alpha(\mathbf{r}_1)\, \frac{1}{r_{12}}\, g_\beta(\mathbf{r}_2) g_\gamma(\mathbf{r}_2)\,, \tag{3.30}$$

where g_α denotes a Gaussian function. The computational effort scales as $\sim N_g^3$, where N_g is the dimension of the Gaussin set, and should be calculated for every incident ($\mathbf{k}_i$) and outgoing ($\mathbf{k}_f$) wave vectors (i.e., all energies and directions). To reduce the computational effort required by these and other integrals, the use of the analytical norm-conserving pseudopotentials of Bachelet, Hamann, and Schlüter [24] in the SMC framework was proposed [25]. In this version of the method (SMCPP), the nuclei and core electrons are replaced by the pseudopotentials, such that only the valence electrons are explicitly described. This may lead to a significant reduction of the numerical effort, especially if heavy atoms are present, for instance, in the halogenated species (bromo-uracil and others) employed in cancer therapy. The efficient use of pseudopotentials in the SMCPP framework required the optimization of Gaussian basis sets, following the procedure of Bettega et al. [26]. Even if pseudopotentials are employed, the description of polarization effects, according to Eq. 3.27, can make the computations very demanding. Fortunately, the SMC and SMCPP methods allow for efficient parallelization. There are parallel versions of the SMC [27] and SMCPP [28] codes based on MPI and OpenMP directives, respectively. The SMCPP approach was recently reviewed [29].

We have so far discussed elastic scattering within the SE and SEP approximations, where the open-channel projector comprises only the ground state, $P = |\Phi_0\rangle\langle\Phi_0|$. In multichannel calculations, the P space should, in principle, be gradually augmented as the collision energy increases, such that all open channels are included in the P operator. For instance, let us assume that the first electronic excitation threshold is found at the energy E_{tr}. For $E < E_{\text{tr}}$, there will be a single channel in the P space, and therefore a single channel in the projected Green's operator, as well as a single energy ($k_i^2/2 = k_f^2/2$) in the numerator of Eq. 3.18. For $E \geq E_{\text{tr}}$, there will be an additional batch of Green's operator matrix elements, arising from the sum over the target states in Eqs. 3.28 and 3.29, and also additional numerator matrix elements for $k_f = [k_i^2 - 2(E - E_{\text{tr}})]^{1/2}$. The same situation will be found in every electronic threshold, and the scattering state expansion may always include virtual target excitations (closed channels) to account for polarization effects. Since we pointed out that elastic scattering with polarization can already be computationally demanding, it should be evident that multichannel computations are very costly. In the early implementations of the SMC and SMCPP codes, electronically excited open states were described with spin-adapted single Slater determinants, often employing improved virtual orbitals as particle orbitals. The SMCPP code has been modified to allow for the description of the excited states within the configuration-interaction with single excitations (CIS) framework. In view of the computational cost, the CIS active space is usually kept as small as possible, defining the minimum orbital basis for single configuration interaction (MOB-SCI) approximation [30].

3.2.4 Applications: Ethylene and 5-Halouracils

We close this section with applications of the SMCPP method to ethylene (C_2H_4) [31], 5-bromouracil (5-BrU), and 5-iodouracil (5-IU) [32]. The details of the calculations are given in the original papers. For the 5-halouracils, we show in Fig. 3.2 elastic (single-channel) A'' partial cross sections obtained in the SE and SEP approximations. The study of halouracils, used as radiosensitizers in cancer therapy, is made possible by the use of pseudopotentials and parallel processing, in view of the relevance of scalar relativistic effects for the heavy halogens and the large number of electrons, 92 and 110, respectively, for 5-BrU and 5-IU (spin–orbit couplings were neglected). For both systems, the A'' symmetry component of the integral cross section (ICS) show three peaks at the SE level, corresponding to π^* shape resonances, labeled π_1^* to π_3^* in order of increasing energy. The improved SEP calculations point out that only the higher-lying π_2^* and π_3^* states are resonances, while π_1^* is actually

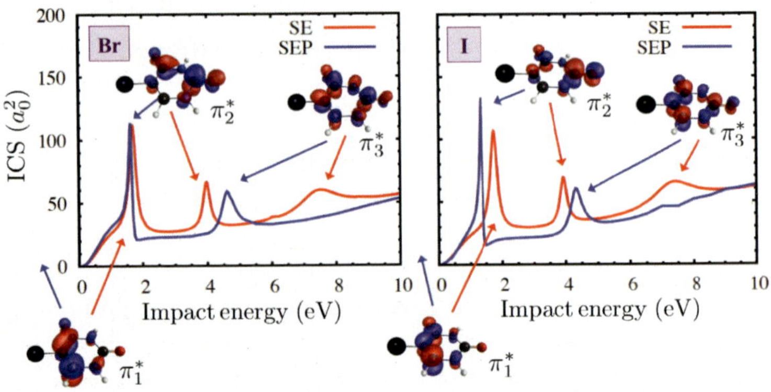

Figure 3.2 A'' symmetry component of the integral cross section (ICS) for elastic electron scattering by 5-BrU (left) and 5-IU (right). The results obtained at the SE (red lines) and SEP (blue lines) approximations are indicated. The amplitudes of compact virtual orbitals are also shown to provide insight into the characters of the anion states. The π^* states are labeled in order of increasing energies, and the valence-bound π_1^* states are obtained as shape resonances in the SE calculations in view of the neglect of correlation-polarization effects. The SEP calculations employed nearly 19,000 CSFs for each molecule.

a valence-bound state. Although not shown here, both systems also have dipole-bound states and σ_{CX}^* resonances, where X = Br or I [32].

Differential cross sections (DCSs) for electron scattering by ethylene are shown in Fig. 3.3 for a few impact energies. Although the latter molecule is considerably smaller than the halouracils, the multichannel calculations include up to 45 open states, depending on the collision energy [31]. For 10 eV, 15 eV, and 30 eV, we show the elastic DCS ($X\,^1A_g \rightarrow X\,^1A_g$) obtained from single-channel and multichannel models. The agreement with experimental data [33–36] is systematically better for the latter, especially at higher collision energies. The single-channel elastic DCS magnitudes are overestimated because the scattering flux into electronic inelastic channels is not accounted for. The cross section for the $X\,^1A_g \rightarrow \tilde{a}\,^3B_{1u}$ excitation is also shown for 18 eV, along with the experimental data [37, 38].

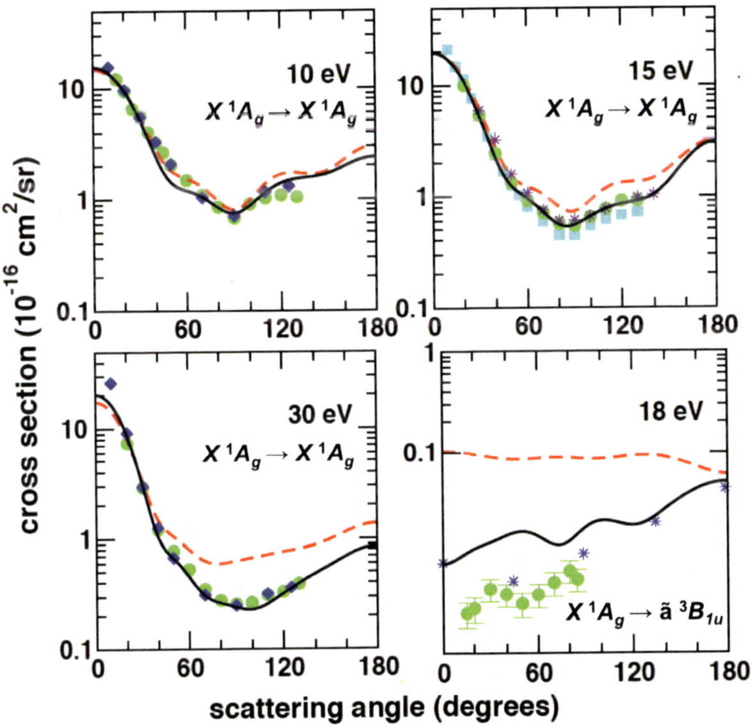

Figure 3.3 Differential cross section (DCS) for elastic ($X\,^1A_g \rightarrow X\,^1A_g$) electron scattering by ethylene molecules for 10 eV, 15 eV, and 30 eV. The results obtained from single-channel (dashed red lines) and multichannel models, with up to 45 states in the latter case (solid black lines), are shown along with the experimental data: [35] (diamonds), [36] (circles), [33] (stars), and [34] (squares). At 18 eV, we show DCSs for the electron impact $X\,^1A_g \rightarrow \tilde{a}\,^3B_{1u}$ excitation obtained from 2-state (red dashed line) and 45-state multichannel (solid black line) models, along with experimental data: [37] (blue stars), and [38] (circles with error bars).

3.3 Discrete Momentum Representation Method

The one-electron nature of the low-energy resonances, together with a simple language of potential scattering theory, led to the development of optical potential methods (one-electron methods). Within these methods the molecular target with a possibly complex electronic structure is replaced by a local or nonlocal interaction potential seen by the impinging electron. The interaction potential is in general a function of nuclear coordinates and thus allows us to incorporate nuclear dynamics initiated by the electron impact.

3.3.1 Electron–Molecule Interaction

3.3.1.1 Static-exchange approximation

In this section we attempt to outline a procedure that leads to the construction of a simple form of the optical potential. Instead of an in-depth derivation we focus on the initial model, and then we emphasize the basic assumptions used during the derivation. The inquisitive readers will be referred to excellent reviews by [39] and [14] for the necessary details. Furthermore, all the following theory will be presented under the Born–Oppenheimer approximation, and the passive dependence of wave functions and electronic operators on the nuclear coordinates will be omitted for clarity reasons.

We start with an $(N+1)$-electron wave function $\Psi(\mathbf{r}_1, \cdots, \mathbf{r}_{N+1})$ describing the continuum state of the composite system of molecule and the scattered electron. This total scattering state is an eigenstate of the full Hamiltonian H

$$[H(\mathbf{r}_1, \cdots, \mathbf{r}_{N+1}) - E]\,\Psi(\mathbf{r}_1, \cdots, \mathbf{r}_{N+1}) = 0\,. \tag{3.31}$$

A typical task of conventional quantum chemistry is to find a minimum energy E (ground state) or a few of higher energies (excited states) together with corresponding wave functions Ψ that satisfy the Shrödinger equation (3.31). In the scattering theory the total energy E of the system is prescribed and fixed. This energy is equal to sum of the kinetic and internal energies of the two colliding bodies, while excluding the mutual interaction. In the present case it is given by the kinetic energy of the impinging electron and the internal energy of the target molecule. According to this the full Hamiltonian H may be divided to three parts as

$$H(\mathbf{r}_1, \cdots, \mathbf{r}_{N+1}) = H_N(\mathbf{r}_1, \cdots, \mathbf{r}_N) - \frac{1}{2}\nabla^2_{N+1} + V(\mathbf{r}_1, \cdots, \mathbf{r}_{N+1})\,, \tag{3.32}$$

where the first term on the right-hand side is the molecular Hamiltonian and the second term is the kinetic energy of the scattered electron. The interaction potential V is basically a

Coulomb interaction of the continuum electron with N target-bound electrons and M molecular nuclei

$$V(\mathbf{r}_1, \cdots, \mathbf{r}_{N+1}) = \sum_{i=1}^{N} \frac{1}{|\mathbf{r}_{N+1} - \mathbf{r}_i|} - \sum_{j=1}^{M} \frac{Z_j}{|\mathbf{r}_{N+1} - \mathbf{R}_j|}, \qquad (3.33)$$

where the vectors $\mathbf{R}_j$ denote positions of the nuclei. The last necessary component for the following projection are the target eigenstates Φ_m, that is, eigenfunctions of the molecular Hamiltonian H_N

$$[H_N(\mathbf{r}_1, \cdots, \mathbf{r}_N) - E_m]\, \Phi_m(\mathbf{r}_1, \cdots, \mathbf{r}_N) = 0\,. \qquad (3.34)$$

The total electronic eigenfunction $\Psi(\mathbf{r}_1, \cdots, \mathbf{r}_{N+1})$ can be then expanded in the complete set of the target eigenstates Φ_m

$$\Psi(\mathbf{r}_1, \cdots, \mathbf{r}_{N+1}) = \mathcal{A} \sum_m \Phi_m(\mathbf{r}_1, \cdots, \mathbf{r}_N)\varphi_m(\mathbf{r}_{N+1})\,, \qquad (3.35)$$

where the operator $\mathcal{A}$ antisymmetrizes the products of N-electron functions Φ_m and one-electron functions φ_m.

One of the simplest forms of the optical potential used in the electron–molecule scattering calculations, the so-called static-exchange (SE) approximation, is obtained by retaining only the ground state in the expansion (3.35), that is,

$$\Psi_{\mathrm{SE}}(\mathbf{r}_1, \cdots, \mathbf{r}_{N+1}) = \mathcal{A}\, \Phi_0(\mathbf{r}_1, \cdots, \mathbf{r}_N)\varphi_0(\mathbf{r}_{N+1})\,. \qquad (3.36)$$

This approximative version of the total electronic wave function Ψ_{SE} may be used in Eq. 3.31, and then the resulting equation is projected onto the target ground state Φ_0 as follows:

$$\int \Phi_0^*(\mathbf{r}_1, \cdots, \mathbf{r}_N)[H(\mathbf{r}_1, \cdots, \mathbf{r}_{N+1}) - E]$$
$$\Psi_{\mathrm{SE}}(\mathbf{r}_1, \cdots, \mathbf{r}_{N+1})d\mathbf{r}_1 \cdots d\mathbf{r}_N = 0\,. \qquad (3.37)$$

The equation above is a one-electron equation with spatial variable $\mathbf{r}_{N+1}$ that will be denoted by $\mathbf{r}$ from now on. Moreover, we also remove the index "0" from wave functions as only the target's electronic ground state will be included in the following model. In the last step we assume we are dealing with a closed-shell molecular target that is described in the HF approximation by a single determinant wave function

$$\Phi(\mathbf{r}_1, \cdots, \mathbf{r}_N) = \frac{1}{\sqrt{N!}} |\phi_1(\mathbf{r}_1) \cdots \phi_N(\mathbf{r}_N)|\,, \qquad (3.38)$$

where ϕ_i are HF spin orbitals of the unperturbed target molecule. After some algebraic manipulations (see [14, 39]) we arrive at the following one-electron equation for the continuum function $\varphi(\mathbf{r})$:

$$\left[-\frac{1}{2}\nabla^2 + V_s + V_x - \frac{k^2}{2}\right]\varphi(\mathbf{r}) = 0\,, \tag{3.39}$$

where $k^2/2 = E - E_0$ and the V_s denotes static interaction of the continuum electron with the nuclei and with the bound electrons

$$V_s(\mathbf{r}) = \sum_{i=1}^{N}\int\frac{|\phi_i(\mathbf{r}')|^2}{|\mathbf{r}-\mathbf{r}'|}\mathrm{d}\mathbf{r}' - \sum_{j=1}^{M}\frac{Z_j}{|\mathbf{r}-\mathbf{R}_j|}\,. \tag{3.40}$$

The HF exchange interaction V_x is nonlocal, and we rather express its action on the wave function φ as

$$\langle\mathbf{r}|V_x|\varphi\rangle = -\sum_{i=1}^{N/2}\phi_i(\mathbf{r})\int\frac{\phi_i^*(\mathbf{r}')\varphi(\mathbf{r}')}{|\mathbf{r}-\mathbf{r}'|}\mathrm{d}\mathbf{r}'\,. \tag{3.41}$$

The SE equations (3.39)–(3.41) played a very important role in first implementations of electron–molecule-scattering methods. However, a lack of the target's orbitals response to the presence of the scattered electron leads to an inaccurate description of the scattering event at lower energies (typically under 10 eV). One of the methods that attempts to amend this shortcoming will be presented in the following section.

3.3.1.2 Polarization potential

A modification of the SE interaction by some local form of a potential accounting for polarization and even a correlation effects of the target's electron density is an old idea. One of the first models [40] employed the spherical long-range form

$$V_p(\mathbf{r}) = -\frac{\alpha_0}{2r^4}\,, \tag{3.42}$$

where α_0 is the spherical polarizability of the neutral target molecule. A cut-off function was used to remove the singularity at the origin. A single parameter of the theory, the cut-off radius, was determined by matching a well-defined property of the calculated results (resonance position) to the experimental data. Nonparametric versions of these models came with a description of the short-range part of the polarization potential by correlation functionals based on density functional theory (DFT). First correlation functionals [41, 42] employed local spin density (LSD) approximation emerging from the field of solid-state physics [43, 44].

Generalized gradient approximation (GGA) by [45] established a framework to further improve the LSD models by incorporating higher gradients of electron spin density into the correlation functional (functionals PW91, PBE). A different way to incorporate the density gradient corrections was used by Lee, Yang, and Parr [46], who turned the correlation energy formula of [47] into a DFT functional form by use of the first and second gradients of electron density (LYP functional).

However, all the correlation functionals mentioned above suffer from incorrect long-range behavior. They decay exponentially outside the molecular charge density instead of having a decrease of the form (3.42). The long-range part of a correlation potential for a general polyatomic molecule has the form (see, for example, [48]):

$$V_{\mathrm{p}}(\mathbf{r}) = -\frac{1}{2r^6}\sum_{ij}^{3} \alpha_{ij} r_i r_j \, , \tag{3.43}$$

where the α_{ij} is a polarizability tensor of the neutral target molecule and $r^2 = r_1^2 + r_2^2 + r_3^2$. There are several ways how to connect an exponentially decaying correlation functional with the anisotropic long-range form (3.43). One of the approaches stems from a mathematical property of expression (3.43) that contains only $l = 0$ and $l = 2$ partial waves for a symmetric tensor α_{ij}. Authors in [49] proposed that all these six components may be smoothly connected to the corresponding $l = 0$ and $l = 2$ components of the short-range correlation potential V_{c}, thus creating a composite correlation polarization potential V_{cp}. The remaining nonconnected angular components of V_{c} are then left to vanish exponentially. Another method of connection was proposed ([50]) in which the authors modified the asymptotic form (3.43) by adding higher-order induced multipoles and achieving the overall smoothness of the V_{cp} potential.

Typical shapes of the composite V_{cp} potentials for different correlation functionals (LSD, PBE, LYP) are displayed in Fig. 3.4.

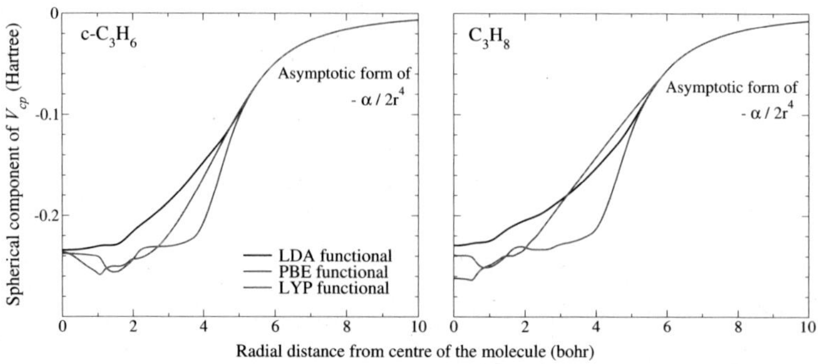

Figure 3.4 Visual comparison of different correlation functionals is used in the present study. For simplicity only the dominant spherical components are shown. The left panel displays the data for the cyclopropane molecule, while the right is for the propane molecule.

The left panel shows the curves for cyclopropane, while the right panel is for propane molecule. Only the spherical components are displayed. The LSD approximation, often denoted as local density approximation (LDA) in the scattering theory literature, is usually smoother than PBE and LYP functionals that contain also first and second density gradient components. One may note that all the functionals smoothly connect to the asymptotic form (3.42) in the spherical component at $\approx$ 5 a.u. in the case of cyclopropane and at $\approx$ 6 a.u. in the case of propane. One also notes that the V_{cp} potentials are attractive as they are supposed to substitute the relaxation of the molecular orbitals in the presence of the scattered electron.

3.3.2 The Scattering Equations

The discrete momentum representation (DMR) method is based on the one-electron Lippmann–Schwinger equation for the transition operator T (original method published in [51])

$$T = V + VG_0(E)T\ , \tag{3.44}$$

where $V = V_{\mathrm{s}} + V_{\mathrm{x}} + V_{\mathrm{cp}}$ describes interaction of the scattered electron with molecular target and G_0 denotes the free Green's function

$$G_0(E) = \frac{1}{E - H_0 + i\varepsilon}\ , \tag{3.45}$$

and H_0 is the free Hamiltonian of the system. In case of fixed-nuclei calculations the above Eq. (3.44) may be expanded in momentum space as follows:

$$\langle \mathbf{k}|T|\mathbf{q}\rangle = \langle \mathbf{k}|U|\mathbf{q}\rangle + \int \frac{\langle \mathbf{k}|U|\mathbf{k}'\rangle\langle \mathbf{k}'|T|\mathbf{q}\rangle}{k^2 - k'^2 + i\varepsilon} \mathrm{d}\mathbf{k}' . \tag{3.46}$$

Here U stands for a double of the interaction potential V, $k^2/2 = E$ and T is twice as large as T-operator in Eq. 3.44. The vectors $\mathbf{k}$ and $\mathbf{q}$ stand for plane-wave functions normalized as follows:

$$\langle \mathbf{x}|\mathbf{k}\rangle = \frac{1}{(2\pi)^{3/2}} e^{i\mathbf{k}.\mathbf{x}} . \tag{3.47}$$

Although a numerical discretization of the integral on the right-hand side of Eq. 3.46 leads to a matrix equation for the T-matrix, the complex integration path around a singularity is not suitable for a numerical implementation. This difficulty is solved in the DMR method via the following steps. By integration over angular coordinates $\hat{\mathbf{k}}'$, the integral on right-hand side of Eq. 3.46 becomes

$$\int_0^\infty k'^2 \frac{f(\mathbf{k}, \mathbf{q}, k')}{k^2 - k'^2 + i\varepsilon} \mathrm{d}k' , \tag{3.48}$$

where the function $f(\mathbf{k}, \mathbf{q}, k')$ is defined as an angular integral

$$f(\mathbf{k}, \mathbf{q}, k') = \int \langle \mathbf{k}|U|\mathbf{k}'\rangle\langle \mathbf{k}'|T|\mathbf{q}\rangle \mathrm{d}\hat{\mathbf{k}}' . \tag{3.49}$$

The integral (3.48) can be evaluated by means of its Cauchy principal value and the residue

$$\int_0^\infty k'^2 \frac{f(\mathbf{k}, \mathbf{q}, k')}{k^2 - k'^2 + i\varepsilon} \mathrm{d}k' = \mathcal{P} \int_0^\infty k'^2 \frac{f(\mathbf{k}, \mathbf{q}, k')}{k^2 - k'^2} \mathrm{d}k' - i\frac{\pi k}{2} f(\mathbf{k}, \mathbf{q}, k) . \tag{3.50}$$

The integration path is now real, but the principal value integral on the right-hand side of Eq. 3.50 cannot be evaluated by a simple numerical quadrature yet. The result is very unstable and strongly depends on how close to the singularity $k' = k$ the integration points happen to be. One of the approaches that results in numerical integration of a smooth function is method based on subtraction of singularity. The simple idea starts with the following identity (the dependence on momentum vectors $\mathbf{k}$ and $\mathbf{q}$ is omitted)

$$\mathcal{P} \int k'^2 \frac{f(k')}{k^2 - k'^2} \mathrm{d}k' = \int k'^2 \frac{f(k') - f(k)}{k^2 - k'^2} \mathrm{d}k' + f(k)\mathcal{P} \int k'^2 \frac{1}{k^2 - k'^2} \mathrm{d}k' . \tag{3.51}$$

Assuming that $f(k')$ is analytic in k, the integrated function in the first integral of the right-hand side of Eq. 3.51 is smooth around k and hence suitable for a numerical integration. The second integral in the above equation is simple and can be evaluated in a closed form.

The fixed-nuclei implementation described above was later extended to include nuclear motion ([52]) in a simple form of harmonic vibrations. After limiting the vibrational space to two channels the Eq. 3.46 becomes

$$\langle\chi_1\mathbf{k}|T|\chi_0\mathbf{q}\rangle = \langle\chi_1\mathbf{k}|U|\chi_0\mathbf{q}\rangle + \sum_{i=0}^{1}\int\frac{\langle\chi_1\mathbf{k}|U|\chi_i\mathbf{k}'\rangle\langle\chi_i\mathbf{k}'|T|\chi_0\mathbf{q}\rangle}{k^2-2E_i-k'^2+i\varepsilon}\mathrm{d}\mathbf{k}', \tag{3.52}$$

where $E_0 = 0$ (for the elastic channel) is the energy of the vibrational ground state and E_1 is the energy of the first excited vibrational state. These two states are described by the vibrational functions χ_0 and χ_1, respectively.

A numerical discretization of the integral on right-hand side of the above equation leads to a set of two coupled matrix equations:

$$\begin{bmatrix}T_{00}\,T_{01}\\T_{10}\,T_{11}\end{bmatrix} = \begin{bmatrix}U_{00}\,U_{01}\\U_{10}\,U_{11}\end{bmatrix} + \begin{bmatrix}U_{00}\,U_{01}\\U_{10}\,U_{11}\end{bmatrix}\begin{bmatrix}G_0 & \\ & G_1\end{bmatrix}\begin{bmatrix}T_{00}\,T_{01}\\T_{10}\,T_{11}\end{bmatrix}, \tag{3.53}$$

with the following interaction matrix elements:

$$\begin{aligned}[U_{00}]_{ij} &= \langle\chi_0\mathbf{k}_i|U|\chi_0\mathbf{q}_j\rangle = [U_{11}]_{ij}\,,\\ [U_{01}]_{ij} &= \langle\chi_0\mathbf{k}_i|U|\chi_1\mathbf{q}_j\rangle = [U_{10}]_{ij}\,.\end{aligned} \tag{3.54}$$

The above identities are based on a first-order expansion of the interaction along the dimensionless normal mode coordinate Q as

$$U(Q) = U(0) + Q\frac{\partial}{\partial Q}U(0) + O(Q^2)\,, \tag{3.55}$$

where $U(0)$ stands for the double of interaction potential at equilibrium geometry. Neglecting the higher-order terms $O(Q^2)$ results in a simple integration over the coordinate Q:

$$\langle\chi_0(Q)\mathbf{k}|U(Q)|\chi_0(Q)\mathbf{q}\rangle = \langle\mathbf{k}|U(0)|\mathbf{q}\rangle\,, \tag{3.56}$$

$$\langle\chi_0(Q)\mathbf{k}|U(Q)|\chi_1(Q)\mathbf{q}\rangle = \frac{1}{\sqrt{2}}\frac{\partial}{\partial Q}\langle\mathbf{k}|U(0)|\mathbf{q}\rangle\,. \tag{3.57}$$

The interaction potential $V(Q) = U(Q)/2$ is generated from target state electron density

$$\rho(Q) = \sum_{\mu\nu} D_{\mu\nu}(Q) g_\mu(Q) g_\nu(Q)\,, \tag{3.58}$$

where $D_{\mu\nu}(Q)$ is a HF density matrix and $g_\mu(Q)$ represents a Gaussian-type basis set. As already described in the previous section the interaction potential $V(Q)$ consists of static, exchange, and correlation polarization contributions

$$V(Q) = V_\mathrm{s}(Q) + V_\mathrm{x}(Q) + V_\mathrm{cp}(Q)\,. \tag{3.59}$$

For simplicity in the following formulas we suppress explicit dependence on the vibrational coordinate Q. The momentum–space matrix elements (3.56) of V_s and V_x are then evaluated through hybrid plane-wave and Gaussian repulsion integrals

$$\langle \mathbf{k}|V_\mathrm{s}|\mathbf{q}\rangle = \sum_{\mu\nu} D_{\mu\nu}\left(g_\mu g_\nu|\mathbf{kq}\right) - \frac{1}{2\pi^2|\mathbf{q}-\mathbf{k}|^2}\sum_{A=1}^{M} Z_A e^{i(\mathbf{q}-\mathbf{k}).\mathbf{R_A}}\,, \tag{3.60}$$

$$\langle \mathbf{k}|V_\mathrm{x}|\mathbf{q}\rangle = -\sum_{\mu\nu} D_{\mu\nu}\left(\mathbf{k}g_\mu|g_\nu\mathbf{q}\right)\,, \tag{3.61}$$

with the second term of Eq. 3.60 representing a nuclear attraction to M nuclei positioned at vectors $\mathbf{R}_A$. The static and exchange hybrid repulsion integrals are defined as follows:

$$\left(g_\mu g_\nu|\mathbf{k}\,\mathbf{q}\right) = \frac{1}{(2\pi)^3}\iint \frac{g_\mu(\mathbf{x}) g_\nu(\mathbf{x}) e^{i(\mathbf{q}-\mathbf{k}).\mathbf{y}}}{|\mathbf{x}-\mathbf{y}|}\mathrm{d}\mathbf{x}\mathrm{d}\mathbf{y}, \tag{3.62}$$

$$\left(\mathbf{k}g_\mu|g_\nu\mathbf{q}\right) = \frac{1}{(2\pi)^3}\iint \frac{g_\mu(\mathbf{x}) e^{-i\mathbf{k}.\mathbf{x}} g_\nu(\mathbf{y}) e^{i\mathbf{q}.\mathbf{y}}}{|\mathbf{x}-\mathbf{y}|}\mathrm{d}\mathbf{x}\mathrm{d}\mathbf{y}. \tag{3.63}$$

In most of the present implementations of a correlation potential in quantum chemistry, the V_cp is calculated as a function of the electron density, ρ, and its first and second spatial gradients, $\nabla\rho$ and $\nabla\nabla\rho$, respectively. In a similar manner the correlation polarization potential V_cp is evaluated here on a rectangular spatial grid with the molecule sitting in its center. The momentum space matrix elements are computed by use of fast Fourier transform (FFT) technique applied on the following integral:

$$\langle \mathbf{k}|V_\mathrm{cp}|\mathbf{q}\rangle = \frac{1}{(2\pi)^3}\int V_\mathrm{cp}(\mathbf{r}) e^{i(\mathbf{q}-\mathbf{k}).\mathbf{r}}\mathrm{d}\mathbf{r}\,. \tag{3.64}$$

Convergence of the above integral with respect to the FFT box size is quite slow, owing to the long-range nature of V_{cp} shown in Eq. 3.43. However, the asymptotic tail of V_{cp} may be removed and evaluated analytically [49], leaving a short-range part of the V_{cp} that is amenable to the numerical evaluation based on FFT. Vibrational coupling elements (3.57) require differentiation of the equations (3.60)–(3.64) with respect to coordinates of the nuclei. The nuclear gradients are implemented in the DMR method in an analytical form [52].

3.3.3 Application to Study Shape Resonances

Using a vibrationally inelastic channel to study shape resonances is, on the one side, profitable because of the weak background cross section, but it also brings a new level of complexity. The advantage of the inelastic channel may be demonstrated with the case of the cyclopropane molecule C_3H_6. The closed C–C–C ring of this molecular system allows the existence of well-known σ^* resonance that has an antibonding character with respect to all the C–C bonds. The resonance was first seen at collision energy of 5.5 eV in the experimental cross section for vibrational excitation of the symmetric C–C ring stretch vibration ν_3 [57]. Its symmetry A_2' was later confirmed computationally ([56] and [58]). However, Fig. 3.5 demonstrates that there is very little of the resonance seen in the fixed-nuclei calculations. On the other hand, the vibrationally inelastic cross sections, shown in Fig. 3.6, exhibit a strong resonant behavior in many modes and at different energies. The 5.5 eV A_2' resonance is computationally reproduced in the excitation cross section of the ν_3 mode. The resonance position determined by theory is slightly shifted to the higher energies as is seen at 5.9 eV. The seemingly surprising result is that this resonance also visibly dominates the cross section for excitation of H–C–H twist mode ν_4. An understanding of this phenomena brings the new level of complexity mentioned earlier. There are several other methods that have been developed to determine the resonant parameters. Among them the two used extensively at present are eigenphase analysis [59] and an analysis using a time delay matrix [60].

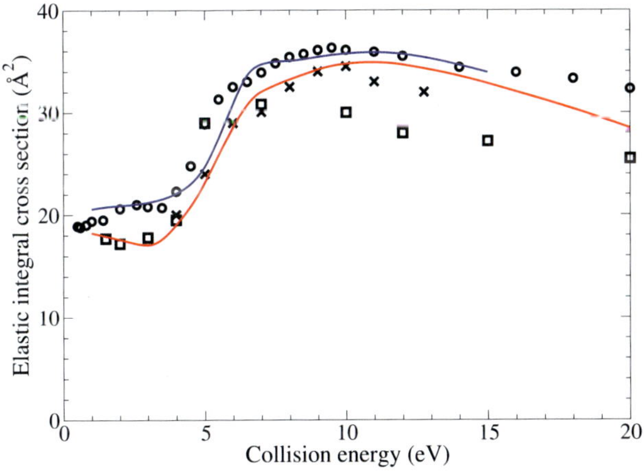

Figure 3.5 Comparison of the elastic integral cross sections for cyclopropane. Experimental data are displayed with crosses [53], squares [54], and circles (total cross sections of [55]). The DMR calculations are shown with the red curve, while the blue curve represents calculations using single-center expansion method [56].

3.3.4 Symmetry of Shape Resonances

The aim of this section is to determine computationally the dominant electronic symmetry components that drive resonant vibrational excitation of cyclopropane. The DMR method does not attempt to reduce computational effort by making use of the symmetry components corresponding to the point group of the molecule. This is because the method is aimed at larger molecular systems that generally have low symmetry and the computational gain would be small. Moreover, even symmetric molecules (like cyclopropane) reduce their symmetry if vibrational motion is included. Therefore, in vibrationally inelastic collisions one may work within two available symmetry schemes: One may reduce the applied point group so that it is conserved for the particular vibrational mode, or one may use the equilibrium point group (D_{3h} in this case) and accept that irreducible representations become coupled by the vibrational motion of less symmetric modes. In the present analysis we have chosen the latter.

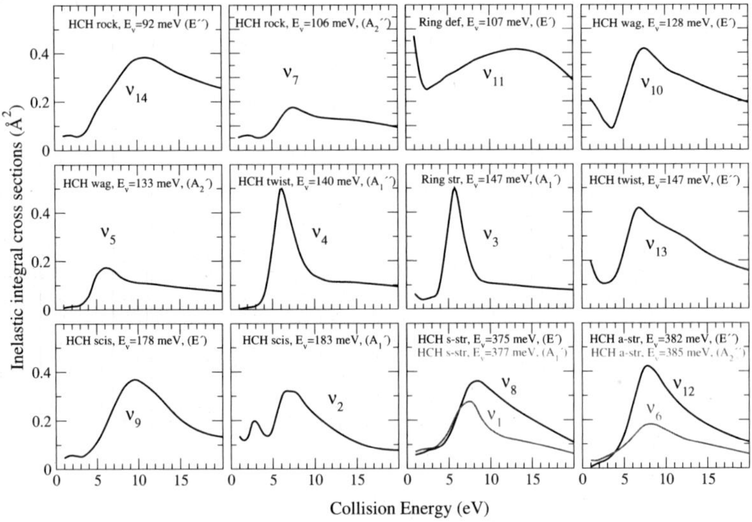

Figure 3.6 Calculated $0 \rightarrow 1$ vibrationally inelastic integral cross sections (full curves) for cyclopropane. The modes are ordered by increasing vibrational frequency. The last two graphs contain two modes each for space-saving purposes.

The inelastic T_{10}-matrix elements are obtained by solving the set of equations (3.53)

$$[T_{10}]_{ij} = \langle \chi_1 \mathbf{k}_i | T | \chi_0 \mathbf{q}_j \rangle \,, \tag{3.65}$$

where the vectors $\mathbf{q}_i$ span the incoming momentum sphere with radius determined by collision energy E and the vectors $\mathbf{k}_j$ span the outgoing momentum sphere with radius determined by energy of the outgoing electron. Since the angular grid on the sphere is described by Lebedev angular quadrature designed to exactly integrate spherical harmonics up to a particular order [61], the partial-wave components of the T_{10} matrix can be obtained by a straightforward technique

$$T_{lm}^{l'm'} = \sum_{ij} w_i w_j S_{lm}(\hat{\mathbf{k}}_i) \, [T_{10}]_{ij} \, S_{l'm'}(\hat{\mathbf{q}}_j) \,, \tag{3.66}$$

where w_i are the weights of the Lebedev quadrature and the vectors $\hat{\mathbf{k}}_i$ and $\hat{\mathbf{q}}_j$ are unit vectors with directions of vectors $\mathbf{k}_i$ and $\mathbf{q}_j$, respectively. The angular functions S_{lm} are chosen as real spherical

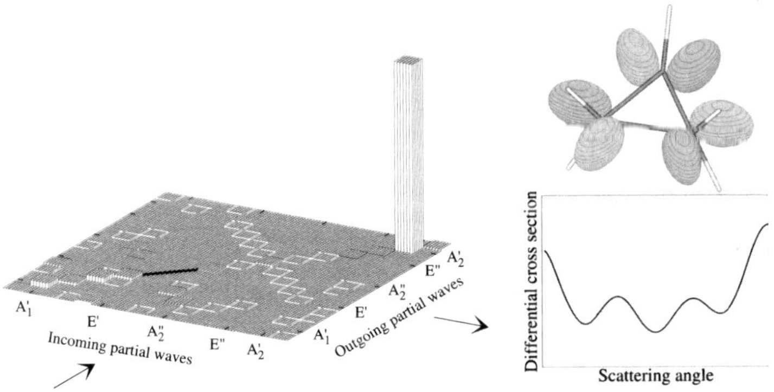

Figure 3.7 Symmetric C–C stretch ν_3. The left panel displays the squared elements $|T_{lm}^{l'm'}|^2$ for vibrational excitation of the ring stretch mode ν_3 calculated via Eq. 3.66. Partial waves up to $l_{\max} = 3$ are separated into irreducible representations of D_{3h} point group. The right panel displays shape of computed differential cross section (lower picture) and shape of a virtual orbital corresponding to an outgoing wave of A_2' symmetry (upper picture).

harmonics. The partial-wave expanded T-matrix (3.66) may have then its ingoing and outgoing channels (partial waves) organized in such a way that they form irreducible representation of the point group of the molecule.

The squared elements $|T_{lm}^{l'm'}|^2$ for the symmetric stretch mode ν_3 are shown as vertical columns for all combinations of the incoming and outgoing partial waves up to $l_{\max} = 3$ in the left panel of Fig. 3.7. The fact that a single column dominates the figure clearly demonstrates that the resonant vibrational excitation of the ν_3 mode is driven by a single partial wave, which is the lowest partial wave of A_2' symmetry. With the carbon ring of the molecule placed in the xy plane and one carbon atom on the y axis we may identify this partial wave as $l = 3$, $m = 3$.This single contribution to the inelastic cross section is diagonal and as such it describes a collision in which an incoming A_2' wave excites the fully symmetric ring stretch mode ν_3 and leaves the system in the same partial wave, A_2'. The right panel of Fig. 3.7 displays the calculated differential cross section and the spatial shape of the corresponding virtual orbital of A_2' symmetry. It is very clear that the angular shape of the differential cross section

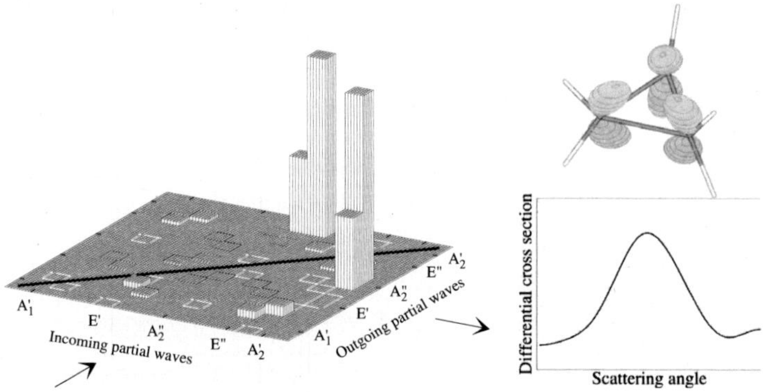

Figure 3.8 H–C–H twist mode ν_4. The left panel displays the squared elements $|T_{lm}^{l'm'}|^2$ for vibrational excitation of the ring stretch mode ν_4 calculated via Eq. 3.66. The right panel displays the shape of the computed differential cross section (lower picture) and the shape of a virtual orbital corresponding to the outgoing wave of the A_2'' symmetry (upper picture).

describes an almost pure f-wave scattering event in agreement with what the partial-wave analysis of the T-matrix shows in the left panel of Fig. 3.7.

A comparison of the squared elements $|T_{lm}^{l'm'}|^2$ for the ν_4 mode (Fig. 3.8) with those for the totally symmetrical mode ν_3 in Fig. 3.7 reveals that the resonant excitation mechanism is quite different. The coupling between incoming and outgoing partial waves is off-diagonal. Figure 3.8 shows that the resonant vibrational excitation of the ν_4 mode is caused by the incoming partial wave ($l = 3$, $m = 3$) of the A_2' symmetry, as in the previous case of the ν_3 mode. However, now the electron causes excitation of the nonsymmetric H–C–H twisting mode ν_4 of the A_1'' symmetry and leaves the molecule dominantly in two different partial waves, $l = 1$, $m = 0$ and $l = 3$, $m = 0$, shown as a pair of adjacent columns in the left panel of Fig. 3.8.

Note that in the angular momentum space limited to $l \leq 3$ used here, and for a given orientation of the molecule, these two partial waves form exactly the irreducible representation A_2''. The individual partial waves are not labeled explicitly in the left panel of Fig. 3.8, but note that always two partial waves belong to one irreducible representation so that each pair of adjacent columns belongs to

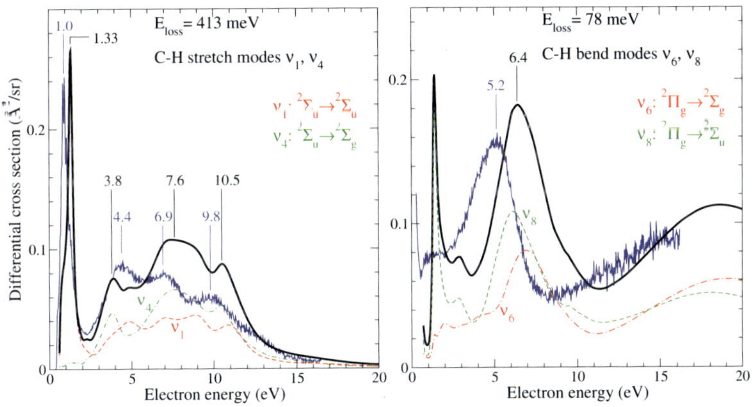

Figure 3.9 Computed (black curves) and measured (blue lines) data for vibrationally inelastic cross sections of diacetylene at the two energy losses indicated. Green and red dashed lines show contributions from the individual vibrational modes. The scattering angle is 180°. The roles of the symmetry components is explained in the text.

one irreducible representation. The fact that there are two pairs of columns, arranged symmetrically across the diagonal, is due to the principle of microscopic reversibility, that is, the symmetries of the incoming and outgoing waves can be switched. What can be observed here is an f-wave resonance of A_2' symmetry that leaks to a p-wave in A_2'' symmetry, while causing resonant vibrational excitation of the nonsymmetric ν_4 mode of A_1'' symmetry.

The mechanism described presents an additional complexity mentioned above. This complexity is not present in fixed-nuclei calculations in which symmetries do not couple. Another example of the symmetry coupling in the resonant vibrational excitation is displayed in Fig. 3.9 for the diacetylene molecule. The left panel of the figure shows four shape resonances of $^2\Sigma_u$ symmetry and their influence on vibrational excitation of the ν_1 and ν_4 modes. An analysis similar to the one carried out before for cyclopropane suggests that the $^2\Sigma_u$ resonances decay via $^2\Sigma_g$ symmetry in the case of the asymmetric C–H stretch mode ν_4. This coupling is even more emphasized in the right panel of Fig. 3.9, where the $^2\Pi_g$ resonance at 5.2 eV decays through the $^2\Sigma_g$ and $^2\Sigma_u$ symmetries for the vibrational excitation of symmetric ν_6 and asymmetric ν_8 C-H bend modes, respectively. More details can be found in [62].

3.4 *R*-Matrix Method

The *R*-matrix method is an approach to solving the time-independent Schrödinger equation. It was initially developed in the 1940s [63] to treat nuclear reactions and adapted to the study of electron–atom collisions in the early 1970s [64]. The first applications to diatomic molecules soon followed [65–67] and by the 1990s implementations to treat electron scattering from polyatomic molecules were available [68, 69].

The basic idea of the method is to solve the problem by splitting configuration space in two. In the case of electron collisions with atoms or molecules, the problem is difficult to solve when the projectile is close to the target and exchange and correlation effects need to be taken into account. However, the projectile–target interaction can be modeled much more easily when the electron is far from the target, and the resulting equations are easier to solve. Combining the solutions in these two regions allows the efficient determination of scattering data (cross sections, resonance parameters, etc.). The main strengths of the method are as follows: (a) The more complex part of the problem can be solved without taking into consideration the kinetic energy of the scattering electron; this makes it possible to generate scattering data for a large number of energies relatively cheaply; and (b) inclusion of electronically excited states is fairly straightforward.

The application of the *R*-matrix method to atomic, molecular, and optical processes has been described in detail in a recent book [70], and a recent article [71] has focused on molecular processes. Here, too, we will concentrate on the method as it applies to electron–molecule scattering.

3.4.1 One-Channel Potential Scattering

To introduce the *R*-matrix method, we will look at the scattering of a spinless particle from a short-range spherically symmetric potential. In this simple case, we aim to solve the time-independent Schrödinger equation for the Hamiltonian $H = -\frac{\nabla^2}{2} + V(r)$:

$$\left[-\frac{\nabla^2}{2} + V(r)\right] \Psi(r, \theta, \phi) - E\Psi(r, 0, \phi) \tag{3.67}$$

with $V(r) = 0$ for $r > r_0$.

It is possible to express $\Psi(r, \theta, \phi)$ as a partial wave expansion:

$$\Psi(r, \theta, \phi) = \sum_l \sum_m \frac{F_l(r)}{r} Y_{l,m}(\theta, \phi) \tag{3.68}$$

Then, the problem of determining $\Psi(r, \theta, \phi)$ is simplified to finding the reduced radial wave functions $F_l(r)$ solution of

$$\left[-\frac{d^2}{dr^2} + 2V(r) + \frac{l(l+1)}{r^2} - 2E\right] F_l(r) = 0. \tag{3.69}$$

To solve these equations, we impose the boundary condition $F_l(0) = 0$.

For $r > r_0$ the solutions of Eq. 3.69 can be expressed in terms of the solutions $j(kr)$ and $\eta(kr)$ of the free particle radial equation [10]:

$$F_l(r) \sim j_l(kr) + K(E)\eta_l(kr) \tag{3.70}$$

Here $k = \sqrt{2E}$. Asymptotically, $j(kr)$ and $\eta(kr)$ reduce to simple trigonometric functions. $K(E)$ contains information on how the potential has affected the particle (and is, in fact, related to the phase shift [10]). Cross sections, resonance information, etc., can be obtained from $K(E)$.

We obtain an expression for $K(E)$ by imposing continuity of the logarithmic derivative of F_l at r_0:

$$\left.\frac{dF_l(r)/dr}{F_l(r)}\right|_{r=r_0} = \left.\frac{d(j_l(kr) + K(E)\eta_l(kr))/dr}{j_l(kr) + K(E)\eta_l(kr)}\right|_{r=r_0} \tag{3.71}$$

The question now arises of how to solve Eq. 3.69 to determine $F_l(r)$ so that $K(E)$ can be calculated.

As mentioned earlier, the R-matrix method is based on dividing the configuration space into an internal and an external region.

These regions are separated by a sphere of radius a that must be chosen to be large enough so that the interaction potential (in this case $V(r)$) takes a simple form for $r > a$. For potential scattering we will take $a = r_0$ and start by finding $F_l(r \leq r_0)$. We use a set of functions $u_i(r)$, which we will call R-matrix basis functions, to expand the F_l:

$$F_l(r \leq r_0) = \sum_{i=0}^{\infty} A_i(E) u_i(r) \tag{3.72}$$

$u_i(r)$ are solutions in the inner region of the time-independent Schrödinger equation for the Hamiltonian $H + L$ with eigenvalues E_i (the operator L, called the Bloch operator, is added to H to make the Hamiltonian hermitian inside the sphere. Details of the form of the operator can be found in [70]).

If we introduce the functions F_l in Eq. 3.69 and solve it for $r < r_0$, we obtain an expression for coefficients $A_i(E)$ in terms of the functions $u_i(r)$ and $F_l(r)$ that, substituted into Eq. 3.72, give us:

$$F_l(a) = \frac{1}{2a} \sum_{i=0}^{\infty} \frac{u_i(a) u_i(a)}{E_i - E} \left(a \frac{dF_l(r)}{dr}\bigg|_a - bF_l(a) \right), \tag{3.73}$$

where b is a parameter of the Bloch operator. We can rewrite this equation as

$$F_l(a) = R(E) \left(a \frac{dF_l(r)}{dr}\bigg|_a - bF_l(a) \right), \tag{3.74}$$

where the R-matrix has been defined as

$$R(E) = \frac{1}{2a} \sum_{i=0}^{\infty} \frac{u_i(a) u_i(a)}{E_i - E}. \tag{3.75}$$

The key point here is that we have found an expression for the radial wave functions $F_l(r)$ at the R-matrix boundary in terms of a quantity determined in the *inner region*: $R(E)$. The R-matrix can be calculated from information obtained solving the problem in the inner region. Knowledge of the R-matrix allows us to determine the scattering information by solving the problem in the outer region. We can now match the solution 3.74 to the known general solution in the outer region given by Eq. 3.70 and obtain an expression for $K(E)$ in terms of the inner region quantity $R(E)$:

$$K(E) = \frac{j_l(ka) - aR(E)j_l'(kr)\Big|_{r=a}}{aR(E)\eta_l'(kr)\Big|_{r=a} - \eta_l(ka)}, \tag{3.76}$$

where the prime denotes the derivative of the function with respect to r. Note that $R(E)$ can be calculated for any energy once the inner region basis set is known. Similarly, the energy-dependent coefficients $A_i(E)$ need not be calculated explicitly.

In the case of electron–molecule scattering, many channels are involved in the description of the process. These channels are defined by the internal state of the target molecule (in the fixed-nuclei approximation, the electronic state) and the partial wave Y_{lm} describing the behavior of the projectile; they define the possible states of the system before or after the collision. In the many-channel case $K(E)$ and $R(E)$ become matrices whose elements, $K_{ij}(E)$ and $R_{ij}(E)$, respectively, correspond to the event in which the system is initially described by channel i and after the collision by channel j. Of course, in the case of an electron interacting with a molecule $V(r > r_0) \neq 0$, even when the molecule has no permanent dipole moment, other moments (e.g., quadrupole) and/or the polarization effects will ensure that, in most cases, the electron–molecule interaction potential cannot be considered zero until the asymptotic region.

3.4.2 *R*-Matrix Method for Electron–Molecule Scattering

In the description below, we assume the nuclei of the molecule are held fixed during the collision, and solve the electronic part of the problem. That is, we work within the fixed-nuclei approximation.

Let us look at the case of an electron colliding with an N-electron molecule. We start by analyzing the inner region. In this case, we will write the energy-dependent wave function describing the collision in terms of the R-matrix basis functions $\Psi_k(\mathbf{r}_1, \ldots \mathbf{r}_{N+1})$:

$$\Psi(E) = \sum_{i=0}^{\infty} A_i(E)\Psi_k(\mathbf{r}_1, \mathbf{r}_2 \ldots \mathbf{r}_{N+1}), \tag{3.77}$$

where $\mathbf{r}_i$ denote the coordinates of electron i. This is the equivalent to Eq. 3.72 for the electron–molecule scattering case. To determine the $\Psi_k(\mathbf{r}_1, \mathbf{r}_2 \ldots \mathbf{r}_{N+1})$, we expand them in the following way:

$$\Psi_k(\mathbf{r}_1, \mathbf{r}_2 \ldots \mathbf{r}_{N+1}) = \mathcal{A} \sum_{mj} a_{mjk} \Phi_m(\mathbf{r}_1 \ldots \mathbf{r}_N) u_{mj}(\mathbf{r}_{N+1}) + \sum_i b_{ik} \chi_i(\mathbf{r}_1, \mathbf{r}_2 \ldots \mathbf{r}_{N+1}) \quad (3.78)$$

The antisymmetrization operator $\mathcal{A}$ ensures the $\Psi_k(\mathbf{r}_1, \mathbf{r}_2 \ldots \mathbf{r}_{N+1})$ are antisymmetric with respect to the exchange of two electrons, u_{mj} are continuum orbitals describing the scattering electron and Φ_m are wave functions describing electronic states of the target molecule with energy ε_m. χ_i are multicenter quadratically integrable functions known as L^2 functions used to represent short-range correlation and polarization effects.

The wave functions Ψ_k are required to be eigenfunctions of the operator $H + L$ in the inner region with eigenvalues E_k (L being the Bloch operator for the $[N + 1]$ particle case). This means that the coefficients a_{mjk} and b_{ik} can be obtained diagonalizing the matrix representing this operator on the basis of $\Phi_m u_{mj}$ and χ_i.

The crucial point in applying this method to electron scattering is to determine what value a, the R-matrix radius, should take. We stated that the interaction potential should be simpler in the outer region. In the case of electron–molecule scattering this translates into ensuring that the nonlocal component of this potential is negligible in the outer region. In other words, the scattering electron must be distinguishable from the target electrons. For this to hold, it is essential that the electronic density associated to the target electronic wave functions Φ_m is negligible outside the R-matrix sphere.[f]

We now need to find an expression for the R-matrix in terms of Ψ_k. To do this we define the reduced radial wave functions at the boundary:

$$F_i(a) = \left\langle \frac{(\Phi_j)_i \Upsilon_{l_i m_i}}{r_{N+1}} \middle| \Psi(E) \right\rangle_{inner} \quad (3.79)$$

[f]In practice, we require that, of the single- and many-particle wave functions in Eq. 3.78 only u_{ij} have nonnegligible amplitude at the boundary between the inner and outer regions. In other words, the χ_i must also have negligible amplitude at the boundary.

and the surface amplitudes:

$$f_{ik}(a) = \left\langle \frac{(\Phi_j)_i \Upsilon_{l_i m_i}}{r_{N+1}} \middle| \Psi_k \right\rangle_{inner} \tag{3.80}$$

The subscript *inner* indicates that the integration implied by the brackets needs to be performed over the inner region; r_{N+1} indicates the radial coordinate of the scattering electron. The product $(\Phi_j)_i \Upsilon_{l_i m_i}$ defines the scattering channel i that describes the target in electronic state j and the scattering electron possessing angular momentum quantum numbers l_i and m_i.

By solving $(H + L)\Psi(E) = E\Psi(E)$ using, for $\Psi(E)$, the expression in Eq. 3.78 and after some operations we arrive at:

$$F_i(a) = \frac{1}{2a} \sum_j \sum_k \frac{f_{ik}(a) f_{jk}(a)}{E_k - E} \left(a \frac{dF_j}{dr_{N+1}}\bigg|_a - bF_j(a) \right), \tag{3.81}$$

where the first summation runs over the channels included in the calculation and the second over all Ψ_k. We can see the similarities with the one-channel derivation in the previous section. The elements of the R-matrix now become

$$R_{ij}(E) = \frac{1}{2a} \sum_k \frac{f_{ik}(a) f_{jk}(a)}{E_k - E}. \tag{3.82}$$

In the same fashion we can now write

$$F_i(a) = \sum_{j=1} R_{ij}(E) \left(a \frac{dF_j}{dr_{N+1}}\bigg|_a - bF_j(a) \right). \tag{3.83}$$

What we now need to do is to be able to link this to results in the outer region. The wave function describing the collision has a simpler form in the outer region:

$$\Psi_k(\mathbf{r}_1, \mathbf{r}_2 \ldots \mathbf{r}_{N+1}) = \sum_i^n \sum_j \Phi_i(\mathbf{r}_1 \ldots \mathbf{r}_N) \frac{F_j(r_{N+1})}{r_{N+1}} \Upsilon_{l_j m_j}(\hat{\mathbf{r}}_{N+1}), \tag{3.84}$$

where n is the number of target states included in the expansion and $\hat{\mathbf{r}}_{N+1}$ indicates the angular coordinates of the scattering electron. As in the single-channel case, we know the asymptotic form of F_j:

$$F_{ij} \overset{r \to \infty}{\longrightarrow} \delta_{ij} \delta_{l_i l_j} \delta_{m_i m_j} R_i^+ + K_{ij} R_j^i, \tag{3.85}$$

where

$$R_i^{\pm} = \frac{1}{\sqrt{k_i}} \begin{Bmatrix} \sin \\ \cos \end{Bmatrix} \left(k_i r - l_i \frac{\pi}{2} + \frac{Z}{k_i} + \Gamma \left(l_i + 1 - i \frac{Z}{k_i} \right) \right). \tag{3.86}$$

Here $k_i = \sqrt{2E - \varepsilon_i}$, Z is the charge of the molecule (if it is an ion), and Γ indicates, as usual, the gamma function.

Therefore, if we determine the F_i and match them to the asymptotic solutions, we obtain K_{ij}. The F_i are, in fact, the reduced radial wave functions that we defined earlier and describe, together with $\Upsilon_{l_i m_i}(\hat{\mathbf{r}}_{N+1})$, the behavior of the scattering electron.

$F_i(r_{N+1})$ are solutions of the following set of differential equations:[g]

$$\left(-\frac{1}{2} \frac{d^2}{dr_{N+1}^2} + \frac{l_i(l_i+1)}{2r_{N+1}^2} \right) F_i(r_{N+1}) + \sum_{j=1} V_{ij}(r_{N+1}) F_j(r_{N+1})$$
$$= (E - \varepsilon_i) F_i(r_{N+1}) \tag{3.87}$$

$V_{ij}(r)$ are coupling potentials; in practice they are expressed as single-center expansions where the coefficients in the linear combination are related to the permanent and transition multipole moments of the target electronic states included in the calculation [71].

The set of Eqs. 3.87 are solved using as boundary conditions the $F_i(a)$ from Eq. 3.83 and employing a propagation method [70]. These $F_i(a)$ are, of course, calculated from the R-matrix determined in the inner region.

Finally, we should note that, in practical applications, symmetry and group theory are used to simplify and reduced the size of the calculations. Since the Ψ_k belonging to the different irreducible representations of the point group the molecule belongs to are orthogonal, the scattering equations can be solved independently for each irreducible representation.

3.4.3 Practical Considerations

The question now arises of how we go about solving the equations presented in the previous section. Several levels of approximation

[g]These equations are obtained by substituting Eq. 3.84 into the Schrödinger equation and projecting on to the channel functions.

can be employed: SE, SEP, and, finally, close-coupling (CC) in which the electronic state of the target molecule is allowed to change. The SE and SEP approximations require only the ground state of the molecule (usually described at the HF level; see Eq. 3.38) to be included in the expansion 3.78. In the CC approximation several target states (described by the Φ_m) are included (additionally, pseudostates can be added to improve the description of the polarization or represent discretized ionization states). Two UK molecular R-matrix suites are currently available: the UKRmol and the more recent UKRmol+. These share a number of programs, with the main difference being the description of the continuum orbitals: the calculation of integrals and manipulation of orbitals has been completely rewritten in UKRmol+ in order to enable the use of a mixed basis set (see below). Both suites use orthogonal orbitals, so an orthogonalization of the continuum and molecular (describing bound electrons) orbitals must be performed. This can be done much more accurately with the UKRmol+ suite because all computations can be performed, when required, in quadruple precision.

A summary of the practical considerations that should be taken into account when using these suites to perform a CC calculation follows (for a description of computational chemistry concepts and methods see, for example, [72]):

- The wave functions Φ_m describing the target states and the functions χ_i are generated as antisymmetrized linear combinations of products of molecular orbitals (i.e., using a configuration interaction [CI] model).
- The molecular orbitals are, in turn, expanded on (standard, but sometimes adapted) the basis of GTOs.
- The molecular orbitals are normally determined in an HF self-consistent field or complete active space self-consistent field (CASSCF) calculation, the latter using standard quantum chemistry software. Externally generated orbitals can be used and are provided via a file in Molden format.
- the L^2 functions χ_i must be carefully built to ensure that the target and $(N+1)$ wave functions are described to a similar level of accuracy [73]. For this reason, complete active space

CI expansions are preferred (to, for example, CISD) for the target wave functions.

- In addition the χ_i must be such that they describe, as accurately as possible, the polarization effects. This can be difficult to achieve: both underrepresentation of this effect and overrepresentation ("overcorrelation" of the calculation) are possible. No accurate way of establishing a priori the right set of χ_i has been established (except, possibly, the use of pseudostates; unfortunately, their use can lead to very big calculations, particularly for larger molecules, so pseudostates have so far only been used in a few calculations with small targets).
- The continuum orbitals are also expanded on the basis of GTOs, either on their own (UKRmol suite) or in combination with B-spline-type orbitals (BTOs), products of a B-spline and a spherical harmonic, UKRmol+ suite. These "continuum" GTOs are centered on the center of mass of the system; the GTO basis is generated by fitting Bessel or Coulomb functions over a finite range [74] and depend on the value of a. Use of BTOs in the UKRmol+ suite makes it possible to perform calculations for higher kinetic energies.
- The choice of R-matrix radius a should be determined by the spatial extent of the wave functions of the electronic target states included in the calculation and of the orbitals used to expand the χ_i. In practice, if UKRmol is used, radii larger than 18 a_0 are not feasible (due to linear dependence problems) so the wave functions included in the CC expansion must be built to ensure they are negligible outside the R-matrix sphere. This implies that when the states are diffuse (in particular of Rydberg character) their description is normally very poor. Since the UKRmol+ suite can be compiled in quadruple precision, the orthogonalization step can be performed more accurately, thus avoiding linear dependence problems and enabling the use of larger a.
- The partial wave expansion is currently limited to $l \leq 6$.
- The "model" (basis set, molecular orbitals, CASSCF active space, χ_i functions) to be used is chosen so as to describe

the following as accurately as possible, while ensuring the size of the calculation is reasonable: ground-state energy, ground-state dipole moment, excitation thresholds for those states included in the calculation, and polarization effects. If the target is nonpolar, the permanent quadrupole moment becomes relevant.

- The main limiting factor in the calculations (apart from a) is the size of the Hamiltonian matrix for the target+electron system that needs to be diagonalized in order to obtain the coefficients a_{ijk} and b_{ik} for the wave functions described by Eq. 3.78. This is given by the number of terms that go into the summations in this equation, and this, in turn, is related to the active space used. In particular, the need to ensure balance between the target and target+electron description means that the number of terms that go into Eq. 3.78 is linked to and much bigger than the number of configurations used in the description of the target states. A very recent improvement to the suites [75] will enable significantly larger Hamiltonians to be diagonalized.

3.4.4 Application: Resonances in Pyrimidine

One of the strengths of the R-matrix method is its ability to describe electronic excitation and two-particle one-hole (i.e., core-excited) resonances. A case in point is the study of pyrimidine ($C_4H_4N_2$). Three nucleobases (cytosine, thymine, and uracil) are pyrimidine derivatives; therefore pyrimidine can be used as a model system to understand electron interaction with them. The advantage of pyrimidine is that its higher symmetry (the molecule belongs to the C_{2v} point group) makes the calculations computationally less demanding than for the nucleobases. Details of the calculations can be found in [76, 77]: the challenge was to find a model that would allow the description of the excited states of the molecule as well as the effects of the polarization. Pyrimidine has a large spherical polarizability (around 50–60 a_0^3) and a significant dipole moment ($\mu = 2.3$ Debye).

Describing the polarization effects accurately is crucial to describing, in particular, shape resonances accurately. We included

28 excited states in the calculation and partial waves up to $l = 4$ (note that this means that these calculations involved a few hundred channels).

Three shape resonances had been clearly identified in pyrimidine, both in electron transmission spectroscopy (ETS) (the first data is from [79]) and theoretically [80]. In addition to these three π^* resonances, our calculations showed the presence of a high-lying (>9.5 eV) broad shape resonance, a larger than expected number of core-excited shape resonances, and at least a couple of very narrow Feshbach resonances. Only one core-excited resonance was identified by Modelli et al. [81] around 5.5 eV in their ETS spectra.

Due to their characteristics many of the core-excited shape resonances are not visible in the elastic cross section, but some of them can be seen in the total electronically inelastic cross section: these are the resonances labeled E3, E4, I5, and E6 in Fig. 3.10

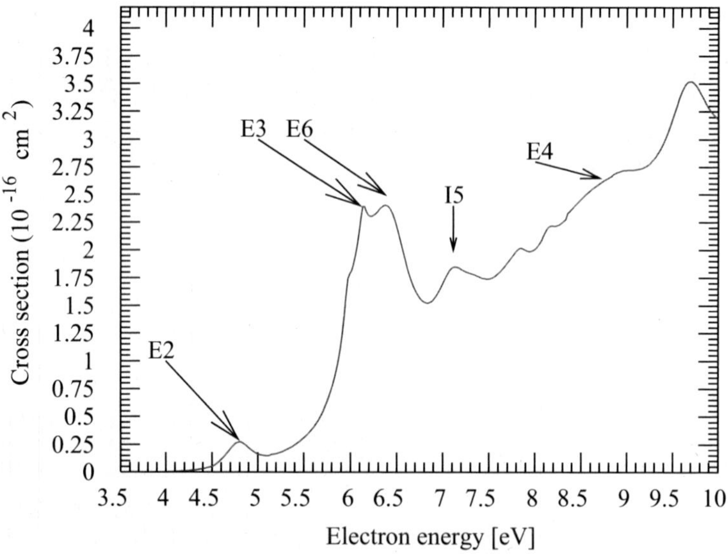

Figure 3.10 Total inelastic cross sections (summed over all electronically excited states included in the calculation) of pyrimidine calculated at the CC level using a cc-pVDZ basis set. Those resonances visible in the cross section are indicated with arrows (the Feshbach resonances are not visible due to the energy resolution chosen for the calculation). A Born correction based on dipole transition moments has been added to the cross sections.

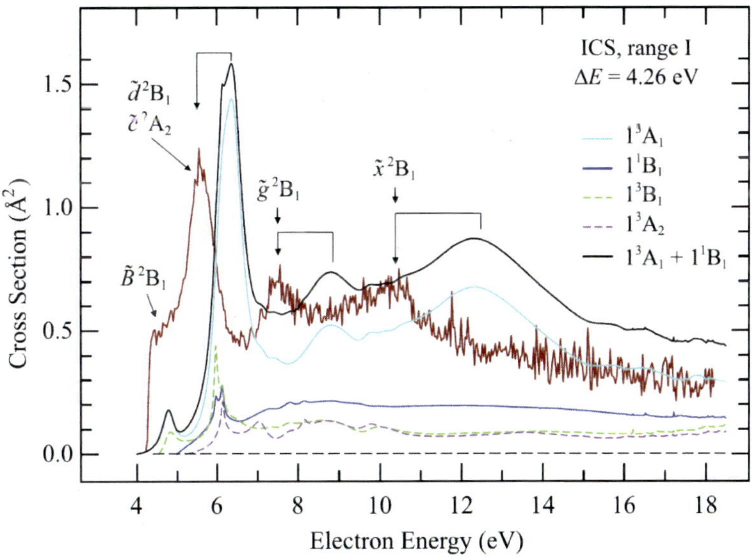

Figure 3.11 Integral cross section for electronic excitation of some low-lying states of pyrimidine [82]. The smooth coloured curves are calculated integral cross sections for each of the four electronic states indicated. The black curve is the sum of the cross sections for exciting the $1\,^3A_1$ and $1\,^1B_1$ states, which contribute to the experimental excitation signal recorded at $\Delta E = 4.26$ eV, shown by the red curve with statistical noise. The $1\,^3B_1$ and $1\,^3A_2$ states are not included in the sum because they lie, respectively, below and above the energy loss of 4.26 eV. The resonances clearly present in both experimental and calculated data are indicated with arrows. Reprinted from Ref. [82], with the permission of AIP Publishing.

(resonance E2 is the third π^* shape resonance that is, in fact, of mixed character). These results suggest that the structure found by Modelli et al. at 5.5 eV corresponds in fact to two overlapping core-excited shape resonances of $^2B_1$ and $^2A_2$ symmetry found by us around 6 eV (labeled E3 and E6 in the figure).

Experimental confirmation of the presence of many of these resonances was provided by electron energy loss spectra [82]. Figure 3.11 shows the cross section for electronic excitation of the lowest singlet and triplet states of pyrimidine: five resonances (four core-excited and the third π^* resonance) are clearly visible in both the measured and calculated cross sections. Excitation cross sections measured for an energy loss of 5.2 eV show the presence of two additional core-excited resonances of $^2B_1$ and $^2A_2$ symmetry. The shift to higher energy of the calculated resonance positions with respect to experiment is well understood, and it is due to a an incomplete description of polarization effects (see [82] for more details). It is worth noting that the measured and calculated excitation cross sections agree very well, demonstrating the quality of the methodologies used.

Similar models have been applied to study electron scattering from DNA bases, while similar as well simpler (SEP) calculations have been performed for amino acids (glycine and alanine, respectively) [83]. Studies of small molecular clusters containing one molecule of biological relevance and up to five water molecules have also been performed in order to understand the effects of microhydration on resonance formation [84, 85]. Finally, use of the UKRmol+ suite (and BTOs for the continuum description) is enabling highly accurate calculations for small molecular systems like BeH [86].

3.5 Summary

We have concisely presented three low-energy methods to treat electron–molecule collisions: the SMC, the DMR method, and the R-matrix method. We have illustrated their application to the study of resonances in furan, cyclopropane, and pyrimidine, highlighting their strengths. The methods and their computational implementations continue to evolve as the need for more and more accurate data for an expanding number of targets and processes increases.

References

1. Condon, E. (1926). A theory of intensity distribution in band systems. *Phys. Rev.*, **28**, pp. 1182–1201.
2. Franck, J. (1926). Elementary processes of photochemical reactions. *Trans. Faraday Soc.*, **21**, pp. 536–542.

3. O'Malley, T. F. (1966). Theory of dissociative attachment. *Phys. Rev.*, **150**, pp. 14–29.
4. Čížek, M. and Houfek, K. (2012). Nonlocal theory of resonance electron-molecule scattering. In *Low-Energy Electron-Molecule Scattering from Molecules, Biomolecules and Surfaces*, Čársky, P. and Čurík R., eds. CRC Press, Boca Raton.
5. Riss, U. V. and Meyer H.-D. (1993). Calculation of resonance energies and widths using the complex absorbing potential method. *J. Phys. B*, **26**, p. 4503.
6. Moiseyev, N. (1998). Quantum theory of resonances: calculating energies, widths and cross sections by complex scaling. *Phys. Rep.*, **302**, pp. 212–293.
7. Simon, B. (1979). The definition of molecular resonance curves by the method of exterior complex scaling. *Phys. Lett. A*, **71**, pp. 211–214.
8. Horáček, J., Mach, P., and Urban, J. (2010). Calculation of S-matrix poles by means of analytic continuation in the coupling constant: application to the $^2\Pi_g$ state of N_2^-, *Phys. Rev. A*, **82**, p. 032713.
9. Jagau, T.-C., Bravaya, K. B., and Krylov, A. I. (2017). Extending quantum chemistry of bound states to electronic resonances. *Annu. Rev. Phys. Chem.*, **68(**1), pp. 525–553.
10. Taylor, J. R. (2006). *Scattering Theory: The Quantum Theory on Nonrelativistic Collisions*. Dover.
11. Smith, F. T. (1960). *Phys. Rev.*, **118**(1), pp. 349–356.
12. Resigno, T. N., McCurdy, C. W., Orel, A. E., and Lengsfield III, B. H. (1995). In *Computational Methods for Electron Molecule Collisions*, Huo, W. M. and Gianturco, F., eds. Plenum Press, New York, pp. 1–44.
13. Schneider, B. I. and Rescigno, T. N. (1988). Complex Kohn variational method: application to low-energy electron-molecule collisions. *Phys. Rev. A*, **37**, pp. 3749–3754.
14. Gianturco, F. A., Thompson D. G., and Jain, A. (1995). Electron-scattering from polyatomic molecules using a single-center-expansion formulation. In *Computational Methods for Electron-Molecule Collisions*, Huo, W. M. and Gianturco, F. A., eds. Plenum Press, New York.
15. Blanco, F. and García, G. (2004). *Phys. Lett. A*, **330**, p. 230.
16. Takatsuka, K. and McKoy, V. (1984). Theory of electronically inelastic scattering of electrons by molecules. *Phys. Rev. A*, **30**(4), pp. 1734–1740.
17. Joachain, C. J. (1975). *Quantum Collisions Theory*. North-Holland.
18. Geltman, S. (1969). *Topics in Atomic Collision Theory*. Academic Press, New York.

19. Lima, M. A. P. and McKoy, V. (1988). Aspects of the Schwinger multichannel variational formulation. *Phys. Rev. A*, **38**(1), pp. 501–503.
20. Kossoski, F. and Bettega, M. H. F. (2013). Low-energy electron scattering from the aza-derivatives of pyrrole, furan, and thiophene. *J. Chem. Phys.*, **138**(23), p. 234311.
21. Winstead, C. and McKoy, V. (1998). Electron collisions with nitrous oxide. *Phys. Rev. A*, **57**(5), pp. 3589–3597.
22. Takatsuka, K. and McKoy, V. (1981). Extension of the Schwinger variational principle beyond the static-exchange approximation. *Phys. Rev. A*, **24**(5), pp. 2473–2480.
23. Lima, M. A. P., Brescansin, L. M., da Silva, A. J. R., Winstead, C., and McKoy, V. (1990). Applications of the Schwinger multichannel method to electron-molecule collisions. *Phys. Rev. A*, **41**(1), pp. 327–332.
24. Bachelet, G. B., Hamann, D. R., and Schlüter, M. (1982). Pseudopotentials that work: from H to Pu. *Phys. Rev. B*, **26**(8), pp. 4199–4228.
25. Bettega, M. H. F., Ferreira, L. G., and Lima, M. A. P. (1993). Transferability of local-density norm-conserving pseudopotentials to electron-molecule-collision calculations. *Phys. Rev. A*, **47**(2), pp. 1111–1118.
26. Bettega, M. H. F., Natalense, A. P. P., Lima, M. A. P., and Ferreira, L. G. (1996). Note on the generation of Gaussian bases for pseudopotential calculations. *Int. J. Quantum Chem.*, **60**(4), pp. 821–824.
27. Winstead, C. and McKoy, V. (1996). Highly parallel computational techniques for electron-molecule collisions. *Adv. Atom., Mol. Opt. Phys.*, **36**, pp. 183–219.
28. dos Santos, J. S., da Costa R. F., and Varella, M. T. do N. (2012). Low-energy electron collisions with glycine. *J. Chem. Phys.*, **136**, p. 084307.
29. da Costa, R. F., Varella, M. T. do N., Bettega, M. H. F., and Lima, M. A. P. (2015). Recent advances in the application of the Schwinger multichannel method with pseudopotentials to electron-molecule collisions. *Eur. Phys. J. D*, **69**, p. 159.
30. da Costa R. F., da Paixão, F. J., and Lima, M. A. P. (2005). Cross sections for electron-impact excitation of the H_2 molecule using the MOB-SCI strategy. *J. Phys. B*, **38**(24), pp. 4363–4378.
31. da Costa, R. F., Bettega, M. H. F., Varella, M. T. do N., de Oliveira, E. M., and Lima, M. A. P. (2014). Electron collisions with ethylene: the role of multichannel-coupling effects. *Phys. Rev. A*. **90**(5), p. 052707.
32. Kossoski, F. and Varella, M. T. do N. (2015). Negative ion states of 5-bromouracil and 5-iodouracil. *Phys. Chem. Chem. Phys.*, **17**(26), pp. 17271–17278.

33. Keane, K., Campbell, C., Guzman, N., and Hazlett, K. (2007). Low energy elastic electron scattering from ethylene. *J. Phys. B*, **40(**18), p. 3601.

34. Kitajima, M., Sakamoto, Y., Gulley, R. J., Hoshino, M., Gibson, J. C., Tanaka, H., and Buckman, S. J (2000). Electron scattering from N_2O: absolute elastic scattering and vibrational excitation. *J. Phys. B*, **33**(9), p. 1687.

35. Mapstone B. and Newell, W. R. (1992). Elastic differential electron scattering from CH_4, C_2H_4 and C_2H_6. *J. Phys. B*, **25**(2), p. 491.

36. Panajotovic, R., Kitajima, M., Tanaka, H., Jelisavcic, M., Lower, J., Campbell, L., Brunger, M. J., and Buckman, S. J. (2003). Electron collisions with ethylene. *J. Phys. B*, **36**(8), p. 1615.

37. Allan, M., Winstead, C., and McKoy, V. (2008). Electron scattering in ethene: excitation of the $\tilde{a}^3B_{1u}$ state, elastic scattering, and vibrational excitation. *Phys. Rev. A*, **77**(4), p. 042715.

38. Do, T. P. T., Nixon, K. L., Fuss, M., García, G., Blanco, F., and Brunger, M. J. (2012). Electron impact excitation of the $\tilde{a}^3B_{1u}$ electronic state in C2H4: an experimentally benchmarked system?. *J. Chem. Phys.*, **136**(18), p. 184313.

39. Lane, N. F. (1980). The theory of electron-molecule collisions. *Rev. Mod. Phys.*, **52**, pp. 29–119.

40. Morrison, M. A. and Collins, L. A. (1978). Exchange in low-energy electron-molecule scattering- free-electron-gas model exchange potentials and applications to e-H_2 and e-N_2 collisions. *Phys. Rev. A*, **17**, pp. 918–938.

41. Connel, O'J. K. and Lane, N. F. (1983). Nonadjustable exchange-correlation model for electron scattering from closed-shell atoms and molecules. *Phys. Rev. A*, **27**, pp. 1893–1903.

42. Padial, N. T. and Norcross, D. W. (1984). Parameter-free model of the correlation polarization potential for electron-molecule collisions. *Phys. Rev. A*, **29**, pp. 1742–1748.

43. Perdew, J. P. and Zunger, A. (1981). Self-interaction correction to density-functional approximations for many-electron systems. *Phys. Rev. B*, **23**, pp. 5048–5079.

44. Vosko, S. H., Wilk, L., and Nusair, M. (1980). Accurate spin-dependent electron liquid correlation energies for local spin-density calculations - a critical analysis. *Can. J. Phys.*, **58**, pp. 1200–1211.

45. Perdew, J. P., Chevary, J. A., Vosko, S. H., Jackson, K. A., Pederson, M. R., Singh, D. J., and Fiolhais, C. (1992). Atoms, molecules, solids, and surfaces - applications of the generalized gradient approximation for exchange and correlation. *Phys. Rev. B*, **46**, pp. 6671–6687.

46. Lee, C. T., Yang, W. T., and Parr, R. G. (1988). Development of the Colle-Salvetti correlation energy formula into a functional of the electron-density. *Phys. Rev. B*, **37**, pp. 785–789.
47. Colle, R. and Salvetti, O. (1975). Approximate calculation of correlation energy for closed shells. *Theor. Chim. Acta*, **37**, pp. 329–334.
48. Umrigar, C. J. and Gonze, X. (1994). Accurate exchange-correlation potentials and total energy components for the helium isoelectronic series. *Phys. Rev. A*, **50**, pp. 3827–3837.
49. Čurík, R. and Šulc, M. (2010). Towards efficient ab initio calculations of electron scattering by polyatomic molecules: III. Modelling correlation-polarization interactions. *J. Phys. B*, **43**, p. 175205.
50. Telega, S., Bodo, E., and Gianturco, F. A. (2004). Rotationally inelastic collisions of electrons with H_2 and N_2 molecules: converged space-frame calculations at low energies. *Eur. Phys. J. D*, **29**, pp. 357–365.
51. Polášek, M., Juřek, M., Ingr, M., Čársky, P., and Horáček, J. (2000). Discrete momentum representation of the Lippmann-Schwinger equation and its application to electron molecule scattering. *Phys. Rev. A*, **61**, p. 032701.
52. Čurík, R. and Čársky, P. (2003). Vibrationally inelastic electron scattering on polyatomic molecules by the discrete momentum representation (DMR) method. *J. Phys. B*, **36**, pp. 2165–2177.
53. Nishimura, H. and Tawara, H. (1991). Some aspects of total scattering cross sections of electrons for simple hydrocarbon molecules. *J. Phys. B*, **24**, p. L363.
54. Makochekanwa, C., Kato, H., Hoshino, M., Tanaka, H., Kubo, H., Lopes, A. R., Lima, M. A. P., and Ferreira, L. G. (2006). Experimental and theoretical elastic cross sections for electron collisions with the isomers. *J. Chem. Phys.*, **124**, p. 024323.
55. Szmytkowski, C. and Kwitnewski, S. (2002). Electron scattering on C_3H_6 isomers. *J. Phys. B*, **35**, pp. 2613–2623.
56. Čurík, R. and Gianturco, F. A. (2002). A computational analysis of low-energy electron scattering from gaseous cyclopropane. *J. Phys. B*, **35**, pp. 717–732.
57. Allan, M. and Andric, L. (1996). σ^* resonances in electron impact-induced vibrational excitation of n-propane, cyclopropane, ethylene oxide, cyclopentane, and cyclohexane. *J. Chem. Phys.*, **105**, pp. 3559–3568.
58. Čurík, R. and Gianturco, F. A. (2002). Quantum calculations for resonant vibrational excitations of cyclopropane by electron impact. *J. Phys. B*, **35**, pp. 1235–1250.

59. Gianturco, F. A. and Thompson, D. G. (1977). Theoretical evidence for short-lived resonances in electron scattering by HCl and HF molecules. *J. Phys. B*, **10**, pp. L21–L26.

60. Stibbe, D. T. and Tennyson, J. (1996). Time-delay matrix analysis of resonances in electron scattering: e^--H_2 and H_2^+. *J. Phys. B*, **29**, pp. 4267–4283.

61. Lebedev, V. I. and Laikov, D. N. (1999). Quadrature formula for the sphere of 131-th algebraic order of accuracy. *Dokl. Akad. Nauk*, **366**, pp. 741–745.

62. Čurík, R., Paidarová, I., Allan, M., and Čársky, P. (2014). Joint experimental and theoretical study on vibrational excitation cross sections for electron collisions with diacetylene. *J. Phys. Chem. A*, **118**, pp. 9734–9744.

63. Wigner, E. P. and Eisenbud, L. (1947). Higher angular momenta and long range interaction in resonance reactions. *Phys. Rev.*, **72**, pp. 29–41.

64. Burke, P. G., Hibbert, A., and Robb, W. D. (1971). Electron scattering by complex atoms. *J. Phys. B: At. Mol. Phys.*, **4**, pp. 153–161.

65. Burke, P. G., Mackey, I., and Shimamura, I. (1977). R-matrix theory of electron-molecule scattering. *J. Phys. B: At. Mol. Phys.*, **10**, pp. 2497–2512.

66. Schneider, B. (1975). R-matrix theory for electron-atom and electron-molecule collisions using analytic basis set expansions. *Chem. Phys. Lett.*, **31**, pp. 237–241.

67. Shimamura, I. (2012). In *Advances in Quantum Chemistry*, Vol. 63, Cleanthes, E. B., Nicolaides, A., and Sabin, J. R., eds. Academic Press, pp. 165–245, ISBN 0065-3276.

68. Morgan, L. A., Gillan, C. J., Tennyson, J., and Chen, X. (1997). *J. Phys. B: At. Mol. Opt. Phys.*, **30**, p. 4087.

69. Pfingst, K., Nestmann, B. M., and Peyerimhoff, S. D. (1994). An R-matrix approach for electron scattering off polyatomic molecules. *J. Phys. B: At. Mol. Opt. Phys.*, **27**, p. 2283.

70. Burke, P. G. (2011). *R-Matrix Theory of Atomic Collisions: Application to Atomic, Molecular and Optical Processes*. Springer.

71. Tennyson, J. (2010). Electron - molecule collision calculations using the R-matrix method. *Phys. Rep.*, **491**, pp. 29–76.

72. Jensen, F. (1999). *Introduction to Computational Chemistry*. Wiley, Chircester.

73. Tennyson, J. (1996). R-matrix calculation of Rydberg states of CO. *J. Phys. B: At. Mol. Phys.*, **29**, pp. 6185–6201.

74. Faure, A., Gorfinkiel, J. D., Morgan, L. A., and Tennyson, J. (2002). GTOBAS: fitting continuum functions with Gaussian-type orbitals. *Comput. Phys. Commun.*, **144**, pp. 224–241.
75. Al-Refaie, A. F. and Tennyson, J. (2017). A parallel algorithm for hamiltonian matrix construction in electron-molecule collision calculations: MPI-SCATCI. *Comput. Phys. Commun.*, **221**, pp. 53–62.
76. Mašín, Z. and Gorfinkiel, J. D. (2012). Shape and core excited resonances in electron collisions with diazines. *J. Chem. Phys.*, **137**(20), p. 204312.
77. Mašín, Z. and Gorfinkiel, J. D. (2014). Resonance formation in low energy electron scattering from uracil. *Eur. Phys. J. D*, **68**(5), p. 112.
78. Norcross, D. W. and Padial, N. T. (1982). The multipole-extracted adiabatic-nuclei approximation for electron-molecule collisions. *Phys. Rev. A*, **25**, pp. 226–338.
79. Nenner, I. and Schulz, G. J. (1975). Temporary negative ions and electron affinities of benzene and N-heterocyclic molecules: pyridine, pyidazine, pyrimidine, pyrazine, and s-triazine. *J. Chem. Phys.*, **62**, p. 1747.
80. Palihawadana, P., Sullivan, J., Brunger, M., Winstead, C., McKoy, V., García, G., Blanco, F., and Buckman, S. (2011). Low-energy elastic electron interactions with pyrimidine. *Phys. Rev. A*, **84**, p. 062702.
81. Modelli, A., Bolognesi, P., and Avaldi, L. (2011). Temporary anion states of pyrimidine and halopyrimidines. *J. Phys. Chem. A*, **115**(39), pp. 10775–10782.
82. Regeta, K., Allan, M., Mašín, Z., and Gorfinkiel, J. D. (2016). Absolute cross sections for electronic excitation of pyrimidine by electron impact. *J. Chem. Phys.*, **144**, p. 2.
83. Gorfinkiel, J. D. and Ptasińska, S. (2017). Electron scattering from molecules and molecular aggregates of biological relevance. *J. Phys. B: At. Mol. Opt. Phys.*, **50**(18), p. 182001.
84. Sieradzka, A. and Gorfinkiel, J. D. (2017a). Theoretical study of resonance formation in microhydrated molecules. I. pyridine-$(H_2O)_n$, n = 1,2,3,5. *J. Chem. Phys.*, **147**(3), p. 034302.
85. Sieradzka, A. and Gorfinkiel, J. D. (2017b). Theoretical study of resonance formation in microhydrated molecules. II. thymine-$(H_2O)_n$, n = 1,2,3,5. *J. Chem. Phys.*, **147**(3), p. 034303.
86. Darby-Lewis, D., Mašín, Z., and Tennyson, J. (2017). R-matrix calculations of electron impact electronic excitation of BeH. *J. Phys. B: At. Mol. Opt. Phys.*, **50**(17), p. 175201.

Chapter 4

Plasmas: Low-Energy Electrons in Action

Nigel J. Mason
School of Physical Sciences, University of Kent, Canterbury, Kent, CT2 7NH, UK
School of Physical Sciences, The Open University, Walton Hall,
Milton Keynes, Bedfordshire, MK7 6AA, UK
n.j.mason@kent.ac.uk

In this chapter the role of low-energy electrons in plasmas is reviewed. Plasmas are ubiquitous; indeed, plasma is the most common form of matter in the universe. Inherent to the formation and sustaining of any plasma are the charged species within it. Low-energy electrons (often below the first ionization energy of the plasma neutral feed gas) are omnipresent in plasmas, and many of the plasma characteristics are derived from these electrons. The role of different electron collision processes in different plasmas is discussed and our state of knowledge of such processes reviewed since, despite over a century of study, our ability to define and hence control such interactions is limited by our lack of knowledge of some of the underpinning physics and chemistry. Particular attention will be given to the role of dissociative electron attachment (DEA), a process common in electronegative plasmas, the understanding of which has led to refinement of current plasma technology and provides intriguing opportunities for next-generation plasma systems and development of nanostructured devices.

Low-Energy Electrons: Fundamentals and Applications
Edited by Oddur Ingólfsson

ISBN 978-981-4800-00-6 (Hardback), 978-0-429-05882-0 (eBook)
www.panstanford.com

4.1 Introduction

4.1.1 Plasma: A Ubiquitous State of Matter

A plasma is composed of almost equal concentrations of positive (usually ions) and negative particles (commonly electrons) such that the overall charge of a plasma is roughly zero. Plasmas have been called the "fourth state of matter" compatible with the more common gas, liquid, and solid forms. Indeed, most of the universe exists in a plasma state with the stars and interstellar medium (ISM) being examples of high- and low-temperature plasmas. Plasma-based technology underpins some of the world's largest industries (Fig. 4.1), producing a significant proportion of the world's global commercial products, including computers, cellphones, automobiles, airplanes, paper, and textiles [1]. Foremost among these is the electronics industry, in which plasma-based processes are indispensable for the manufacture of ultralarge-scale integrated microelectronic circuits. According to the Semiconductor Industry Association, worldwide sales of microelectronic components were $338.9 billion in 2016 up from $247.7 billion in 2006, with worldwide semiconductor wafer-level manufacturing equipment (WFE) revenue totaling $37.4 billion in 2016, an 11.3% increase from 2015, with no evidence of a reduction in growth (indeed, in July 2017 global sales increased by 24% over 2016, with US companies reporting a record 36% increase).

Plasma processing and coating of materials is a key technology in the aerospace, automotive, paper, and textile industries. Since 70%–80% of the light on the planet is produced by plasmas, plasma technology also underpins the global lighting industry. With the growing consensus that anthropogenic climate change is now inevitable, plasma technology is set to play a key role in the development of a cleaner and more environmentally conscious world, for example, in toxic waste management, air pollution detection and removal, and the development of fusion as a clean, sustainable energy source. Plasma processing will also play an essential role in the emerging technologies, including carbon-based electronics, solar cell manufacture, and hydrogen storage. The role of plasmas in biomedicine, where they may be used both as a

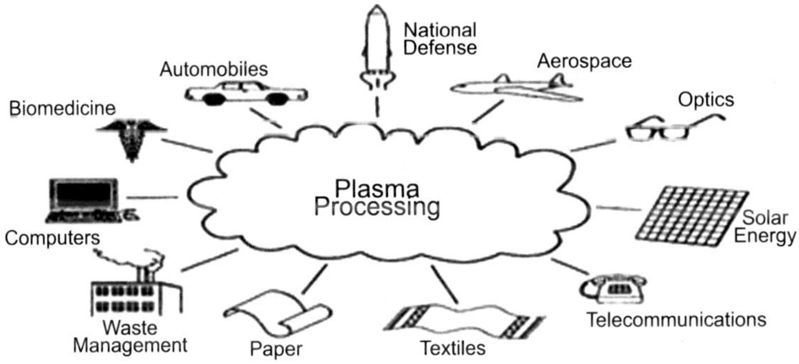

Figure 4.1 Plasma processing is a critical technology to many worldwide industries. Reprinted with permission from Ref. [1], (1991) by the National Academy of Sciences, Courtesy of the National Academies Press, Washington, D.C.

sterilization agent and as a possible surgical tool, with minimum tissue damage, is also making rapid progress. Finally, and crucially, plasma technology currently still provides the only truly viable methodology for further reductions in the size of structures in integrated circuits (e.g., fabrication of carbon nanotubes) and hence the introduction of nanoelectronics into the marketplace.

Underpinning all plasma technology is the formation and persistence of the plasma itself, which is dependent upon the fundamental collision processes that generate the plasma and the species generated therein that subsequently control the chemistry induced by the plasma in its local environment. Amongst the most important collision processes are those involving low-energy electrons. In this chapter we will explore the role of electrons in plasmas and how they influence the physical and chemical properties of the plasma.

4.1.2 Plasma Properties

A plasma is characterized by a series of parameters; these include traditional and well-known terms such as density and temperature but also plasma-specific terms such as degree of ionization, plasma frequency, and the Debye length. For a full description of these terms and plasma characteristics, the reader is referred to classic plasma textbooks [2–4].

4.1.2.1 Plasma density

Plasma density is defined as the number of charge carriers per unit volume, traditionally quoted as particles per cubic centimeter. The density may vary considerably depending upon the type of plasma, ranging from $<10^{-5}/\text{cm}^3$ in space to $>10^{20}/\text{cm}^3$ for laser-induced plasma and stellar atmospheres. Since plasmas are quasi-neutral the density of positive and negative species over the whole plasma are similar, $n_i \approx n_e$, where n_i and n_e are the densities of ions and electrons, respectively, but in different regions of the plasma the density for any one species may vary significantly. However, in any plasma where multiply charged ions are common we have to modify this slightly such that if z is the charge number of a positive ion and n_z is the density of z times charged ions, the condition for quasi-neutrality will be when $n_e = \Sigma_z \times n_z$. Moreover when considering the chemistry within the plasma it is also important to recall the neutral species density.

4.1.2.2 Plasma temperature

Plasma temperature is a measure of the thermal kinetic energy per particle, so it is commonly reported in electron volts as well as traditional Kelvin: 1 K corresponds to $8.6173324(78) \times 10^{-5}$ eV; this factor is the ratio of the Boltzmann's constant to the elementary charge and thus 4000 K is 0.35 eV. Often, the plasma temperature referenced in literature is the *electron temperature* and is a direct measure of the energy of the electrons in the plasma. Due to the large difference in mass, the electrons reach thermodynamic equilibrium amongst themselves much faster than when they come into equilibrium with ions or neutral atoms. For this reason, the ion temperature may be very different from (usually lower than) the electron temperature. This is especially common in weakly ionized technological plasmas, where the ions are often near the ambient neutral gas temperature. The definition of an electron temperature should not be seen as a "monochromatic" electron energy within the

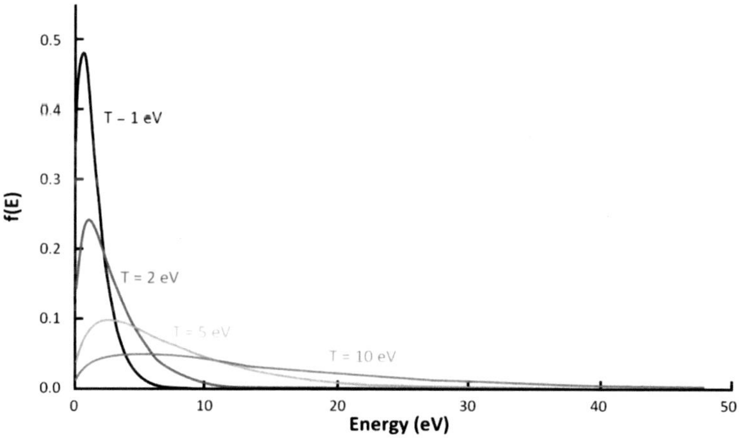

Figure 4.2 Relationship between energy distribution in a plasma and the electron energy distribution function $f(E)$ for different electron temperatures. Note the high-energy tail in the distribution, which sustains ionization in the plasma even at low electron temperatures.

plasma; rather the electron temperature is a measure of the electron energy distribution function (EEDF) (Fig. 4.2). If the velocities of a group of electrons follow a Maxwell–Boltzmann distribution, then the electron temperature is well defined as the temperature of that distribution; for other distributions, two-thirds of the average energy is often referred to as the temperature, since for a Maxwell–Boltzmann distribution with 3 degrees of freedom,

$$\langle E \rangle = (3/2)\langle k_{\rm B}T \rangle. \tag{4.1}$$

Very high temperatures are usually needed to sustain ionization of a plasma, so the kinetic energy of the electrons must be high. A plasma with $T < 10^5$ K is considered a low-temperature plasma (typical of most industrial plasmas), and a plasma with $T \approx 10^6$–10^8 K or higher is regarded as a high-temperature plasma (e.g., laser-induced, fusion, and stellar plasmas). Although the electron temperature may be low (i.e., the mean or peak of the energy distribution is low) and beneath the ionization energy of the neutral species of the plasma, there must be sufficient high energy electrons in the "tail" to sustain the plasma.

4.1.2.3 Plasma frequency

Separation of charges in the plasma leads to the creation of an electrostatic field E, which, in turn, leads to movement of the charges generating a restoring force $F = eE$, leading to oscillatory motion of the particles. The plasma frequency, ω_p, is a measure of timescales of such oscillations in the plasma, defined as $(ne^2/\varepsilon_0 m)^{1/2}$, where e and m are the charge and mass of the oscillating electrons or ions, respectively; n is their number density; and ε_0 is the permittivity of empty space. Clearly, there is a different plasma frequency for each charged species; however, the relatively fast electron frequency is the most important and the "plasma frequency," as stated in most texts, refers to the electron plasma frequency $\omega_{pe} = 2\pi \times 8.9 \times \sqrt{n_e m^{-3}}$. A more detailed analysis of the plasma shows that, instead of these oscillations, different types of acoustic waves can propagate in plasmas; however, the electron and ion plasma frequencies are still commonly used as important parameters for characterizing the plasma.

4.1.2.4 Debye length

The Debye length, also commonly called *Debye radius*, is the measure of the effective range over which the electrostatic effect of any charge carrier in a plasma is felt by neighboring particles. A Debye sphere is then defined as the volume whose radius is the Debye length; within the sphere the electrostatic fields influence targets, and outside the sphere all charges in the sphere are electrically screened. The Debye length may cover several orders of magnitude. In space plasmas such as in the magnetosphere, the solar wind, and the ISM, the electron and ion densities are relatively low, so the Debye length may reach 10 m, whereas in a dense plasma such as a gas discharge or a fusion plasma, $\lambda_D = 10^{-4}$ m. Again, the Debye length is often referenced to the electrons in the plasma with

$$\lambda_D = \sqrt{\frac{\varepsilon_0 k T_e}{n_e q_e^2}} \tag{4.2}$$

where ε_0 is the permittivity of free space, k_B is Boltzmann's constant, q_e is the charge on an electron, T_e is the electron temperature, and n_e is the density of electrons.

4.1.2.5 Degree of ionization

The ratio of the number of ionized atoms to the total number of atoms per unit volume of the plasma is called the plasma's degree of ionization, a. Depending on the magnitude of a, we characterize plasmas as weakly, strongly, or completely ionized plasmas. Most technological plasmas are only weakly ionized: $a = 10^{-6}$ to 10^{-4}. Examples of nearly completely ionized plasmas ($a \approx 1$) include the solar wind (interplanetary medium), stellar interiors (the sun's core), and fusion plasmas. It is important to remember that even in highly ionized plasmas, the number density of neutral species remains high, and in most plasmas the number of ions/electrons per unit volume remains low; hence there is a high probability that the ions/electrons will interact with the neutral species initiating chemistry. Hence in studying any plasma the role of plasma-induced chemistry is important, if not dominant, and since electrons are a major charged species in the plasma, electron-induced chemistry may be play dominant role in the plasma chemistry.

4.1.3 Electron Processes in Plasma Systems

This chapter does not aim to provide a complete description of the physical and chemical processes in plasmas—this is the subject of many textbooks and reviews [2–4]. Rather, in this chapter we will review the role of electrons in plasmas and in particular processes involving low-energy electrons. There are many electron collisional processes in any plasma; these include elastic scattering, inelastic scattering and ionization.

4.1.3.1 Elastic scattering

$$e^- + AB \longrightarrow e^- + AB$$

In this equation, no kinetic energy is exchanged between the incident electron and the target. However, the electron is scattered such that the momentum transfer cross section is nonzero. Elastic cross sections (Q_{Elastic}) are usually measured at specific energies and angles. These data are used to determine the integral elastic cross section by integrating over the angular range, which may involve extrapolation or use of theoretical models to extend the range of measurements. The elastic differential cross section (elastic DCS) therefore provides a more stringent test of theory than simple comparison with the integral cross sections. This is particularly true of collisions with polar molecules, which show a strong DCS at small angles (i.e., strong forward scattering) where empirical integral cross sections are often strongly dependent upon the method used to represent low-angle scattering.

The integral elastic cross section at a given electron impact energy E is given by

$$Q_{\text{INT,elastic}}(E) = 2\pi \int_0^{\pi} \sigma(E,\theta) \cdot \sin\theta \cdot d\theta \tag{4.3}$$

where $\sigma\ (E,\theta)$ is a DCS.

In many plasmas the effective gas temperature is such that the molecular species composing the plasma feed gas are rotationally and even vibrationally excited, while many electronic states may be excited within the plasma (the decay of which leads to the optical transitions by which most plasmas are observed). In this case effective elastic cross sections are required, which can be referred to as rotationally unresolved, vibrationally unresolved, electronically unresolved, etc.

The elastic momentum transfer cross section $Q_{\text{MT,elastic}}$ is the $(1 - \cos\theta)$ weighted, angle-integrated DCS for elastic scattering, where θ is the scattering angle.

$$Q_{\text{MT,elastic}}(E) = 2\pi \int_0^{\pi} \sigma(E,\theta) \cdot (1-\cos\theta) \cdot \sin\theta \cdot d\theta \tag{4.4}$$

This can differ significantly from the elastic integral cross section $Q_{\text{Inelastic}}$, obtained by direct integration of the corresponding DCS over all angles. This cross section may also be rotationally unresolved, vibrationally unresolved, electronically unresolved, etc.

In electron transport modeling commonly used in plasmas, input to the two-term electron Boltzmann's equation requires both $Q_{\text{MT,elastic}}$ and the *total* momentum transfer cross section $Q_{\text{MT,tot}}$, which includes the contribution of elastic, excitation, and ionization processes such that $Q_{\text{MT,tot}} = Q_{\text{MT,elastic}} + Q_{\text{MT,exc}} + Q_{\text{MT,ion}}$, where

$Q_{\text{MT,exc}}$ is the excitation of the state with the threshold ε_i:

$$Q_{\text{MT,exc\_i}}(E) = 2\pi \int_0^{\pi} \sigma(E, \theta) \cdot \left[1 - \left(1 - \frac{\varepsilon_i}{E}\right)^{1/2} \cos\theta\right] \cdot \sin\theta \cdot d\theta \qquad (4.5)$$

These cross sections appear naturally in the collision term in Boltzmann's equation when the angular dependence of the velocity is expanded in Legendre polynomials. However, anisotropies in the inelastic cross sections are often neglected, introducing the practice that an approximation $Q_{\text{MT,tot}} \approx Q_{\text{MT,elastic}} + Q_{\text{exc}} + Q_{\text{ion}}$ is used.

4.1.3.2 Inelastic scattering

$$\text{e}^- + \text{AB} \longrightarrow \text{e}^- + \text{AB}^*$$

Here AB* denotes some excited state of the target (in the case of a molecule this may be a rotational, vibrational, or electronically excited state). Inelastic scattering is the process by which kinetic energy from the electron is transferred to the target and hence is pivotal to the energy balance and chemistry induced by the plasma. Cross sections for each excitation process may be evaluated or measured (Q_{rot}, Q_{vib}, and $Q_{\text{electronic}}$) and are the core input for any model of the plasma. It should be noted that if an electron scatters from an excited target, it may actually *gain* energy by de-exciting the target; such a process is called *superelastic scattering* and is one way by which higher-energy electrons may be formed in the plasma, which itself may have a low electron temperature.

In its pure form a rotational excitation cross section should be defined as state to state (and is described as such in theoretical calculations), but experiments normally do not have sufficient resolution to allow such precise cross sections to be measured, and so the term "rotationally elastic" is often used, which means the (measured) elastic cross sections contains unknown contributions from rotational states.

Similarly, in its pure form a vibrational excitation cross section should be defined as state to state (and is again described as such in theoretical calculations). Some electron energy loss experiments have sufficient resolution to allow these cross sections to be measured, but in many cases two or more vibrational transitions are overlapping such that theory may be used to deconvolute individual cross sections.

Electronic excitation can be strictly defined as $e^- + AB \longrightarrow e^- + AB^*$, where AB^* is an electronically excited state of the target. However, in many cases it may actually be described as a two-step process: the initial excitation of the target (step 1) and the resultant products/decay pathways of that excitation (photons, metastables, or dissociation products) (step 2).

$$\begin{array}{ccc} & \text{Step 1} & \text{Step 2} \\ e^- + AB \longrightarrow & e + AB^* \longrightarrow & e + AB^{*\prime} + h\nu \quad \text{or} \quad e^- + A + B \end{array}$$

Many of these fragments may be reactive and induce chemistry in the plasma; this is particularly true when the target is a polyatomic molecule and the fragments may be radicals as is the case (by choice) in many technological plasmas, such as those developed for plasma etching. Ideally any description of an electron excitation process would report (a) initial cross sections (for step 1) and then (b) branching ratios and lifetimes for the subsequent dynamical process(es) (for step 2). Most applications usually require cross sections for step 2 since the cross sections describing the product yields are used in subsequent chemistry.

Electron energy loss spectroscopy (EELS) measures and theory generally calculates the step 1 process and is described as the *inelastic cross section for electronic excitation*. However, other experiments measure the final products (step 2), the yields of which may be only a partial cross section or complicated by other processes (e.g., cascade); this is particularly true when monitoring metastable and fluorescence yields from a target.

4.1.3.3 Ionization

The most important processes in a plasma are those that produce the ions that characterize the plasma. The most fundamental electron processes in a plasma are therefore those that lead to the formation of long-lived ions; these include electron impact ionization cross sections $e^- + AB \longrightarrow e^- + AB^+$, with the parent cation often subsequently decaying $AB^+ \longrightarrow A^+ + B$, a process called *dissociation ionization*. The ionization process is defined by the total ionization cross section, which is the sum of all partial ionization cross sections (each cross section defining a pathway for a particular cation). The total ionization cross section is the cross section as defined in many semi-empirical calculations, but experiments may report both total and partial cross sections.

4.1.3.4 Dissociative electron attachment

At low energies (below the ionization potential) low-energy electrons may be temporarily captured by the target to form a temporary negative ion (TNI), which lives for only a femtosecond or picosecond before decaying by autodetachment, hence performing delayed elastic scattering.

$$e^- + AB \longrightarrow AB^- \longrightarrow AB + e^-$$

Such a process may lead to the observation of phenomena in the elastic scattering cross section, for example, sharp, often narrow structures; hence the common name for a TNI is *resonance* [5] (Fig. 4.3). Similar structures may be observed in the vibrational excitation cross section; indeed, for many polyatomics at low energies, such resonance phenomena dominate the cross section, with direct electron-induced vibrational excitation being 1 or 2 orders of magnitude lower.

An alternative route of TNI decay leads to the fragmentation of the TNI to form an anionic product $AB^- \longrightarrow A + B^-$, a process called *dissociative electron attachment* (DEA). It is a process common to most molecules, though the probability (cross section) is often low ($<10^{-18}$ cm^2) [6]. Since the anionic product is charged it can be readily detected (e.g., by conventional mass spectrometry), and therefore DEA has been the subject of extensive studies. DEA cross sections are strongest for targets containing halogen atoms (since they are electronegative) but may also be large in compounds containing psuedo-halogen chemical groups such as –CN. DEA occurs naturally in Earth's ionosphere, where O^- anions are formed by DEA to molecular oxygen and in many industrial plasmas; indeed, in several plasmas the anion yield may be dominant, and hence such plasmas are called electronegative plasmas. The specific nature of

Figure 4.7 Aurorae are perhaps the most beautiful of Earth's atmospheric plasma. From https://pl.pinterest.com/pin/312859505342271390/?lp=true.

strongly in the UV, where electrons excite electronic states of nitrogen. Important electron collisions are both direct electron impact excitation and DR of molecular ions of oxygen and nitrogen.

Ionospheres are common to other planets, with those of Venus and Mars being particularly well studied, while that of Jupiter revealed a rich aurora arising from the H_3^+ ions. The recent Cassini–Huygens mission to the Saturnian system has revealed a rich ionospheric chemistry on Titan, including the observation of anions [15]. Titan is the only atmosphere outside Earth to show a rich anion chemistry, which arises due to the DEA of nitriles (molecules containing pseudo-halogen –CN groups) formed in Titan's atmosphere. Recent work has also shown, for the first time in a natural plasma, the importance of ion pair formation, with ion pair formation contributing up to 40% of the anion yield from ethylene in Titan's atmosphere [7].

Atmospheric discharges are increasingly used in industry since they are easily prepared and require no complex (and expensive) vacuum systems. Today several different types of atmospheric discharges are used in industry:

- Direct current (DC) arc
- Atmospheric pressure glow discharge (APGD)
- Dielectric barrier discharge (DBD)

Atmospheric pressure plasmas have a range of applications from surface treatment of materials [16] (e.g., to provide hydrophobic textiles) to medicine, indeed the latter has seen the largest growth over the past decade, with atmospheric plasma discharges now also being used in dentistry (to replace the old-style dentist drill) and surgery. Clinical applications have grown to include use of atmospheric pressure plasmas for blood coagulation and skin sterilization, skin cancer treatment, and treatment of cornea lesions, creating a new field of plasma medicine [17].

Atmospheric pressure plasmas are very efficient sources of highly reactive oxygen and nitrogen species and UV radiation, all of which have biological effects. The current challenge is to quantify the important species and processes in the plasma and correlate these with different biological mechanisms; this, in turn, requires accurate knowledge of the plasma parameters. Any atmospheric pressure plasma using air as the feed gas will contain water as well as the oxygen and nitrogen. A discharge may also form copious amounts of ozone (recognized by its distinct sweet smell); thus a comprehensive data set of electron interactions with these molecules is needed. Furthermore, operating at STP and in the presence of water engenders the formation of clusters often larger hydrated clusters [18]; however, to date our knowledge of electron interactions with such clusters is at best rudimentary. Elastic electron scattering from clusters is expected to have a high cross section due to geometry effects, but the cross sections for ionization, DEA, and fragmentation may also be high. However, whether electron cluster cross sections can be simply scaled from the monomers is a subject of much current study (both experimental and theoretical).

4.2.2 Rare Gas Plasmas

Some of earliest and still most commonly used plasmas are those using rare gases as feed gas, most particularly helium and argon. Most extraterrestrial helium is found in a plasma state, with properties quite different from those of atomic helium. In a plasma helium cations have very high electrical conductivity, even when the gas is only partially ionized; thus helium is often added to other plasmas to increase their conductivity and so stabilize them.

Atmospheric pressure discharges of helium are being developed for medical use and thus helium jets have recently been commercialized for such use [19].

Argon plasmas are routinely used in plasma etching and surface cleaning because argon removes all kinds of contamination via ion bombardment (also known as sputtering) without causing a chemical reaction or oxidation on the surface of the substrate. An argon plasma can remove about 100 Å or more from the bulk surface of a substrate in about 30 s, thus removing any contaminants from the surface of the substrate, and, as a gaseous effluent, can be extracted from the processing chamber before redeposition can occur. In addition to contamination removal, argon plasmas are also used to roughen the substrate and can activate chemical bonds on the surface in order to improve bonding and adhesion and (as can helium) increase the wettability of the substrate surface.

The electron density and temperature of atmospheric helium and argon plasmas operated under similar experimental conditions may be compared. It is found that a helium plasma has a higher electron temperature and a lower electron density than an equi-operational argon plasma, that is, an argon plasma that is operated at the same external conditions [20]. This is mainly caused by the higher excitation potential of the first excited state and the lower mass of helium. Helium and argon plasmas, however, both support the excitation of long-lived excited states (metastable states); indeed they are central to, for example, the plasma that sustains the helium-neon laser. The presence of such metastables (which may be as much as 1% of the total neutral density), excited by direct electron impact, leads to two additional collisional processes in such plasmas: *penning ionization*, where the large internal energy of the metastable (A*) may be transferred to a molecule of lower ionization potential $A^* + XY \longrightarrow X^+ + Y$, and *superelastic scattering*, where the scattered electron gains the energy of the metastable state as it de-excites, generating more electrons with energies sufficient to ionize other molecules in the plasma.

Electron impact ionization or stepwise excitation of the metastable state to higher-lying electronic states of the neutral atoms provides additional pathways in the plasma, removing low-energy electron flux from the plasma since, typically, the energy

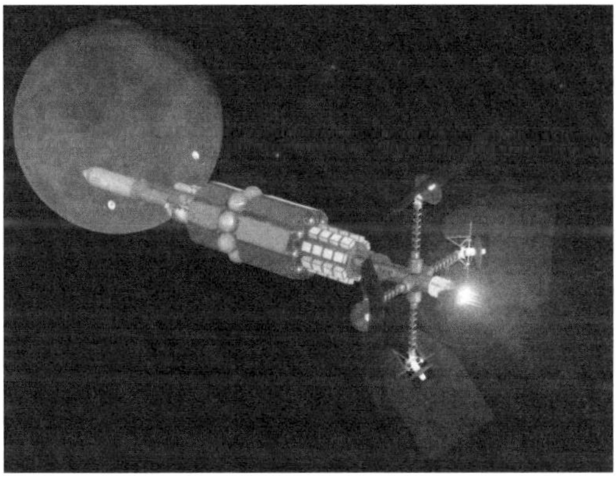

Figure 4.8 Representation of a spacecraft powered by plasma thruster technology for interplanetary space travel. Human Mission to Mars with 10 MW NEP-VASIMR® reprinted with permission. Image used under license with Ad Astra Rocket Company.

difference between the metastable excitation energy and the ionization energy is <5 eV. Neutral metastable species may also drift out of the plasma and upon striking the wall of the chamber create a new source of electrons that may re-enter the plasma.

Recently the excitation of metastable states in such rare gas plasmas has been utilized as a viable plasma propulsion methodology for long-distance space travel, potentially cutting interplanetary travel times from years to months (Fig. 4.8) [21].

4.2.3 Fluorocarbon Plasmas

Fluorocarbon plasmas have dominated the study of technological plasmas for more than three decades since they are the principle plasma source used in plasma etching to fabricate microstructures in chip technology [22, 23]. Much of the technological progress in modern electronics and computational power has been due to advances in our understanding and exploitation of fluorocarbon plasmas. Indeed, the ability of industry to match Moore's law has been due to these improvements. Today some of the most detailed and realistic simulations of any technological process involve the simulation of plasma etching building microstructures

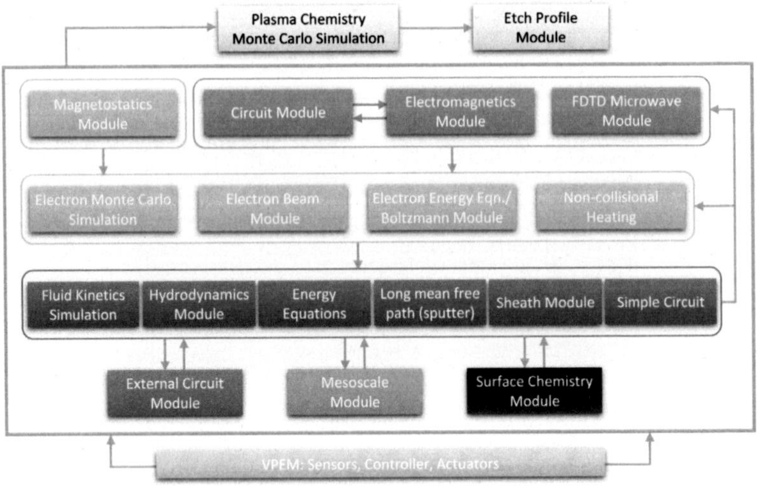

Figure 4.9 A virtual factory is a computerised simulation of a fabrication facility.

using fluorocarbon plasmas. Such models have been called *virtual factories* (Fig. 4.9) since they are sufficiently robust that they may be used to test new methodologies for chip manufacture before expensive technical plants are constructed, and since they may test dozens of different scenarios, they provide a high cost benefit to the industry. This has become particularly important when the choice of feed gas is increasingly restricted due to environmental concerns (see below) [2].

Typical fluorocarbons include SF_6, CF_4, C_2F_4, and variants thereof; accordingly the electron interactions with such species have been extensively studied and a detailed (but still not complete) database of such interactions has been compiled) [23]. It should be noted that all halogenated species have strong cross sections for electron attachment and hence fluorocarbon plasmas are electronegative plasmas, with the anions playing a strong role in subsequent plasma chemistry. Indeed, since DEA is a resonant-driven process, changing the electron temperature may alter the relevant DEA pathways, allowing the chemistry in the plasma to be tuned (see Section 4.3).

Unfortunately all of the commonly used fluorocarbon feed gases have high global warming potentials (GWPs) [24]; indeed, SF_6 has the highest GWP (=23,000) of any known compounds, that is, one molecule of SF_6 is equivalent to 23,000 CO_2 molecules. Accordingly, with increasing awareness of the role of the potential effect of global warming, protocols have been put in place to ban the use of such compounds within a decade. Such protocols provided the semiconductor industry with its greatest challenge since it attacked the very basis of the industry. Major research programmes were established to explore alternatives but to date the alternatives (e.g., CF_3I, NF_3) are not as efficient as those gases they replace, or may have other drawbacks (e.g., toxicity or reactivity). An alternative is to use nonfluorocarbon species such as NH_3, but the reproducibility and stability of the microstructures such plasmas create remains uncertain, in part due to the lack of a comprehensive database to develop a virtual factory using such feed gases against which to test fabrication conditions.

In all such plasmas a core electron-induced process is the formation of the etching radicals (CF_x, SF_x, or NH_x, where x = 1 to 3). Subsequent to their formation such radicals may experience electron collisions, including excitation, ionization, and DEA; however, to date experiments on such species are limited to only a few cases since it is difficult to produce sufficient concentrations of such radicals for current experimental techniques [25]. Theoretical studies of such species are also limited both by the ability to create accurate representations of such scattering targets and the complex and computational expensive models needed to study electron-induced dissociative processes. Thus despite the great progress in electron-scattering theory and creation of virtual factories, a full description of electron-induced processes and subsequent chemistry in fluorocarbon plasmas remains to be completed.

4.2.4 Fusion Plasmas

The largest plasma technology project in human history is the International Thermonuclear Experimental Reactor (ITER) project aimed at establishing fusion as a viable future energy source. The

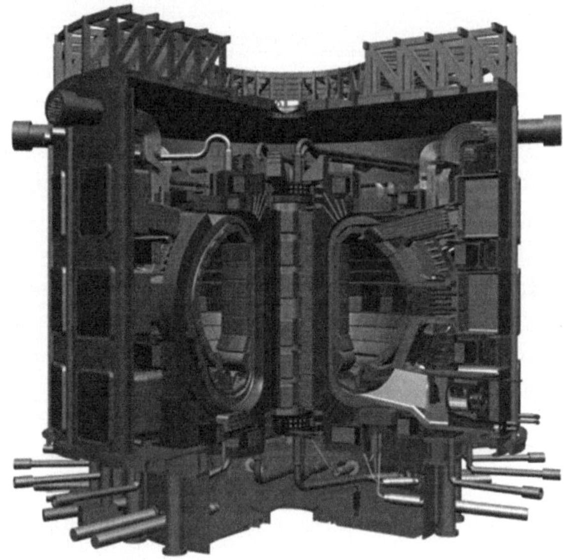

Figure 4.10 Schematic representation of the ITER, the largest technological plasma in the world. Under the CC-BY version 4.0 license (https://creativecommons.org/licenses/by/4.0/).

ITER tokamak (Fig. 4.10) will measure 24 m in height and 30 m in width. It will be smaller than a conventional power station but is expected to produce up to 500 MW of thermal power in a toroidal fusion plasma of 800 m^3 volume. The ITER experiment is developed using the concept of the tokamak, which was a device first conceived in Moscow back in the 1960s. A tokamak, is a torus, or doughnut-shaped, continuous tube, surrounded by coils that produce a magnetic cage to confine the high-energy plasma. In the ITER, all the magnetic coils are superconducting, and it is expected to bring together all necessary technology for a future fusion power station.

The ITER is 30 times more powerful than the previous largest tokamak, the Joint European Torus (JET), and will produce 10 times more power than is necessary to maintain plasma at fusion temperatures such that it can demonstrate the physics of a burning plasma—a plasma that is heated by internal fusion reactions rather than external heating. Engineers will then be able to test the heating, control, diagnostic, and remote maintenance systems that will be needed for a real fusion-based power station.

Atomic collision processes play a large role in both magnetic and inertial confinement fusion research. There are four principal areas of fusion research where atomic processes are important: (i) the hot central plasma where the fusion reactions are expected to occur, (ii) the plasma edge where the plasma interacts with the external environment, (iii) plasma heating methods used to produce the hot plasma, and (iv) diagnostic techniques used to measure the physical properties of the plasma. In all cases electron collisions play an important role, and of particular relevance are electron interactions with hydrogen (atomic and molecular) and its isotopes deuterium and tritium. The high temperatures within the plasma are such that molecular hydrogen is present in highly vibrationally excited states, and hence electron-scattering cross sections (and rate constants) for electron scattering from ro-vibrationally excited $H_2(\upsilon)$ are required. Such cross sections may be significantly different from those of the ground state [26]; indeed, DEA leading to the formation of $H + H^-$ may change by orders of magnitude (Fig. 4.11), while there are also strong isotope effects, with the formation of anions via the lowest (4 eV) resonance in D_2 being 1/100 of that of corresponding H_2, a process that is important in developing neutral beam injectors. Experiments using radioactive tritium are of course sparse such that data sets are mainly theoretical; however, even for such a simple molecular system as H_2 calculations of cross sections provide severe challenges (e.g., the cross section for excitation and subsequent dissociation of the lowest excited state resulting in two ground-state atoms). Due to the high degree of ionization in the fusion plasma DR is an important process, but cross sections (particularly for ro-vibrationally excited cations) remain debatable as do the fluorescent yields of H atoms excited in the plasma, which is important in calibrating several spectroscopic diagnostic techniques.

In the plasma edge hydrocarbon species become important as do beryllium hydrides (beryllium being the material chosen to line the tokamak wall facing the plasma). Electron interactions with many of these species (particularly CH_x radicals and beryllium hydrides) are largely unquantified and often rely on theoretical calculations

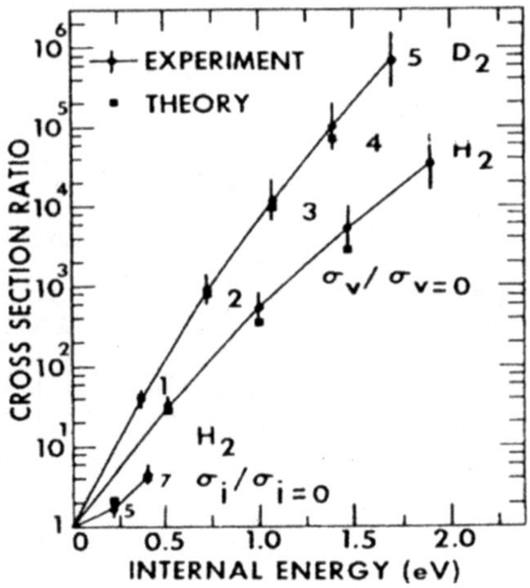

Figure 4.11 Strong vibrational energy and isotope dependence on the DEA cross section in H_2 and D_2. Reprinted (figure) with permission from Ref. [26]. Copyright (1978) by the American Physical Society.

that are poorly benchmarked. Semi-empirical calculations [27, 28] of electron impact ionization cross sections appear, in general, to be reliable, though they do not giving branching ratios and recent generally available codes (e.g., Quantemol [29] and the independent atom approximation [30]) are shown to give good estimates of elastic and summed inelastic cross sections over a wider energy range, but excitation cross sections (ro-vibrational and electronic) and dissociative pathways (DEA, etc.) remain largely uncharacterized and will form a research area concurrent with the development of an ITER facility.

4.2.5 Plasmas for Environmental Remediation

Environmental concerns have assumed major political and societal importance in the past two decades. The need to clean the spoils of industrialization has been highlighted. Plasma technology has been harnessed to provide methodologies for the removal of unwanted and harmful waste from effluent gases and water waste (see, for example, Fig. 4.12).

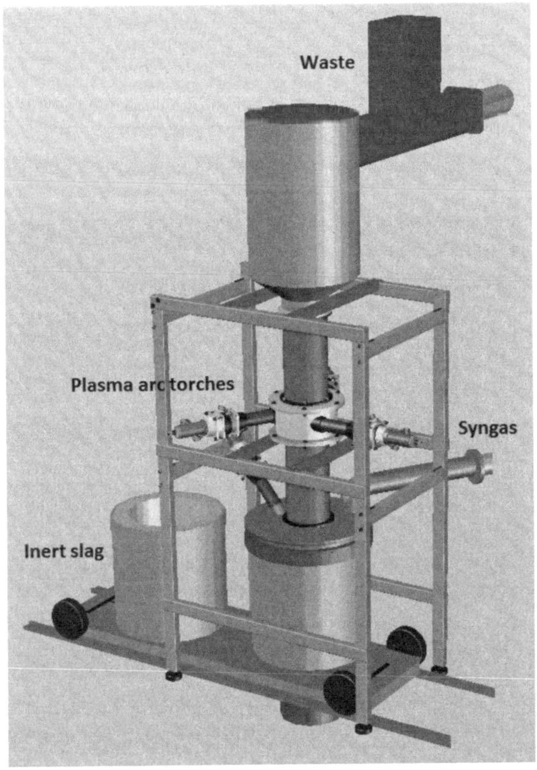

Figure 4.12 The Plasma Pyrolysis Reactor PR-100 commercially developed by High Temperature Technologies Corporation in Canada is used as a basis for the system that treats gaseous, liquid, and disperse medical, toxic, and chemical wastes. The device is based on plasma torches fed by air, it handles up to 100 kg waste per hour, and the power used by the system is up to 150 kW. Courtesy of High Temperature Technologies Corporation (http://www.httcanada.com/furnacesh.html).

Often (but not exclusively) atmospheric plasmas are used, since they are cheap and readily prepared in most environments. However, the interaction of plasmas (and with it electrons) with the many chemicals provides a challenge since how such interactions influence (and hence could be controlled) such waste remediation remains largely unknown. Furthermore such interactions are not just in the gas phase but in liquid (water) and solid phases; indeed, there appears to be cumulative effect if a plasma is combined with a catalyst in many remedial techniques [31]. The study of electron collisions of surfaces and how these are perturbed by the presence of the surface (e.g., the position and lifetime of a TNI formed in electron collisions) is discussed elsewhere in this volume. Unfortunately our knowledge of how such collisions are modified in the liquid phase is almost totally unexplored due to the difficulty in developing experimental techniques for such studies. Nevertheless the importance of electron interactions with water clusters is highlighted (see Section 4.4 later), and recent international research coordination projects have begun to explore models on plasma–liquid interactions and the databases (including electron interactions) needed for such modeling [32]. It should noted that there is a major overlap between this and the study of plasma medicine where plasma interacts with skin and blood (with high concentrations of water) for which models have been developed [33].

4.3 Controlling Plasmas with Electrons

One of the major goals of any technology is to be able to control the product of that technology and hence be able to design, refine, and commercialise that product. Accordingly in most technology-based plasmas the aim of controlling specific parameters of the plasma in order to control the effect and hence the product of that plasma has been central to the development of the plasma industry. In this section two examples will be given of how knowledge of electron collisions in the plasma can lead to improvement in technologies.

4.3.1 Improving Ozone Generation

A special case of plasma used for environmental control is the generation of plasmas that produce ozone. Ozone is a chemically reactive species that has strong sterilizing properties as well as ability to cure materials. The study of ozone-forming discharges is thus over 150 years old since the invention of the first ozonizer in

1855. Ozone is formed in a simple three-body reaction

$$O + O_2 + M \longrightarrow O_3 + M,$$

where M is a third body (atom, molecule, or surface) necessary to stabilize nascent ozone formed in the excited state. O atoms are readily formed by electron impact dissociation of O_2 in an oxygen-rich discharge (though ozone is also formed in other discharges with feed gases capable of generating O atoms, and thence O_2, for example, CO_2). Ozone is, however, also readily destroyed by the reaction

$$O + O_3 \longrightarrow 2O_2.$$

Hence for ozone concentrations to stabilize, the production and destruction mechanisms must balance. This balance is achieved in Earth's stratosphere where there is sufficient molecular oxygen to create O atoms by photodissociation and sufficient density to provide enough concentrations of M to allow three-body collisions; the stratosphere is also sufficiently dry to prevent catalytic destruction by the OH radical that would otherwise dominate destruction over formation.

$$OH + O_3 \longrightarrow HO_2 + O_2$$

$$HO_2 + O_3 \longrightarrow OH + 2O_2$$

$$\text{Net: } O + O_3 \longrightarrow 2O_2$$

Hence ozone yields are very sensitive to humidity, which explains the very different yields experienced in operating commercial ozonizers (Fig. 4.13) feed by nascent air in different parts of the world (ozone yields from commercial ozonizers fed by nascent air being higher in dry Gulf states than in the humid tropics).

Whilst the best ozone yields are generated with pure oxygen feed gas practicality requires the use of air as a feed gas. While nitrogen does not affect the ozone chemistry (indeed, it acts as M) the nitrogen oxides that can be formed in such plasmas (NO_x) act as catalytic destruction agents of ozone in same way as OH; thus yields of air-fed ozonizers are (per O_2 molecule) less than those fed by pure oxygen.

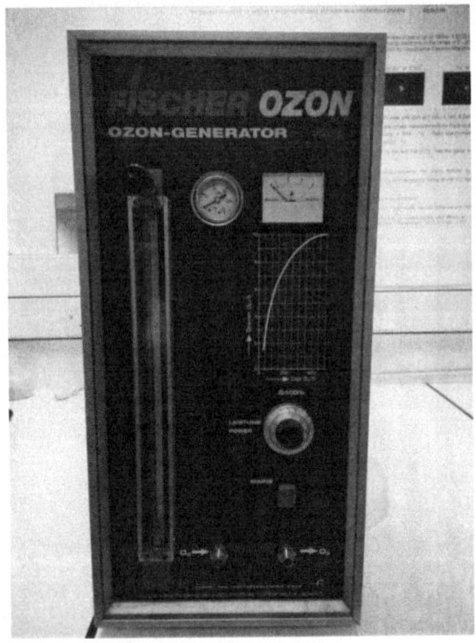

Figure 4.13 Commercial ozonizer used to produce ozone for laboratory-based experiments.

The use of ozone in, for example, water and clinical tool sterilization and cleaning (including deactivating microbes) has led to considerable research on the design of ozonizers with the aim of increasing the yield of ozone per kilowatt-hour. Several models of the ozone-generating discharge have been developed, including ones that allow for surface- as well as gas-phase reactions [34]. Key to these models has been understanding the role of low-energy electrons in the discharge and in particular the interaction of low-energy electrons with ozone itself [35]. Since ozone is an unstable, highly reactive species few experimental studies had been made before the author's group in the 1990s; furthermore ozone is a particularly hard target to represent, having a myriad of low-lying (<2 eV) electronically excited states such that, to date, few theoretical studies of electron–ozone scattering have been made.

The author and coworkers' study of low-energy electron scattering from ozone revealed an unexpected feature, namely a strong low-energy DEA cross section (Fig. 4.14) with several overlapping resonances coupled to the low-lying excited states [36]. Although the original zero-energy DEA peak now appears to have been an artifact

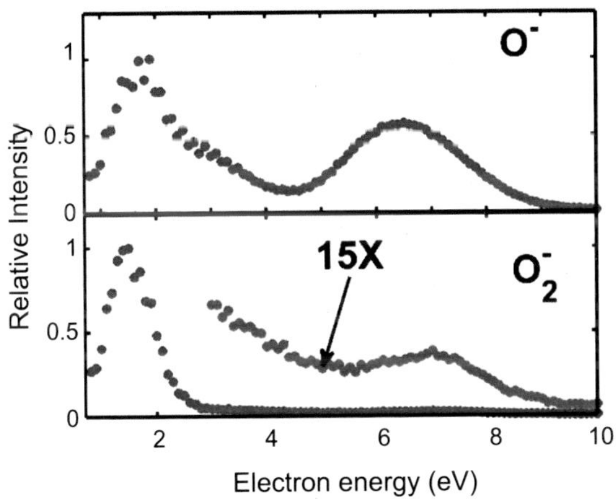

Figure 4.14 DEA in ozone showing a high cross section at low energy (<4 eV).

the DEA cross section is strong from ~0.5 to 4 eV. This has important consequences. In the plasma discharge creating the O atoms that are needed for the formation of the nascent ozone, many low-energy electrons are created. These low-energy electrons (<5 eV) will initiate DEA with the ozone product, destroying it and hence reducing the ozone yields. Thus to reduce the destruction of ozone such low-energy electrons should be removed from the discharge. Accordingly several groups tried adding "electron scavengers" to the feed gas to change the electron distribution (raise the electron temperature) in the plasma and remove electrons with energies less than 5 eV. SF_6 and CCl_4 are known electron scavengers (removing low-energy electron by DEA), but the neutral products also proved to be catalytic agents for ozone [37]. An alternative method is to redesign the ozonizer such that the discharge creating the O atoms is separated from the high-pressure region where O atoms react with O_2 to form the ozone. Grids placed at the exit of the discharge region further ensure that no electrons are passed from the discharge region into the high-pressure ozone-forming region. Such a technique has led to more efficient ozone generation. This is an excellent example of how knowledge of the electron processes in the discharge can lead to improved performance of the plasma tool.

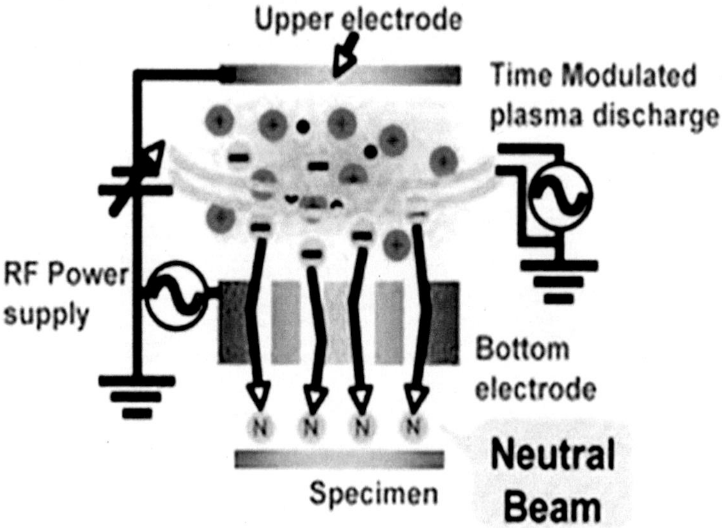

Figure 4.15 Schematic of a modulated plasma reactor that exploits electron-induced chemistry to produce anions, which are withdrawn from the plasma, charge-exchanged with a carbon surface, and then used to make a neutral beam for surface etching [38, 39]. Reprinted with permission from Ref. [39]. Copyright (2002), American Vacuum Society.

4.3.2 Controlling Electron Temperature in Fluorocarbon Plasma for Nanotechnology

As discussed before the need to control and refine fluorocarbon plasmas and even substitute the feed gases for the use in the microelectronics industry is one major example of attempts to control plasma.

In seeking to tune a plasma one of the physical or chemical characteristics of that plasma must be selected as the probe for effect of the change induced. While in any etching plasma it is the surface modification/structure that is the final required product, it is impractical to use these as an online tool; thus concentrations of specific molecular species in the plasma are commonly used as a probe of the plasma. Concentrations of such species may be measured using in situ probes (e.g., a Langmuir probe to measure electron density) or derived from spectroscopic observations, the latter having the advantage of easy, real-time monitoring.

Such diagnostic probes are often dependent upon the electron energy (temperature) of the plasma. Should it be possible to control the electron temperature it may be possible to control the energetics and hence the chemistry of the plasma. The electron energy (temperature) in a plasma can be controlled by the voltages applied to the plasma-generating electrodes. In most commercial plasma systems the voltages on such electrodes are modulated (Figs. 4.15 and 4.16) such that the electron and ion concentrations are also modulated; however, by modulating the electrode voltage the electron temperature is also modulated. Samukawa et al. [38, 39] have demonstrated that by pulsing the electrode voltages generating the plasma in an industrial etching reactor, the electron temperature may be lowered to such an extent that the electron temperature is reduced to a mean energy centred where DEA has a high cross section and the corresponding density of anions in the plasma is increased (Fig. 4.16). These anions may be drawn out of the plasma and, by passing through a carbon filter, converted to the neutral radicals of the anionic species. Thus anions may be converted to neutral radicals, which may then be used to etch the surface without the disadvantages of charging of the surface by charge carriers in the plasma or UV damage from the plasma. Such neutral beam etching has been shown to produce well-defined 20–30 nm structures (Fig. 4.17) [40].

4.4 The Future

4.4.1 Controlling Plasma Chemistry

The two previous examples reveal how our growing knowledge of the role of different electron processes in plasmas is delivering new methodologies and advancing the technology. However, in the future it may be possible to further develop these ideas and control the chemistry in the plasma.

DEA is a bond-specific process such that each resonance (TNI) while decaying may cleave a single bond in the target. Hence by

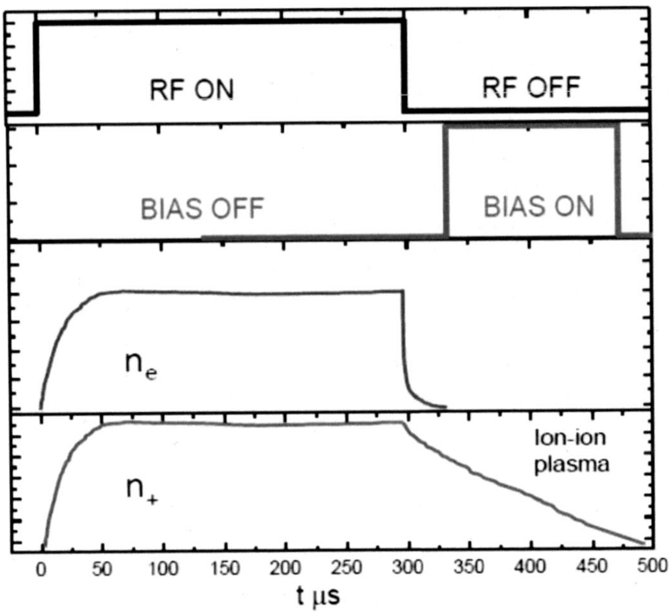

Figure 4.16 Time modulation of the upper electrode in Fig. 4.15 showing a change in the electron and anion flux. Reprinted with permission from Ref. [39]. Copyright (2002), American Vacuum Society.

controlling the energy of the incident electron it is possible to control the molecular fragmentation pathways—the holy grail of chemistry for more than a century.

Figure 4.18 illustrates this electron-induced chemical control. Very low-energy electrons (~0 eV) incident upon $CFCl_3$ are able to cleave the C–Cl bond exclusively (producing a Cl^- anion and $CFCl_2$ neutral), while electrons at ~3.2 eV cleave the C–F bond, with the electron being shared by the F and CCl_3 fragments [41]. Thus by tuning the electron temperature (e.g., by the time modulation method described earlier) the ratio of different anions and radicals can be altered. Different radicals have different etching rates and produce different etch profiles; thus by changing the population of the radicals in the plasma further control of the etching rate and profiles may be possible. Alternatively using the neutral beam method developed by Samukawa et al. [38–40] it may be possible to create different anions, which can be drawn from the plasma to form

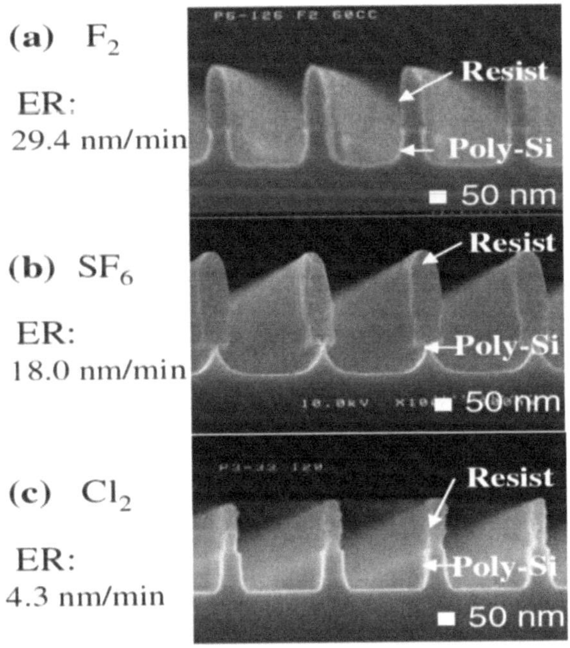

Figure 4.17 Nanostructures in polysilicon created using the neutral beam etching technique described in Figs. 4.15 and 4.16 with F_2, SF_6, and Cl_2 as plasma feed gases (all of which have strong DEA cross sections). Reprinted from Ref. [40]. Copyright 2005 The Japan Society of Applied Physics.

neutral beams. As our knowledge of DEA pathways for plasma feed gases improves, the range of radicals and anions we can selectively produce grows. Indeed, mixing two feed gases delivering different radicals may be possible, for example, CF_3I decays under DEA to liberate CF_3 radicals, while C_2F_4 produces CF_2 radicals. CF_3 and CF_2 have different etching rates and produce different profiles; hence using a mixture of these two feed gases allows in situ control of the nanofabrication.

Should this sound overly complex it should be noted that many commercial plasma prepared nanostructures are formed through a series of etching steps controlled by computer with plasma diagnostics (spectroscopy, mass spectrometry, and electron temperature) monitored in situ during each step.

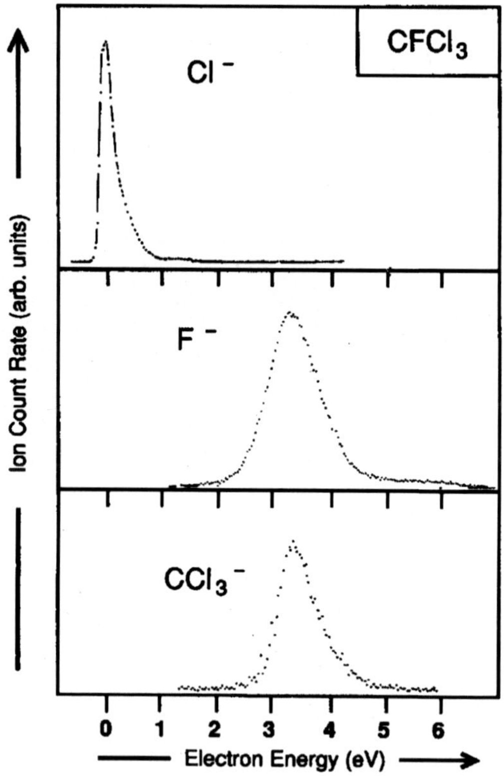

Figure 4.18 Demonstration of the ability of low-energy electrons to selectivity cleave molecular bonds. At low energies (~0 eV) electrons cleave the C–Cl bond in CFCl3, while at ~3 eV they cleave the C–F bond. Reprinted from Ref. [41], Copyright (1989), with permission from Elsevier.

4.4.2 Understanding the Plasma: Radicals and Excited Species

It is now recognized that to understand the physical and chemical properties of a plasma a better understanding of the role of minor species in the plasma is needed. The key minor species are radicals and excited species, both of which may drive the local environmental chemistry. As stated before radical species play an important role in etching plasmas determining the profile of the surface structure, while electron scattering from long-lived (metastable) excited states can open new scattering channels (Penning ionization and superelastic scattering). Therefore current research is exploring new methods for both detecting such species in plasmas and characterizing their electron cross sections.

Spectroscopy provides a standard method for detecting species in a plasma, with optical emissions being routinely used to identify constituent species. Emissions from excited CF_x radicals has been used to probe their concentrations and spatial profile in technological plasmas, but many radicals are produced their ground state and so are dark. Laser-induced fluorescence is one method by which such dark species may be observed [42], and subsequently the dependence of the CF_2 radical density on the electron density may be probed [43].

Plasma simulations require knowledge of the cross sections for electron interactions with such radicals; however, with only a few exceptions [44, 45] it is not proven practical to measure such cross sections, since it is difficult to prepare sufficiently intense radical beam sources for such experiments. Thus most of our knowledge of such cross sections is based on theoretical evaluations (benchmarked against evaluation of scattering from a stable precursor, for example, CF_4 for CF_x species). However, the complex representation of radical targets is still a challenge for modern theory and the derived cross sections are probably only a guide (with large [50%] errors).

Similarly, scattering from excited (metastable) molecular targets is largely unexplored either experimentally or theoretically, while scattering from common metastable rare gas species (He and Ar) reveals discrepancies between theory and experiment. The role of such species in a plasma may nevertheless be high. For example, the DEA cross sections for scattering from an excited state may be orders of magnitude higher than for the ground state [46]. Furthermore the reaction rates of radicals and metastables with other plasma constituents are largely ill-defined.

In electronegative plasmas anion species play a critical role in establishing the plasma characteristics and chemistry. The formation rates of such anions require knowledge of absolute DEA cross sections (and where DEA cross sections are low, ion pair formation). However, to date most DEA cross sections are only relative values. Such relative data may be normalized to a standard but the usual standard is the zero-energy peak in SF_6 or CCl_4, a peak that is dependent upon the resolution of the incident electron beam. However, recently Allan et al. developed a system for measuring absolute DEA cross sections [47], and combined with developing theoretical methods [48] such data may be forthcoming. However, rate constants for anion neutral reactions in any plasma remain, at present, unknown. The adoption of ion trap technology to study such reactions [49] and the use of ion storage rings (such as Desiree [50]) may, in the next decade, provide the first authoritative data on such anionic reactions.

Thus in the next decade we may expect significant progress in our knowledge of electron scattering from radicals and excited species in the plasma, which, in turn, may provide a better understanding of the physical and chemical characteristics of plasma, allowing us to control and tune it in the future.

4.4.3 The Role of Macrospecies in Atmospheric Plasma Clusters

As discussed before a major topic of modern plasma research is to develop atmospheric pressure plasmas [51]. At such high pressures the plasma is dominated by ion/neutral chemistry, the key species of which are formed in the plasma. For example, consider an air-based plasma (as used, for example, in surface treatment of textiles or sterilization of materials). In such atmospheric pressure plasmas there is evidence of cluster formation, often arising from the presence of trace quantities of water [52, 53]. Such clusters demonstrate that within the discharge complex ion–molecule reactions occur, reactions for which we have little data and that will be pressure (number density) as well as temperature dependent. Figure 4.19 shows an illustrative anion cluster spectrum formed in a coronal discharge operating at atmospheric pressure; the $O_3^- \cdot (H_2O)_n$ ions are postulated to be formed by one of the following processes. The three-body process

$$O^- + O_2 + M \longrightarrow O_3{}^- + M \quad k = 3.3 \times 10^{-31} \text{cm}^6/\text{s}$$

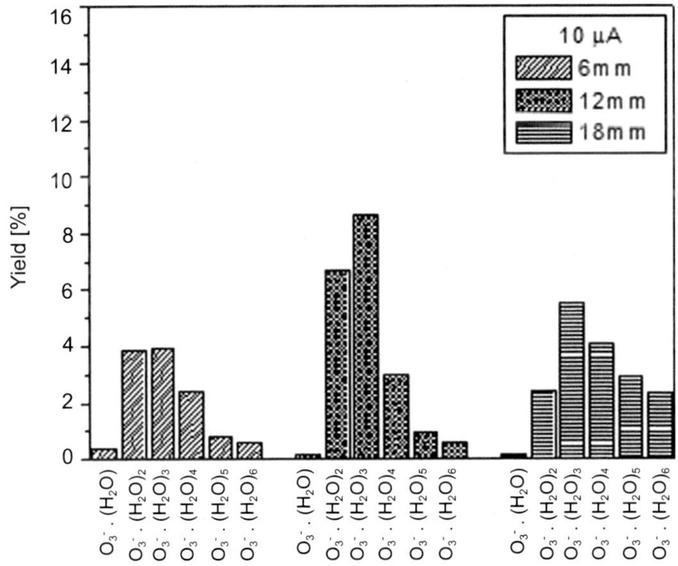

Figure 4.19 Illustrative anion cluster spectrum formed in a coronal discharge operating at atmospheric pressure; the $O_3^-.(H_2O)_n$ ions are postulated to be formed by one of the following processes: the three-body process or the charge transfer process. Reprinted from Ref. [41], Copyright (1979), with permission from Elsevier.

or the charge transfer process

$$O^- + O_3 \longrightarrow O_3{}^- + O \quad k = 5.3 \times 10^{-10} cm^3/s$$

At atmospheric pressure $M = 2.68 \times 10^{19}$ molecule.cm^{-3}; it therefore follows that at atmospheric pressure the ozone concentration [O_3] must be higher than 3300 ppm if the charge transfer is to be the main process by which O_3^- is produced. This value is unlikely in such experiments. Another channel for formation of O_3^- ions is the ion reaction

$$O_2{}^- + O_3 \longrightarrow O_3{}^- + O_2 \quad k = 3 \times 10^{-10} cm^3/s$$

At atmospheric pressure this will only be an important channel when ozone concentrations exceed 5800 ppm, which is not a realistic value in such an experiment. Hence the dominant primary source of negative ozone ions is the three-body process.

The observed $O_3^-.(H_2O)_n$ clusters are then formed in a series of consecutive reactions $O_3^-. (H_2O)_n + H_2O + M \longrightarrow O_3^-. (H_2O)_{n+1} + M$, where $n = 0, 1, 2, \ldots$ and $k \approx 1 \times 10^{-28}$ cm^6/s, and it should be noted no O_3^- monomers were observed.

Such observations highlight the need for the study of electron interactions with larger macromolecular systems such as water clusters $(H_2O)_n$ and solvated aeronomic clusters, for example, $(M.H_2O)_n$. However, to date there are very few studies of electron interactions with clusters, and it is unclear how or whether the cross sections may scale with size or how resonances in one molecule are modified by clustering (in the solid state, for example, in a surface film, both resonance lifetime and energy are altered). Thus a rich field of study in the future is expected to be studies of electron scattering from molecular clusters, and pioneering work by Gorfinkiel et al., exploiting the R-matrix method to explore electron interactions with dimers and trimmers, has recently been reported [54, 55].

Similarly a simple discharge in a methane/nitrogen atmosphere (as representative of Titan's planetary atmosphere) shows a wide variety of anions (Fig. 4.20), the formation rates of which are poorly known since the database for electron scattering from nitriles yielding CN^- anion chemistry is much less well established than that for oxygen/ozone anions [56].

The database for even more exotic ionic species as used in industrial plasmas (arising from halocarbon, silane, and other species) is even more limited, but the role of electron cluster and ion molecule reactions is still expected to be significant (even dominant). Indeed, inter- and intracluster chemistry may occur, opening up new dissociation pathways that are not possible with monomers. For example, consider a plasma comprising two common etching gases, NF_3 and CH_3Cl. It is known that the DEA cross section for CH_3Cl is negligible and no Cl^- anions are formed, but if the two compounds are clustered $(NF_3)_n$ $(CH_3Cl)_m$ then Cl^- anions are observed (Fig. 4.21). This is ascribed to an intracluster reaction where F^- is liberated from NF_3 by DEA and then reacts with CH_3Cl to liberate Cl^- and form CH_3F via $F^- + CH_3Cl \longrightarrow CH_3F + Cl^-$, a reaction that may also occur in a surface film of such species [57].

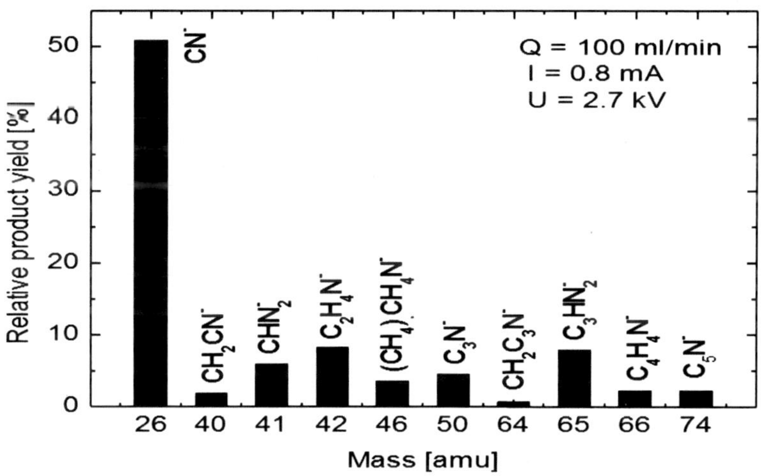

Figure 4.20 Relative abundances of the most dominant anions formed in a 350 torr negative corona discharge of methane (4%) and nitrogen (88%) with 8% argon [56].

New experimental methods to explore electron interactions with clusters are being developed but are rare, in part because they are both complex and expensive. Clusters are prepared in supersonic expansion jets and form an ensemble of sizes that cannot be reduced to a single well-defined cluster. It is also difficult to evaluate the number density of such a cluster source, and thus, to date, it is proven impractical to define cross sections to any observed process.

4.4.4 Plasmas in Other Media

Plasmas are not confined to operating in the gas phase. Plasma–surface interactions are integral to any plasma system. Plasmas are usually produced in a reactor and in many instances are developed specifically to interact with surfaces (e.g., for plasma etching, surface fictionalization or sterilization). There is often a complex interplay between the plasma interaction with surfaces and the chemical composition of the plasma. For example, molecules in the plasma feed gas may be absorbed on the walls, forming a layer (or more often multilayers). The probability of such absorption is characterized by the sticking coefficient, a value that

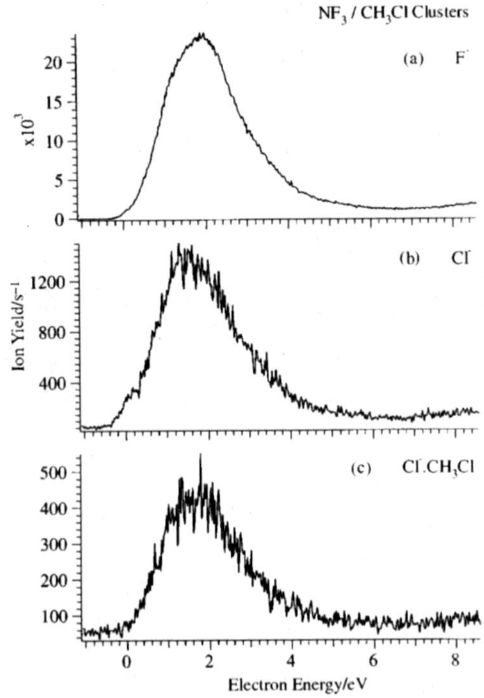

Figure 4.21 DEA observed in a source of $(NF_3)_n\,(CH_3Cl)_m$ showing evidence of intracluster chemistry $F^- + CH_3Cl \longrightarrow CH_3F + Cl^-$. Reproduced from Ref. [57] with permission of the PCCP Owner Societies.

is highly variable depending upon the surface material, the surface temperature, and, of course, the gas itself [58, 59]. Thus there is no set parameter (cross section) one can use, since it will certainly change as a function of time (as coverage of the surface changes). Therefore an average sticking coefficient is often assumed and this may have a significant effect on the model.

It is again necessary to determine the role of electrons in the plasma–surface interaction. For example, if we wish to determine the flux of a single radical, R, in the plasma we must know the electron impact dissociation cross section of a parent XR, and since R may then be absorbed on the walls, we must know the sticking coefficient γ. The total radical flux in the plasma may then be a balance of the production rate by electron impact and loss rate by sticking; hence the same final radical density can be achieved by doubling the sticking coefficient or halving the production cross section. However, the subsequent chemistry is quite different, a lower electron impact cross section having a greater influence on the plasma gas chemistry, while a larger sticking coefficient (and thence larger surface concentrations) enhances the heterogeneous chemistry.

Once the surface absorbate is established the surfaces will experience bombardment by energetic ions, electrons, neutrals, and photons. The detail in which we understand the effects of such bombardment varies widely depending upon the particular process. Examples of the data needed are cross sections and rate constants for energy transfer, reaction, emission, surface diffusion, implantation, reflection, disordering, and recombination. However, such cross sections will strongly depend upon the morphology and structure of the surface. Indeed, the very idea of a unique cross section may not be tenable for such surfaces and for a standard electron–molecule gas-phase process such as elastic scattering may have no single meaning. Furthermore, experiments show that electron scattering from molecules absorbed on a surface maybe very different from that in the gas phase. Resonances are usually lowered in energy and their lifetime changed [60], and electronic excitation cross sections (leading to dissociation) are altered (Rydberg states are hindered and maybe altogether suppressed in the solid phase) [61]. The state of the database(s) for such phenomena is therefore hard to assess, and a new review of the data used for assessing surface processes in existing models would be very valuable. Particular attention should be given to how different models estimate the production rates of secondary electrons and ions emitted from the surface back into the plasma.

Plasmas may also be induced in liquids and used to provide new coatings and surface preparation, while plasma–liquid interactions are important in plasma medicine (where the liquids are body fluids) [17]. The study of discharges in liquids has been the subject of a recent COST Action [32] demonstrating the vibrancy of the field. However, studies of electron interactions in the liquid phase are even less well defined than those in the solid state, with no methodology for exploring discrete electron-scattering processes in the liquid phase having been developed. Theoretical methods have been developed describing a liquid as a dense gas and modeling the transport of electrons through the liquid using a solution of Boltzmann's equation together with ab initio liquid phase electron-scattering cross sections calculated using the Dirac–Fock scattering equations [62], but the results reveal the inadequacy of local treatments of exchange that are commonly used in liquid and cluster phase cross-section calculations. Hence understanding electron scattering from targets in the liquid phase remains a challenge for the next generation of experimentalists and theorists.

4.5 Conclusions

In this chapter the role of low-energy electrons in both natural and technological plasmas has been discussed. Indeed, plasmas are in general characterized by the properties of the electrons within them and many plasma descriptors are, by default, those of the electrons in the plasma.

Over the past decade our knowledge and appreciation of the role of low-energy electrons on plasmas have greatly increased, with the role of DEA in electronegative plasmas now being understood and used to develop new plasma techniques and methodologies. Nevertheless the DEA database remains both scattered and limited, though the recent construction of the IDEADB is a great step forward [63]. DEA studies also suffer from a lack of absolute cross-section measurements, though Allan et al. have demonstrated how these can be measured with an accuracy of 10%–20% [47].

The role of excited species in the plasma remains controversial and again lacks both experimental and theoretical study. Earlier work suggested that even minor concentrations of excited species may alter the plasma characteristics significantly. Similarly although many radicals are formed in plasmas we still know remarkably little of how they interact with the nascent species while anion-neutral chemistry is in its infancy.

Therefore the study of low-energy electron interactions will remain of great topical interest to the plasma community since it provides opportunities for developing the field. The two communities should therefore continue to collaborate and develop synergistic research programs.

Acknowledgments

The author wishes to recognize support from the UK EPSRC and generous support from the EU Framework programme (FP6, FP7, and H2020), which has allowed him to develop many collaborations within and across the electron and plasma physics communities. While there are too many collaborators to list, the author wishes to particularly acknowledge late Prof. (Dusan) Skalny of Comenius University, Slovakia, whose support and guidance led the author to develop many of his own projects and introduced him to many of his now established collaborators.

References

1. National Research Council (1991). Plasma Processing of Materials: Scientific Opportunities and Technological Challenges. National Academy Press, Washington, DC.
2. Makabe, T. and Lj Petrovic, Z. (2014). *Plasma Electronics: Applications in Microelectronic Device Fabrication*, 2nd ed. CRC Press, Boca Raton.
3. Bittencourt, J. A. (2010). *Fundamentals of Plasma Physics*, 3rd ed. Springer, NY, USA.
4. Inan, U. S. and Gołkowski, M. (2010). *Principles of Plasma Physics for Engineers and Scientists*. Cambridge University Press, Cambridge.
5. Schulz, G. J. (1973). *Rev. Mod. Phys.*, **45**, p. 378.
6. Christophorou, L., ed. (1984). *Electron-Molecule Interactions and Their Applications*, Vol. 1. Chapter 5, Academic Press, New York.
7. Szymańska, E., Mason, N. J., Krishnakumar, E., Matias, C., Mauracher, A., Scheier, P., and Denifl, S. (2014). *Int. J. Mass Spectrom.*, **365–366**, pp. 356–364.
8. Larsson, M. and Orel, A. E. (2012). *Dissociative Recombination of Molecular Ions*. Cambridge University Press, Cambridge.
9. Curik, R. and Carsky, P. (2012). Low-Energy Electron Scattering from Molecules, Biomolecules and Surfaces. CRC Press, Boca Raton.

10. Franz, G. (2009). *Collisions and Cross Sections*. Springer-Verlag, Berlin, Heidelberg.
11. Laporta, V., Celiberto, R., and Tennyson, J. (2013). *Plasma Sources Sci. Technol.*, **22**, p. 025001.
12. Mason, B. J. and Mason, N. J. (2003). The physics of a thunderstorm. *Eur. J. Phys.*, **24**, p. S99.
13. Smirnov, V. V. (1987). *Phys. Rep.*, **152**(4), pp. 177–226.
14. Ebert, U. and Sentman, D. D. (2008). *J. Phys. D: Appl. Phys.*, **41**, p. 230301.
15. Vuitton, V., Lavvas, P., Yelle, R. V., Galand, M., Wellbrock, A., Lewis, G. R., Coates, A. J., and Wahlund, J.-E. (2007). *Planet. Space Sci.*, **57**, pp. 1558–1572.
16. Zille, A., Oliveira, F. R., and Souto, A. P. (2015). Plasma treatment in textile industry. *Plasma Processes Polym.*, **12**, pp. 98–131.
17. Kong, M. G., Kroesen, G., Morfill, G., Nosenko, T., Shimizu, T., van Dijk, J., and Zimmermann, J. L. (2009). *Plasma medicine: an introductory review. New J. Phys.*, **11**, p. 115012.
18. Skalny, J. D., Orszagh, J., Mason, N. J., Rees, J. A., Aranda-Gonzalvo, Y., and Whitmore, T. D. (2008). *Int. J. Mass Spectrom.*, **272**, pp. 12–21.
19. Hoffmann, C., Berganza, C., and Zhang, J. (2013). *Med Gas Res.*, **3**, p. 21.
20. Jonkers, J., van de Sande, M., Sola, A., Gamero, A., and van der Mullen, J. (2003). *Plasma Sources Sci. Technol.*, **12**, p. 30.
21. Yamamoto, N., Tomita, K., Sugita, K., Kurita, T., Nakashima, H., and Uchino, K. (2012). *Rev. Sci. Instrum.*, **83**, p. 073106.
22. Posseme, N. (2015). Plasma Etching Processes for Interconnect Realization in VLSI. Elsevier Science & Technology.
23. Adamovic, I., et al. (2017). The 2017 plasma roadmap. *J. Phys. D: Appl. Phys.*, **50**, p. 323001.
24. Shine, K. P., Fuglestvedt, J. S., Hailemariam, K., and Stuber, N. (2005). *Clim. Change*, **68**, pp. 281–302.
25. Rozum, I., Limão-Vieira, P., Eden, S., Mason, N. J., and Tennyson, J. (2006). *J. Chem. Phys. Ref. Data*, **35**, pp. 267–284.
26. Allan, M. and Wong, S. F. (1978). *Phys. Rev. Lett.*, **41**, p. 1791.
27. Vinodkumar, M., Joshipura, K. N., Limbachiya, C. G., and Antony, B. K. (2006). *Eur. Phys. J. D*, **37**, pp. 67–74.
28. Kim, Y. K. and Rudd, M. E. (1994). *Phys. Rev. A*, **50**, p. 3954.
29. http://www.quantemol.com/products/quantemol-n/
30. Blanco, F. and García, G. (2003). *Phys. Rev. A*, **67**, p. 022701.

31. Kang, W. S., Lee, D. H., Lee, J. O., Hur, M., and Song, Y.-H. (2013). *Environ. Sci. Technol.*, **47**, p. 11358.
32. http://www.cost-plasma-liquids.eu/
33. Babaeva, N. Yu., Ning, N., Graves, D. B., and Kushner, M. J. (2012). *J. Phys. D: Appl. Phys.*, **45**, p. 115203.
34. Yanallah, K., Pontiga, F., Fernández-Rueda, A., and Castellanos, A. (2009). *J. Phys. D: Appl. Phys.*, **42**, p. 065202.
35. Mason, N. J. (2003). *Phys. Scr.*, **68**, p. C37.
36. Senn, G., Skalny, J. D., Stamatovic, A., Mason, N. J., Scheier, P., and Mark, T. D. (1999). *Phys. Rev. Lett.*, **82**, pp. 5028–5031.
37. Skalny, J. D., Mikoviny, T., Mason, N. J., and Sobek, V. (2002). *Ozone Sci. Technol.*, **24**, pp. 29–37.
38. Samukawa, S., Sakamoto, K., and Ichiki, K. (2001). *Jpn. J. Appl. Phys.*, **40**, p. L997.
39. Samukawa, S., Sakamoto, K., and Ichiki, K. (2002). *J. Vac. Sci. Technol., A*, **20**, p. 1566.
40. Samukawa, S. (2005). *Jpn. J. Appl. Phys.*, **45**, p. 2395.
41. Illenberger, E., Scheunemann, H. U., and Baumgärtel, H. (1999). *Chem. Phys.*, **37**, pp. 21–31.
42. Haverlag, M., Stoffels, W. W., Stoffels, E., Kroesen, G. M. W., and de Hoog, F. J. (1996). *J. Vac. Sci. Technol., A*, **14**, p. 384.
43. Kim, J.-H., Chung, K.-H. and Yoo, Y.-S. (2006). *J. Korean Phys. Soc.*, **47**, p. 249.
44. Francis-Staite, J. R., Maddern, T. M., Brunger, M. J., Buckman, S. J., Winstead, C., McKoy, V., Bolorizadeh, M. A., and Cho, H. (2009). *Phys. Rev. A*, **79**, p. 052705.
45. Tarnovsky, V., Levin, A., Deutsch, H., and Becker, K. (1995). *J. Chem. Phys.*, **102**, p. 770.
46. Christophorou, L. G., Pinnaduwage, L. A., and Datskos, P. G. (1993). In *Linking the Gaseous and Condensed Phases of Matter: The Behavior of Slow Electrons*. Christophorou, L. G., Illenberger, E., and Schmidt, W. F., eds., Plenum, New York, p. 412.
47. Fedor, J., May, O., and Allan, M. (2008). *Phys. Rev. A*, **78**, p. 032701.
48. Fabrikant, I. (2010). *J. Phys. Conf. Ser.*, **204**, p. 012004.
49. Deiglmayr, J., Göritz, A., Best, T., Weidemüller, M., Wester, R. (2012). *Phys. Rev. A*, **86**, p. 043438.
50. Gatchell, M., et al. (2014). *J. Phys. Conf. Ser.*, **488**, p. 012040.

51. Bardos, L. and Barankova, H. (2009). *Vacuum*, **83**, p. 522.
52. Mikoviny, T., Skalny, J. D., Orszagh, J., and Mason, N. J. (2007). *J. Phys. D*, **40**, p. 6646.
53. Skalny, J. D., Orszagh, J., Mason, N. J., Rees, J. A., Aranda-Gonzalvo, Y., and Whitmore, T. D. (2008). *Int. J. Mass Spectrom.*, **272**, p. 12.
54. Bouchiha, D., Gorfinkiel, J. D., Caron, L. G., and Sanche, L. (2008). *J. Phys. B*, **41**, p. 045204.
55. Fabrikant, I., Caprasecca, S., Gallup, G. A., and Gorfinkiel, J. (2012). *J. Chem. Phys.*, **136**, p. 184301.
56. Horvath, G., Aranda-Gonzalvo, Y., Mason, N. J., Zahoran, M., and Matejcik, S. (2009). *EPJ Appl. Phys.*, **49**, p. 13105.
57. Langer, J., Matejcik, S., and Illenberger, E. (2000). *Phys. Chem. Chem. Phys.*, **2**, p. 1001.
58. Leroy, W. P., Mahieu, S., Persoons, R., and Depla, D. (2009). *Plasma Processes Polym.*, **6**, p. S342.
59. Kokkoris, G., Goodyear, A., Cooke, M., and Gogolides, E. (2008). *J. Phys. D: Appl. Phys.*, **41**, p. 195211.
60. Palmer, R. E. and Rous, P. J. (1992). *Rev. Mod. Phys.*, **64**, p. 383.
61. Mason, N. J., Dawes, A., Holtom, P. D., Mukerji, R. J., Davis, M. P., Sivaraman, B., Kaiser, R. I., Hoffmann, S. V., and Shaw, D. A. (2006). *Faraday Discuss.*, **133**, p. 311.
62. Boyle, G. J., McEachran, R. P., Cocks, D. G., and White, R. D. (2015). *J. Chem. Phys.*, **142**, p. 154507.
63. http://ideadb.uibk.ac.at/

Chapter 5

Focused Electron Beam–Induced Processing

D. Howard Fairbrother,[a] Samantha G. Rosenberg,[a] Cornelis W. Hagen,[b] I. Utke,[c] and Petra Swiderek[d]

[a]*Department of Chemistry, Johns Hopkins University, Baltimore, MD 21218, USA*
[b]*Delft University of Technology, Department of Imaging Physics, Charged Particle Optics group, Lorentzweg 1, 2628CJ Delft, The Netherlands,*
[c]*EMPA, Swiss Federal Laboratories for Materials Science and Technology, Feuerwerkerstrasse 39, 3602, Thun, Switzerland*
[d]*University of Bremen, Institute for Applied and Physical Chemistry, Fachbereich 2 (Chemie/Biologie), Leobener Straße / NW 2, Postfach 330440, D-28334 Bremen, Germany*
howardf@jhu.edu, c.w.hagen@tudelft.nl, ivo.utke@empa.ch, swiderek@uni-bremen.de

5.1 Introduction

Chemical reactions induced by electron irradiation are central to electron beam–induced processing, an application at the forefront of nowadays' nanotechnology [1]. Using tightly focused high-energy electron beams, such processes are capable of fabricating structures with arbitrary shapes and dimensions down to a few nanometers [2]. Initially being an unwanted reaction in electron microscopy involving mostly contaminant gases like pump oil, such focused

Low-Energy Electrons: Fundamentals and Applications
Edited by Oddur Ingólfsson

ISBN 978-981-4800-00-6 (Hardback), 978-0-429-05882-0 (eBook)
www.panstanford.com

electron beam–induced processing (FEBIP) has developed into a maskless technique that writes nanostructures directly onto a surface. This is achieved by dosing volatile precursor molecules onto the surface. These precursor molecules are decomposed by the impinging electron beam. Depending on the nature of the resulting fragments, material can then either be deposited or removed, the resulting processes being termed "focused electron beam–induced deposition" (FEBID) and "focused electron beam–induced etching" (FEBIE) [1].

While electron beam processing applies electrons with energies in the kiloelectron volt regime, low-energy electrons make significant contributions to the processes that occur when a high-energy electron from the focused beam enters the solid material on which a nanostructured deposit shall be fabricated [1]. Following the initial impact on the surface, the primary high-energy electron may be back-scattered to the surface to again become involved in precursor decomposition. However, along its track, a large number of the secondary electrons (SEs) with energies below 50 eV are released [1–4]. When exiting from the solid, they initiate, depending on their actual energy, precursor fragmentation via dissociative electron attachment (DEA), neutral dissociation (ND), or dissociative ionization (DI). It is these processes together with the energy and spatial distribution of the back-scattered electrons (BSEs) and SEs that must be known to fully understand the nanostructure formation in the different FEBIP processes. As the BSEs and SEs can exit the surface outside of the primary beam focus and thus induce precursor decomposition and deposit formation over a wider surface area, knowledge of their spatial distribution is important to understand the shape of the deposits. Simulation methods help to obtain such insight but also to predict the energy distribution of the electrons involved in FEBIP. The latter then determines the mechanisms by which a given precursor will fragment, that is, the chemistry underlying the different FEBIP processes and, in particular, the nature of the material produced by FEBID.

FEBID applies volatile precursor molecules that contain atoms of the desired deposit material and side groups, also called ligands, which provide the molecule with sufficient volatility. In an ideal FEBID process, the ligands dissociate completely from the central atom upon electron impact and are volatile enough to desorb. A solid deposit consisting of a pure material is thus left behind. However, in reality the deposits often incorporate significant quantities of impurities [5]. This can, in part, be traced back to residual gases in the vacuum chamber because the FEBID process is typically performed under high vacuum (HV) only [1, 6, 7]. On the other hand, the unwanted material can also result from incomplete dissociation of the precursors [8–14] or secondary reactions of molecular fragments that do not desorb fast enough after dissociation of the precursor [15].

To improve this situation and develop better precursors, a solid understanding of the electron-induced reactions of FEBID precursors, that is, their fragmentation via DEA, ND, and DI, is essential. This can be achieved by a combination of gas-phase studies on the electron-initiated precursor fragmentation [8, 10, 12] and surface science studies performed in ultrahigh vacuum (UHV), that is, under well-defined conditions and in the absence of residual gases [16, 17]. Using this approach [14], the chemical reaction mechanisms underlying the FEBID process can be unraveled. This has become particularly evident through studies on the precursors trimethyl (methylcyclopentadienyl) platinum(IV) ($MeCpPtMe_3$) [8, 9], $Pt(PF_3)_4$ [10, 11], and $Co(CO)_3NO$ [12, 13]. Also, quantitative information on the cross sections of the reactions can be obtained, in particular, through gas-phase experiments. The cross sections can be fed back into simulations to model deposit growth [1, 3, 4], while the understanding of reaction mechanisms helps us to learn how the structure of a precursor must be modified to achieve a complete dissociation under electron impact.

This chapter focuses on the electron-induced chemistry underlying FEBID processes and the approaches of current state-of-the-art work in this area. It provides a basic knowledge of the technique itself, the physics behind the production of low-energy SEs under the focused high-energy electron beam, and experiments unraveling the electron-driven chemistry leading to nanostructure formation. Section 5.2 thus gives an introduction to FEBIP processes and some of its applications. It is followed by a summary of the fundamentals of electron scattering in solid materials, the resulting release of SEs, and its consequences in terms of spatial resolution of FEBIP processes (Section 5.3). Reactive electron–precursor interactions

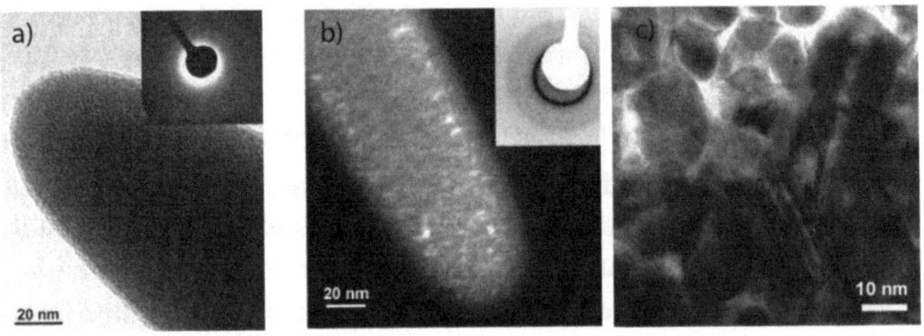

Figure 5.3 Transmission electron microscope images of FEBID material obtained from different precursors. Typical electron beam conditions in FEBID were 25 keV, 100 pA, and an electron flux of about 10^{19} electrons/($cm^2 \cdot s$). (a) Si–O–C tip obtained from a tetraethoxysilane precursor; the diffraction pattern in the inset shows that this material is amorphous. (b) Nanocomposite tip obtained from a $Co_2(CO)_8$ precursor. The bright contrast shows the cobalt nanocrystals embedded in an amorphous carbonaceous matrix; the diffraction inset confirms the presence of cobalt nanocrystals. (c) Polycrystalline pure gold deposit obtained from a $AuClPF_3$ precursor without an embedding matrix. Reprinted with permission from Ref. [1]. Copyright (2008), American Vacuum Society.

[29]. Water co-injection with Me_2Au(tfac) resulted in clean gold deposits [30]. Also, efficient post-FEBID cleaning processes were developed to remove the carbon content from platinum- and cobalt-containing deposits. In this case, the reactive gases are injected after the FEBID process with the organometallic compound and react with the carbon of the deposit. Following this strategy, FEBIE postgrowth treatments with water and oxygen of thin films and 3D structures of as grown Pt—C FEBID material were demonstrated [31, 32]. Hydrogen and substrate heating to 300° C was used for purifying as-grown Co—C material [33]. Changing the composition and percolation of the metal-embedding carbonaceous matrix can also be achieved by postgrowth irradiation without gas at room temperature for Pt—C material [34], just with O_2 and heating by relying on catalytic reactions of Pt in the carbonaceous [35], or by simple vacuum annealing of Co—C material, which reduces the oxygen content inside the matrix [36]. Further studies will be needed to verify whether co-injection or postinjection of reactive gases can yield pure material for a larger variety of metals and thicker films then a few tens of nanometers. H_2O, O_2, and XeF_2

etch carbon efficiently but can also readily oxidize or fluorinate metals that are not noble. A conceivable alternative could be to use hydrogen as an option for a reducing agent forming CH_x species with the carbon matrix and leaving the metal intact.

5.2.4 Applications of FEBID and FEBIE

The functional materials that were obtained by FEBID so far can be classified into (i) pure nanocrystalline metals, (ii) pure amorphous oxides, (iii) carbon, (iv) nanocomposites with metal nanocrystals embedded in a carbonaceous matrix, and (v) alloys. Each of these FEBID materials has already found attractive applications either due to the minimally invasive, direct-write, high-lateral-resolution nature of FEBID or due to the novel material properties involved in FEBID material. A few examples of FEBIP applications are shown in Fig. 5.4.

Pure-metal and high-metal-content FEBID materials serve, for example, as customized electrodes to nanowires or carbon nanotubes (Fig. 5.4b) [37], as components of future magnetic logic circuits [38–41], or as artificially top-down-built materials with applications in plasmonics [42], photonics (Fig. 5.4d), or spintronics [43, 44], just to name a few examples. High-resolution FEBID of pure silicon dioxide and etching (FEBIE) of quartz and molybdenum silicide multilayer materials became an indispensable technological platform for industrial repair of masks for standard UV and EUV photolithography [22, 45]. FEBID of robust, highly dense, diamond-like carbon material serves for tailoring functional scanning probe tips with applications in metrology, biology, and nanoindentation [46]. Carbon FEBID material postannealed by lasers to yield graphite was recently used to contact multiwalled carbon nanotubes at all their shells for lowest contact resistance [47, 48].

Nanocomposite FEBID materials (Fig. 5.3b) have interesting applications in nanoelectronics and nanophotonics due to their specific electronic or photonic coupling mechanisms between the quantum dot–size metal nanocrystals and the dielectric matrix [49, 50]. So far granular Co—C-based highly-flux-sensitive magnetic Hall sensors [51, 52] (Fig. 5.4a), granular W—C-based sensitive piezoresistive strain sensors [53], granular Au—C-based polarization

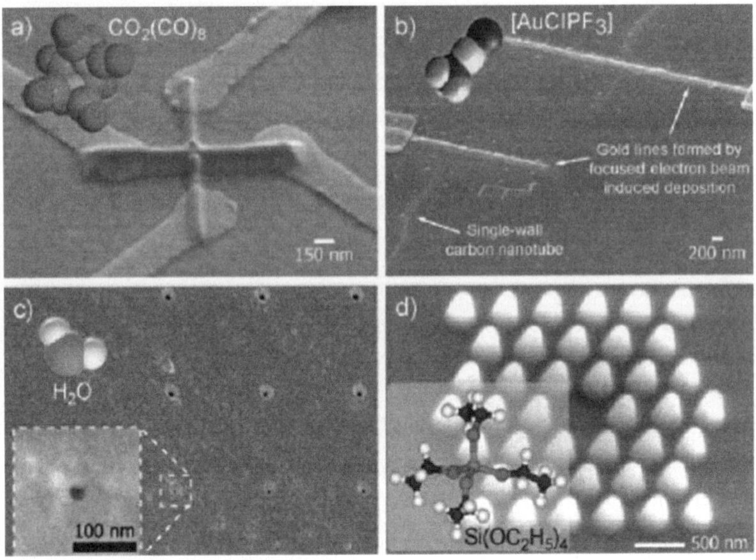

Figure 5.4 Examples of FEBIP applications: (a) A Co—C granular magnetic Hall sensor obtained from FEBID with $Co_2(CO)_8$. (b) Gold connections to a single-walled carbon nanotube written with $AuClPF_3$. (c) Etching of holes in a carbon membrane using water. (d) A photonic crystal structure deposited with tetraethoxysilane. Reproduced from Ref. [19] with permission. Copyright (2010) by John Wiley and Sons.

gratings for mode control in vertical cavity surface-emitting lasers [54], granular Pt—C humidity sensors [55], and large-current field emission tips [56] were realized with this novel material.

FEBID alloys that recently started to attract attention can be fabricated using dual GISs and offer new possibilities to tune material properties [3, 6, 57a]. For more applications of FEBID materials the reader is referred to Refs. [20e, 57b, 58a].

Apart from the industrial application of FEBIE as a mask repair process [22], FEBIE was recently developed as a soft-etch process in the field of quantum photonics to enable photonic crystal cavities into 2D hexagonal boron nitride using water molecules [58b] as well as to nanopattern diamond with water or oxygen [58c], leaving the atomistic structure of the substrates close to the pattern fully intact. This feature is crucial and makes FEBIE unique to reactive ion etching or focused ion beam milling.

5.3 Generation of Low-Energy Secondary Electrons in Electron Beam–Induced Processing

In FEBIP an electron beam is focused onto a substrate at which precursor molecules are adsorbed. As a result of the interaction of the electron beam with the substrate and precursor molecules the latter are dissociated. Because the interaction gives rise to scattered electrons with energies other than the primary electron (PE) beam, the dissociation process is much more difficult to understand than electron-induced dissociation of molecules in the gas phase. Therefore it is useful to provide here some insight into the interaction of electrons with matter. Regarding the scope of this book, it is also important to understand the generation of low-energy SEs, which play a major role in driving the precursor chemistry underlying FEBIP processes.

FEBIP is usually done in scanning electron microscopes where the energy of the PE beam can be varied between ~1 and 30 keV. When a fast electron of energy E hits a solid substrate it will be scattered by the substrate atoms, with the scattering process being either elastic or inelastic. These scattering processes have been studied extensively over the past four or five decades. It is not our intention to rigorously treat the theory here but rather to give an insight into the physics involved and present the ingredients that are frequently used in electron-scattering simulations. Much of the following is based on the books by David C. Joy [59] and N. Silvis Cividjian et al. [60]. For more in-depth information the reader is referred to a book by L. Reimer [61] or the paper by M. Cailler and J. P. Ganachaud [62] and references therein.

5.3.1 Elastic Electron Scattering

In an elastic scattering process the initial electron does not lose energy, but its momentum and thus its direction changes. Typical scattering angles range from ~5° to 180°. This means that part of the electrons, the so-called BSEs, returns from the material. The underlying physics of this type of scattering is the Coulomb

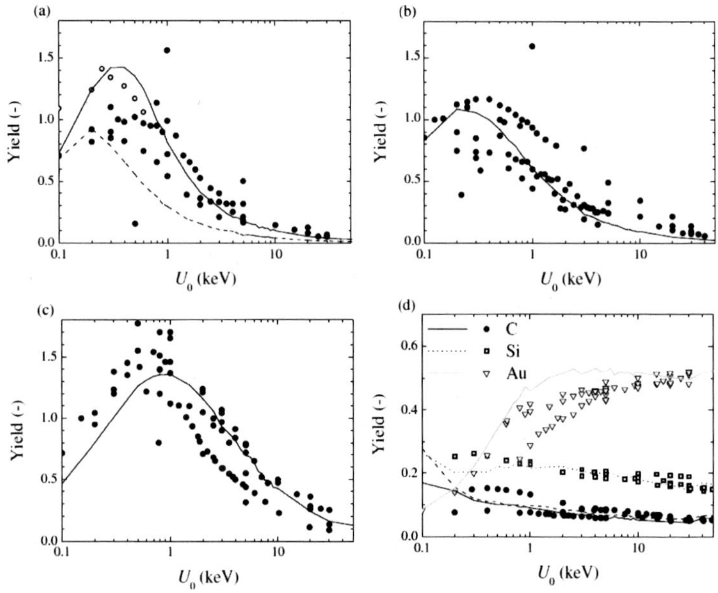

Figure 5.8 SE yields of carbon (a), silicon (b), and gold (c) and BSE yields for the same materials (d) as a function of primary energy U_0. The experimental data from the database by Joy [75] are indicated by dots. The circles in (a) correspond to the HOPG data from Corbridge et al. [76]. The solid lines correspond to the simulation results by Kieft and Bosch [74]. The dashed lines for carbon in (a) and (d) are for glassy carbon; the solid lines correspond to HOPG. From Ref. [74].

the two. The solid lines in Fig. 5.8 are the result of Monte Carlo simulations, to be described in the next section, and the data points are experimentally measured values. As the SE2s come from a much larger area (of radius r_b) than the SE1s, one may wonder whether the spatial resolution of scanning electron microscopy is not determined by the size of that area. The BSEs are expected to generate more emitted SEs than the PE because their average energy is reduced with respect to the PE, and they approach the surface under large angles with respect to the surface normal, which also enhances the SE yield. However, the BSE yield, being smaller than 1, reduces that enhancement, resulting in roughly the same number of SE2s and SE1s [61]. The low escape depth of the SE1s, combined with the enhanced emission from edges and tilted surface elements, allows for 1–10 nm resolution SE imaging.

5.3.4 Monte Carlo Simulation of Electron Scattering

Predictions of the spatial resolution of FEBIP require knowledge of the spatial distribution of the electrons that cross the substrate surface. In addition, it is important to know their energy, as the precursor dissociation probability is energy dependent. The best way to obtain this information is to use Monte Carlo simulations, in which the trajectory of a PE, incident on a substrate, is traced until it has come to rest. In this model, PEs can suffer two types of possible scattering events: elastic or inelastic. The PE trajectory in the target is a chain of linear segments, each describing the electron's motion between two successive interactions. The length of the linear segments (i.e., the step between the events) is Poisson distributed and can be sampled using a uniformly distributed random number RND between 0 and 1 as [77]

$$step = -\lambda_{\mathrm{t}} \ln(\mathrm{RND}), \tag{5.12}$$

where λ_{t} is the total MFP, which is related to the corresponding elastic (λ_{e}) and inelastic (λ_{i}) MFPs through

$$\frac{1}{\lambda_{\mathrm{t}}} = \frac{1}{\lambda_{\mathrm{e}}} + \frac{1}{\lambda_{\mathrm{i}}}. \tag{5.13}$$

The interaction sampling (i.e., the type of event [elastic/inelastic]) will also be determined using a random number in the following way. If RND is a random number between 0 and 1, the interaction will be elastic if RND$< \frac{\lambda_{\mathrm{t}}}{\lambda_{\mathrm{e}}}$, and inelastic otherwise.

Equation 5.12 derives from the expression for the probability, calculated here from the Poisson distribution $P(step) = \lambda_{\mathrm{t}}^{-1} e^{-step/\lambda_{\mathrm{t}}}$, that a segment length has a value between 0 and *step*. This probability and thus the segment length *step* is then related to RND so that the inverse relation yields Eq. 5.12. In electron trajectory simulations, this type of sampling is also applied to make other decisions such as the change in angle or electron energy. In the case of an elastic event a new direction of the electron is calculated using again a random number RND between 0 and 1. The polar angle θ is obtained from

$$\mathrm{RND} = \int_{\Omega} \frac{d\sigma_{\mathrm{e}}(\theta)/d\Omega}{\sigma_{\mathrm{e}}} d\Omega, \tag{5.14}$$

where the numerator, that is, the differential cross section, in the integrand, is given by Eq. 5.3 and the denominator, that is, the total elastic cross section, by Eq. 5.5. The integral represents the probability than an electron is scattered within a solid angle Ω. The polar angle θ is obtained by integration of Eq. 5.14 [64]

$$\cos\theta = 1 - \frac{2\alpha\mathrm{RND}}{(1+\alpha-\mathrm{RND})}. \tag{5.15}$$

The azimuthal scattering angle is assumed to be isotropic for both elastic and inelastic scattering.

In the case of an inelastic scattering event the incident electron loses an amount of energy that is assigned to a newly created SE. As explained before, the determination of the energy loss is a complex problem. Good results are obtained using the dielectric function approach using optical data. Both the incident electron, with a lower energy now, and the newly created SE continue their paths to the next scattering event, their direction being determined from energy and momentum conservation. When an electron arrives at the substrate surface it is either reflected or emitted, depending on its energy with respect to the surface energy barrier related to the material's work function.

In Fig. 5.9 the scattering process is sketched. The process ends when the PE has lost all its energy and when all generated SEs have come to rest or have been emitted. This calculation is done for a large number of PEs and the exit position and energy of all emitted electrons is stored. Such a Monte Carlo simulation code was developed by Kieft and Bosch [74] (hereafter labeled as KB code), as an extension of the GEANT4 [78] simulation platform. For the elastic scattering events this approach used the angular differential Mott cross sections provided by [65] for energies larger than 100 eV and the cross sections for acoustic phonon scattering [79, 80] for energies lower than 100 eV. The energy loss function can be constructed from any sufficiently complete and accurate set of optical data. Useful compilations of optical data can be found in the handbooks edited by Palik [81–83] and on the CXRO

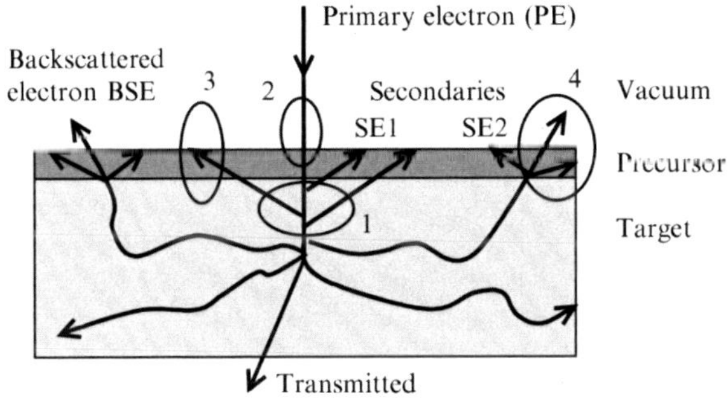

Figure 5.9 Sketch of the scattering process in a target with a layer of adsorbed precursor molecules. The primary beam (PE, zone 2) directly creates SEs in the target (zone 1), which may exit as SE1s (zone 3). The PEs also create BSEs, which, in turn, may create SE2s (zone 4). All electrons that enter or exit the surface may contribute to the dissociation of the adsorbed precursor molecules.

website [84]. To determine the binding energy of a SE, the code uses the ionization cross sections provided by [85], and at the lower energies the bandgap energy was taken or user-defined values. The boundary crossing of electrons from one material to another involves a quantum mechanical transmission and reflection calculation, as described in Ref. [77]. As seen in Fig. 5.8, in which experimental yields are compared to simulation results from the KB code, the simulation provides correct values of the SE yields, unlike most other Monte Carlo simulations of electron–matter scattering. Accurate Monte Carlo simulations are known to be notoriously slow. However, recently the KB code was critically assessed by T. Verduin [86] and reprogrammed using graphics processing units (GPUs) instead of a central processing unit (CPU). The resulting code is 2 orders of magnitude faster than the CPU-based KB code, reducing typical scanning electron microscopy (SEM) image simulations to minutes rather than many hours.

Using such Monte Carlo simulations, the locations at the substrate surface where precursor molecules are dissociated can be simulated, provided that the energy-dependent cross section for the dissociation of the adsorbed precursor molecules is known. In other words, one can simulate where material is deposited. The deposition rate $R(r)$ at a distance r from the point of incidence of the PE beam of energy E_p is given by [87] as

$$R(r) = \int_0^{E_p} f(r, E)\, \sigma_{\text{diss}}(E)\, N dE, \text{ [in molecules/unit area/s]} \tag{5.16}$$

where $f(r, E)$ is the flux of electrons crossing the substrate surface as a function of position r and energy E (in electrons/unit area/s), $\sigma_{\text{diss}}(E)$ is the electron-induced dissociation cross section of the precursor molecule (in unit area) as a function of energy, and N is the surface density of adsorbed precursor molecules (molecules/unit area). The size of the deposit is then convolved with the Gaussian current distribution in the PE beam to obtain the actual deposit size. It is noted that this is the deposit size at the very first instance of the growth. Later, the incident beam also interacts with the deposit, generating SEs, BSEs, and forward-scattered electrons in the deposit as well. To model this requires a dynamic Monte Carlo simulation, which has not been done with the KB code yet. Such simulations have been performed using a Monte Carlo code based on a simpler physical model [88], but this code does not necessarily give the correct yields and consequently can be used only to predict relative trends.

Figure 5.10 shows the electron distribution as obtained from the Monte Carlo simulation using the KB code, for a 20 keV electron beam incident on a 10-nm-thick carbon membrane. The plot shows the radial distribution $N(r)$, that is, the number of electrons that cross the surface in a ring between r and $r + dr$, the radial intensity $I(r)$, that is, the number of electrons per unit area emitted from a ring between r and r + dr, that is, $I(r) = N(r)/(2\pi r dr)$, and the total number of electrons $\Sigma N(r)$ emitted from within a circle with radius r. The simulation was done with 10^6 PEs and a zero-diameter beam. Although the radial intensity $I(r)$ shows a very narrow peak with a full width at half maximum (FWHM) of 1.1 nm, it makes more sense to determine the FW50%, that is, the diameter from within 50% of the electrons are emitted. This value is taken from the $\Sigma N(r)$ curve as 2.6 nm.

To obtain the actual deposit size, the electron distribution has to be convolved with the dissociation cross section. However, dissociation cross sections for adsorbed precursor molecules are hard to find. There are cross sections available for electron-induced

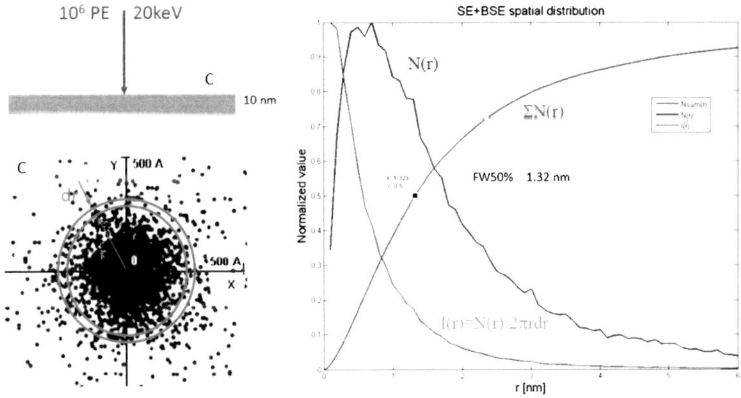

Figure 5.10 (Upper left) 10^6 20 keV electrons are incident on a 10-nm-thick C membrane. (Lower left) Simulated electron exit points on the surface of the membrane. (Right) Radial distribution $N(r)$, radial intensity $I(r)$, and total number of electrons $\Sigma N(r)$ emitted from within a circle with radius r.

dissociation of molecules in the gas phase. Here a general equation, published by Alman [89], for the electron impact dissociation cross section of C_xH_y in the gas phase is used to illustrate the approach of simulating FEBIP growth:

$$\sigma_{\text{diss}}(E) = \begin{cases} 0,\ E \le E_{\text{th}} \\ \sigma_{\max}\left(1 - \frac{(E_{\max}-E)^2}{(E_{\max}-E_{\text{th}})^2}\right),\ E_{\text{th}} < E < E_{\max} \\ \sigma_{\max} \cdot e^{-\frac{E-E_{\max}}{\lambda}},\ E \ge E_{\max} \end{cases} \tag{5.17}$$

This equation expresses the dissociation cross section of a hydrocarbon molecule in terms of four parameters E_{th}, $E_{\max}$, $\sigma_{\max}$, and λ, making it very useful and easy to implement in a computer program. $\sigma_{\max}$, the maximum cross section, scales with the number of carbon (#C) and hydrogen (#H) atoms present in the molecule, according to

$$\sigma_{\max} = (1.89\#C + 0.330\#H - 0.505) \times 10^{-2}.\quad [\text{nm}^2] \tag{5.18}$$

$E_{\max}$, the energy at which the cross section peaks, and λ, a constant that determines the rate of decay of the cross section beyond $E_{\max}$, are given by Alman et al. [89] for many dissociation channels.

For instance, for $C_2H_5 \rightarrow C_2H_4 + H$, $E_{\max} = 25$ eV and $\lambda = 77$ eV; for $C_2H_5 \rightarrow C_2H_3 + 2\ H$, $E_{\max} = 18$ eV and $\lambda = 11.4$ eV. C_2H_5 is one of the hydrocarbon molecules that Yoshimura et al. [90] found

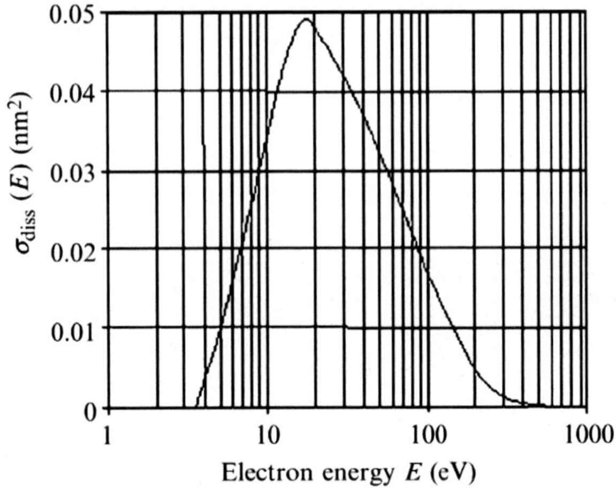

Figure 5.11 Dissociation cross section according to Eq. 5.17 for C_2H_5. The threshold has been shifted from 10 eV down to 3.5 eV.

in electron microscopes. E_{th} is the threshold energy for molecular dissociation and is always around 10 eV for gas-phase hydrocarbons. When the gas molecule is adsorbed on a surface, the dissociation threshold energy is shifted to values lower than that in the gas phase. Experiments of deposition induced by low-energy electrons in a scanning tunneling microscope (STM) by Uesugi and Yao [91] showed that the dissociation threshold for adsorbed hydrocarbon molecules lies at 3–4 eV. We used this information to adjust the C_2H_5 dissociation cross section for the adsorbed phase. Figure 5.11 shows a plot of $\sigma_{diss}(E)$ for adsorbed C_2H_5 (E_{th} = 3.5 eV) with an E_{max} of 25 eV and a λ of 77 eV.

In Fig. 5.12 the distribution of dissociated molecules is shown together with the electron distribution of Fig. 5.10. It is seen that the convolution with the cross section has a narrowing effect on the distribution of dissociated molecules. The FW50% is reduced to 1.4 nm. This demonstrates the very high spatial resolution of FEBIP.

One may wonder what the effect of the SE2s is on the FEBIP process compared to the effect of the SE1s. The ratio of the surface density of SE1s to SE2s is determined by the ratio of the emission areas. Assuming a radius for the SE1 emission area of the order of

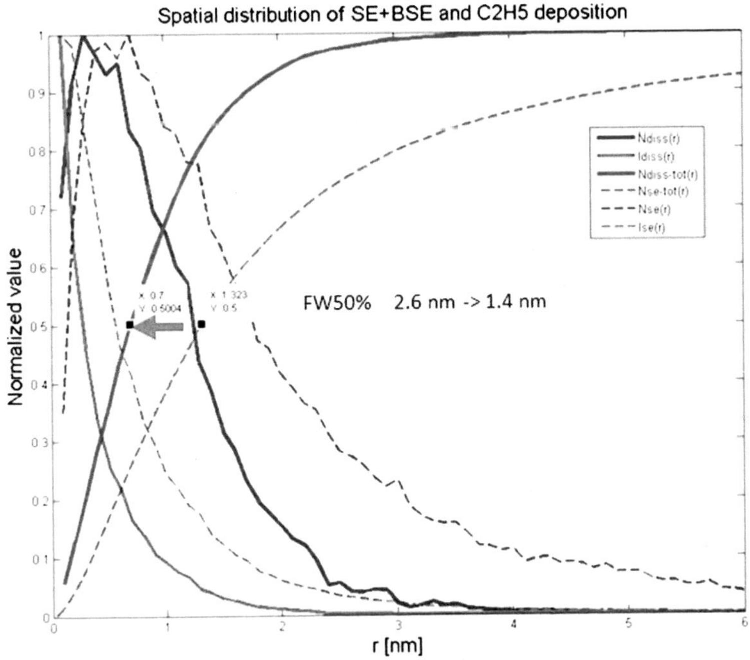

Figure 5.12 Spatial distribution of dissociated molecules (solid lines) compared to spatial distribution of electron exit points, as shown in Fig. 5.10 (dashed lines). $N_{\text{diss}}(r)$, $I_{\text{diss}}(r)$, and $N_{\text{diss-tot}}(r)$ are the radial distribution, radial intensity, and total number of molecules dissociated within a circle with radius r, respectively. $N_{\text{se}}(r)$, $I_{\text{se}}(r)$, and $N_{\text{se-tot}}(r)$ are the radial distribution, radial intensity, and total number of electrons emitted from within a circle with radius r, respectively.

the inelastic MFP, that is, ~2 nm, and $r_b = 5.3$ μm (for Si at 30 keV, from Eq. 5.9, the SE2 surface density is 1.4×10^{-7} times smaller than the SE1 surface density. At 1 keV, where $r_b = 18$ nm and assuming the same inelastic MFP, the SE2 surface density is still only 1% of the SE1 surface density. So, precursor dissociation will predominantly occur by the SE1 in the very small area around the incident PE beam. The SE2s will only cause a very thin deposit spread out over a much larger area. To demonstrate the ultrahigh spatial resolution of FEBIP, Fig. 5.13 shows two experimental results of very-high-resolution deposits on a thin membrane, prepared in a 300 keV scanning transmission electron microscope [92] (Fig. 5.13a), and

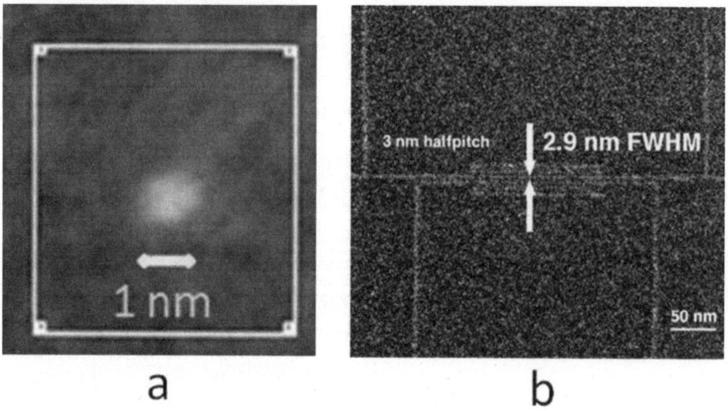

Figure 5.13 (a) HAADF image of a 1 nm W/C dot deposited on a thin C membrane in a 300 keV STEM using $W(CO)_6$ as a precursor. From Ref. [92]. © IOP Publishing. Reproduced with permission. All rights reserved. (b) SEM image of 3 nm Pt/C lines and spaces as written in a 30 keV SEM using $MeCpPtMe_3$ as a precursor. Reprinted with permission from Ref. [93]. Copyright (2011), American Vacuum Society.

on bulk Si, prepared in a 30 keV scanning electron microscope [93] (Fig. 5.13b).

To summarize this, the FEBIP spatial resolution is seen to be determined by, first of all, the beam size of the focused electron beam but, second, by the spatial and energy distribution of the scattered electrons that pass the substrate surface from below. The energy-dependent dissociation cross section of the precursor molecule determines which electrons are most effective in the deposition process. The lack of knowledge on the electron-induced dissociation cross sections of adsorbed precursor molecules hinders successful simulation of the FEBIP process. The dynamic process of the electron scattering in a growing deposit has not been dealt with here. Furthermore, there may very well be an influence of surface diffusion of precursor molecules and precursor fragments on the FEBIP spatial resolution, which has also not been treated here. And finally, an intriguing question, which has not been raised yet, is whether SEs that reach the substrate surface and do not have sufficient energy to be emitted still have enough energy to dissociate a molecule on the surface. Some molecules show dissociative resonant behavior upon attachment of a very low energetic electron. To approach these questions, detailed studies on the electron-induced dissociation of precursor molecules are required, as reviewed in the next section.

5.4 Elucidating the Role of Low-Energy Electrons in the FEBID Precursor Decomposition Process

5.4.1 The Surface Science Approach to FEBID Studies

The typical FEBID approach to creating nanostructures is shown in Fig. 5.14a. Deposition is performed in an electron microscope with a focused, high-energy electron beam (energies between 5 and 200 keV) in a low-vacuum (10^{-5} to 10^{-3} Torr) environment [1, 3, 94, 95]. A GIS allows for a suitably volatile organometallic precursor to be directed toward the substrate [25]. Nanostructures are created when transiently adsorbed precursor molecules decompose as a consequence of electron-stimulated reactions in a localized region defined by the point of intersection between the incident PE beam and the substrate. The volatile fragments produced during deposition are pumped away by the vacuum, while the remaining, nonvolatile fragments become incorporated into the nanostructure.

FEBID nanodeposits created in electron microscopes are typically characterized by SEM, transmission electron microscopy (TEM), and/or energy-dispersive X-ray spectroscopy (EDX). Although these postdeposition analytical techniques (Fig. 5.14a) can yield information on the structure and composition of a deposit, they do not provide any direct molecular-level insights into the mechanisms of precursor decomposition, including the role of the low-energy SEs generated by the interaction of the PE beam with the substrate (see Section 5.3). Developing a more detailed understanding of the molecular events that characterize FEBID is of more than just academic importance because FEBID structures are typically created from precursors designed for thermal processes, such as chemical vapour deposition (CVD). However, ligands that dissociate cleanly during CVD are often decomposed by electron-

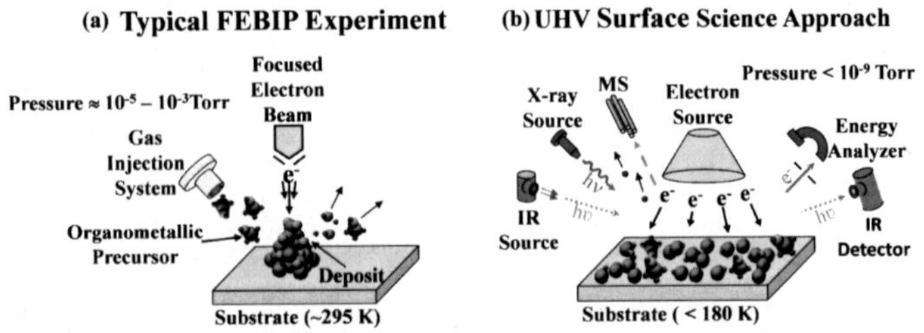

	(a)	(b)
Pressure Regime (Torr)	•$10^{-5} - 10^{-3}$	•$< 10^{-9}$
Electron Energy (keV)	•20 – 200	•.04 - 5
Power Density (mW/mm²)	•$\approx 10^2$	•$\approx 10^{-2}$
Temperature (K)	•≈ 300	•< 160
Precursor Surface Coverage (ML)	•< 1	•1 - 3
Size Scale of Deposits	•$\approx$ nm³	•$\approx$ cm²
Analytical Techniques	•SEM, (S)TEM, EDX	•XPS, RAIRS, MS, TPD
Information Obtained	•Composition and Dimensions of Deposited Structure	•Kinetic and Mechanistic Details of Precursor Decomposition

Figure 5.14 Schematic and table comparing the differences between (a) a typical FEBID experiment conducted in an electron microscope versus (b) surface science experiments conducted under ultrahigh vacuum (UHV) conditions. Reprinted from Ref. [15], Copyright (2011), with permission from Elsevier.

induced processes, resulting in high levels of organic contamination, which negatively impacts the properties and potential applications of FEBID nanostructures. Consequently, it is crucially important to understand how electron-stimulated decomposition occurs and to develop structure–activity relationships that describe the electron-stimulated reactions of surface-bound organometallic precursors. This information will be centrally important in the design of new precursors specifically for FEBID that can generate nanostructures with improved metal contents.

This kind of fundamental, molecular-level information on the reactions of adsorbed precursors with electrons can be extracted using an UHV surface science approach. This approach relies on studying the effect of electron irradiation on nanometer-thick films of precursor molecules adsorbed onto chemically inert substrates at low temperatures (Fig. 5.14b) [15]. In contrast to typical FEBID experiments a broad-beam electron source is needed to facilitate the use of surface analytical techniques that cannot acquire data on nanometer-scale features [96]. The requirement to use a broad-beam electron source (flood gun) means that the incident energy and electron flux are attenuated in these UHV studies. However, the incident electron energy (500–2000 eV) is still sufficient to mimic the higher energies of PEs used in typical FEBID studies ($>$5 keV) because it still generates a plume of lower-energy SEs and BSEs as a result of its interaction with the substrate [1, 92]. Precursor molecules must also be adsorbed at low temperatures, which serves to simplify the analysis by ensuring that the results are a reflection of the purely electron-stimulated reactions. In the UHV surface science approach, well-known surface analytical techniques, such as X-ray photoelectron spectroscopy (XPS) and reflection absorption infrared spectroscopy (RAIRS), can then be used in situ to follow changes to the surface composition and bonding environment of the various elements within the precursor molecule, while mass spectrometry (MS) identifies the volatile species ejected from the film as a consequence of the electron-stimulated reactions. A comparison of the operating conditions and information that can be obtained from a typical FEBID experiment and from a UHV surface science study is tabulated in Fig. 5.14 [15].

A particularly noteworthy advantage of XPS in FEBID studies is that it can often be used to extract information on the kinetics of electron-stimulated deposition by tracking the evolution of the metal XPS region as a function of the electron dose: electron dose = (electron flux $\times$ time)/surface area (e^-/cm^2) [98]. For example, Fig. 5.15a shows the effect that increasing electron dose has on the Au(4f) XPS region of adsorbed dimethyl-(acetylacetonate) gold(III) ($Au^{III}(acac)Me_2$) molecules. The parent $Au^{III}(acac)Me_2$ molecules give rise to the expected Au($4f_{7/2}/4f_{5/2}$) doublet due to spin–orbit splitting. With the onset of electron irradiation the Au($4f_{7/2}/4f_{5/2}$) doublet associated with the parent molecules decreases, accompanied by the appearance of a new Au($4f_{7/2}/4f_{5/2}$) doublet at lower binding energy due to the more metallic Au atoms produced by the electron-induced precursor decomposition.

The change in the formal oxidation state that the transition metal experiences when the organometallic complex is reduced by electron irradiation (in the case of Fig. 5.15a, from Au^{III} to metallic Au) is often large enough to facilitate accurate spectral deconvolution. As shown in Fig. 5.15b, the rate of Au^{III} reduction as

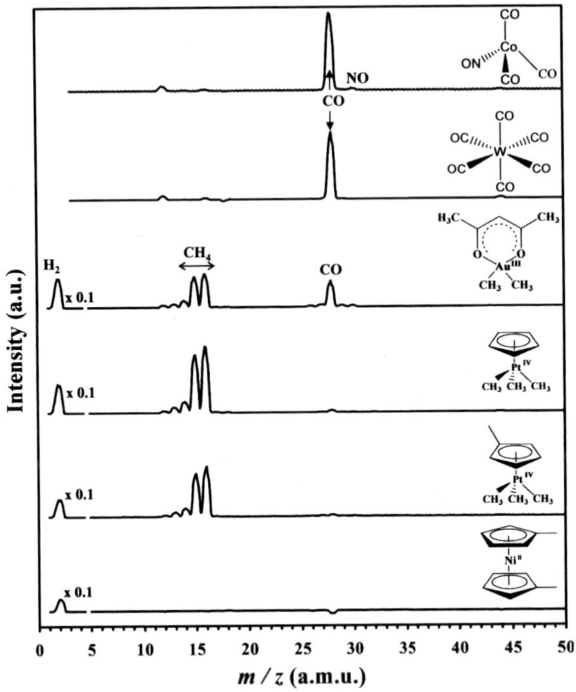

Figure 5.16 Mass spectra (0–50 amu) of the volatile neutral species produced when six different organometallic precursors, adsorbed onto a gold substrate at 100 K, were electron-irradiated. Reprinted by permission from Springer Nature Customer Service Centre GmbH: Springer Nature, *Applied Physics A: Materials Science & Processing*, Ref. [16], copyright (2014).

secondary reactions. These can be either thermal reactions or post-deposition electron beam processing [107–110]. Other processes, such as diffusion and the contribution from electron-stimulated reactions of co-adsorbed contaminant molecules (principally water and hydrocarbons), must also be considered when deposition is performed in an electron microscope due to its environment. A notable advantage of the low-temperature UHV surface science experiments is the ability to prevent contaminant molecules from adsorbing by controlling the substrate temperature. For example, by adsorbing at a substrate temperature held at ~200K we can adsorb the organometallic precursor without co-adsorbing any water or hydrocarbon species. The initial coverage of precursor molecules can also be varied by controlling the substrate's initial exposure to the organometallic precursor. The subsequent electron dose can also be controlled and quantified by measuring the target current. Moreover, since the deposition step is the first step in the FEBID process, and the electron dose provides a measure of the extent of reaction, information obtained in these well-controlled surface science experiments can often be analyzed to determine the nature of the initial deposition step [99].

In FEBID, adsorbed precursor molecules will be exposed to electrons with a disparate range of energies because the interaction of the incident PE beam with the substrate produces, through inelastic scattering, a plume of lower energy BSEs and SEs that collectively encompass a wide range of energies [1, 97]. However, although the interactions of electrons with solids are relatively well understood, the fundamental electron-stimulated interactions with surface-bound precursor molecules that initiate deposition in FEBID are significantly less clear. This is in large part because a number of electron-induced dissociation mechanisms are possible, depending on the electron energy. These include DI, DEA, and electron excitation [111].

Determining the nature of the initial decomposition/fragmentation step has been greatly facilitated by the availability of data from recent gas-phase studies conducted on electron interactions with FEBID precursors (e.g., $W(CO)_6$, $Co(CO)_3(NO)$, $MeCpPtMe_3$, and $Pt(PF_3)_4$ [8, 112–114]). In the isolated gas-phase medium, the incident electron energies can be well defined and products unambiguously identified, as can energy-dependent cross sections and resonance peak positions. A particular advantage of gas-phase studies is that the incident electron energy can be well controlled. This allows low-energy (<10 eV) decomposition processes, such as DEA,

$$ML_n + e^- \rightarrow [ML_n]^- \rightarrow \text{fragmentation products (DEA process)} \tag{5.21}$$

to be studied separately from DI,

$$ML_n + e^- \rightarrow [ML_n]^+ + 2e^- \rightarrow \text{fragmentation products (DI process)}, \tag{5.22}$$

which dominates at higher incident electron energies (typically >50 eV). Results from gas-phase studies can then be contrasted with the fundamental bond-breaking processes and reaction products observed when the same precursor molecules are adsorbed on surfaces and exposed to a broad range of electron energies. By combining results from gas-phase and surface science studies, mechanistic insights into the dominant electron-stimulated processes occurring during FEBID can be acquired.

The remainder of this chapter focuses on three case studies that shed light on the detailed nature of the initial fragmentation/ decomposition step ($ML_{n(ads)} + e^- \rightarrow MY_{(s)} + Z_{(g)}\uparrow$). The three case studies involve two platinum-containing precursors ($MeCpPtMe_3$ and $Pt(PF_3)_4$) [9, 11] and one cobalt-containing precursor ($Co(CO)_3NO$) [13]. In two of these case studies the comparison between surface science and gas-phase experiments points conclusively toward an initial decomposition step that occurs through a DEA-type process involving extremely low-energy (<10 eV) electrons, while the third case study is most consistent with contributions from both DEA and DI processes. In a typical FEBID process conducted in an electron microscope, the electrons responsible for the precursor's fragmentation (either DI or DEA) will be generated by the initial interaction of the primary beam with the substrate (see Section 5.3).

5.4.3 The Case of $MeCpPtMe_3$

Trimethyl(methylcyclopentadienyl) platinum(IV) ($MeCpPtMe_3$) is a volatile precursor (vapour pressure of 0.053 Torr at 23°C) widely used for Pt deposition, not only in FEBID, but also in focused ion beam–induced deposition (FIBID), laser-induced chemical processing, CVD, and ALD. Within the FEBID community, $MeCpPtMe_3$ is one of the most common precursors used to create conductive nanostructures/nanowires [11, 34, 92, 95, 99, 115–122]. However, the presence of significant carbon contamination (typically 80%–90% carbon content) negatively impacts the electronic properties of the deposits. For example, the resistivity of Pt wires created from $MeCpPtMe_3$ by FEBID is typically >1 Ω·cm, severely limiting their ability to serve as nanoelectrodes or nanowires [121].

Figure 5.17 shows the structure of an adsorbed $MeCpPtMe_3$. molecule being electron irradiated. XPS data from UHV surface science experiments shown in Fig. 5.18a demonstrate that when

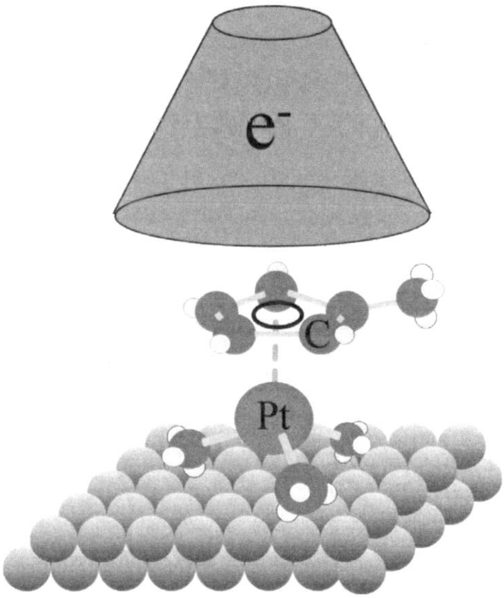

Figure 5.17 Adsorbed $MeCpPtMe_3$ precursor irradiated with a broad electron beam.

a nanometer-thick film of $MeCpPtMe_3$ adsorbed onto a gold surface is exposed to an electron dose (incident electron energy 500 eV) sufficient to reduce all of the native Pt species, and thereby decompose all of the parent $MeCpPtMe_3$ molecules, the C:Pt ratio decreases by approximately 11% of its initial value, creating films that contain 89% carbon and 11% platinum [9]. This change in stoichiometry can be accurately determined by integrating the area of the Pt(4f) and C(1s) peaks observed by XPS before and after electron irradiation (not shown). This highlights a notable advantage of the low-temperature UHV surface science approach, specifically the ability to directly measure the film's composition before and after electron irradiation, facilitating an accurate determination of even relatively small changes in film composition. The significance of this decrease in the ratio of carbon to platinum atoms (C:Pt) is that since the parent molecule contains one central Pt atom and 9 C atoms; a decrease of ~11% corresponds to the loss of exactly one carbon atom from the parent molecule as a result of the electron-induced deposition step [9]. Complementary MS data (Fig. 5.18b) reveals that this carbon atom is released from

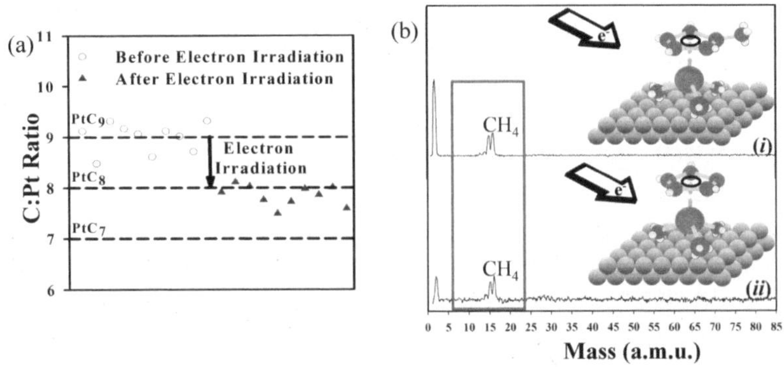

Figure 5.18 (a) Influence of electron beam irradiation on the chemical composition of a nanometer-thick film of MeCpPt(IV)Me_3 determined by XPS analysis of the Pt(4f) and C(1s) regions. Dashed horizontal lines are shown to indicate the C:Pt ratio expected for films with a stoichiometry of PtC_9, PtC_8, and PtC_7. (b) Mass spectrum (0–85 amu) of neutral volatile species produced during the electron irradiation of adsorbed (i) $MeCpPtMe_3$ and (ii) $CpPtMe_3$ molecules. Adapted from Ref. [9]. Copyright (2009) by the American Chemical Society.

the adsorbed layer as methane during irradiation, irrespective of whether $MeCpPtMe_3$ or $CpPtMe_3$, its structural analog, is used. Collectively, the XPS and MS results point to an initial electron-stimulated deposition process that proceeds through the cleavage of one of the Pt–CH_3 bonds, with the remaining carbon atoms becoming trapped in the deposit that forms [9]. It should be noted that in the mass spectrometer (Fig. 5.18b) the methyl fragment is detected as methane, most likely due to the abstraction of a hydrogen atom from the chamber walls prior to being ionized.

Gas-phase studies of $MeCpPtMe_3$ have been conducted in the low-energy (0–10 eV) regime where DEA dominates and at an electron energy of 100 eV where DI dominates [8]. Analysis of the findings from these studies, summarized in Table 5.1, indicate that cleavage of a single Pt–CH_3 bond to produce $[MeCpPtMe_2]^-$ is by far the most dominant reaction channel observed in the 0–10 eV energy regime, accounting for over 87% of the fragments. In contrast, when gas-phase $MeCpPtMe_3$ molecules are exposed to 100 eV electrons, the dominant fragments produced by DI are $[MeCpPtMe]^+ -$ nH and $[MeCpPt]^+ -$ nH, corresponding to the loss

Table 5.1 Comparison of the fragments detected when gas-phase $MeCpPtMe_3$ is irradiated by electrons with (left) <10 eV energy and (right) 100 eV

Incident electron energy <10 eV		Incident electron energy 100 eV	
Fragment	**Rel. intensity (CPS)**	**Fragment**	**Rel. intensity (CPS)**
$[MeCpPtMe_2]^-$	1	$[MeCpPtMe]^+$-nH	1
$[MeCp]^-$	7.6E-2	$[MeCpPt]^+$-nH	5.3E-1
$[C_6H_5Pt]^-$	3.2E-2	$[MeCpPtMe_3]^+$	1.0E-1
$[C_7H_5Pt]^-$	1.4E-2	$[MeCpPtMe_2]^+$-nH	5.8E-2
$[PtCH_3]^-$	1.1E-2	$PtC_4H_n^+$	4.3E-2
$[MeCpPtCH_3]^-$	5.0E-3	$PtC_2H_n^+$	3.2E-2
$[MeCpPtMeH]^-$	4.0E-3	$MeCp^+$-nH	2.9E-2
Pt^-	1.5E-3	$PtC_3H_n^+$	2.1E-2

Source: Reproduced from Ref. [8] with permission from the PCCP Owner Societies.

of two and three methyl groups, respectively. The $[MeCpPtMe_2]^+$–nH fragment, corresponding to the loss of one methyl group, is a relatively minor fragment accounting for less than 3% of the fragments observed. Thus, a comparison of the surface science and gas-phase studies clearly points toward deposition occurring via DEA through the formation of $[MeCpPtMe_3]^-$, which subsequently decomposes via a single Pt–CH_3 bond cleavage event [8, 9]. The small, volatile CH_3 fragment is released while the much heavier platinum-containing fragment remains surface bound [9]. The low-energy electrons (<10 eV) required to initiate DEA in the surface science studies are generated by the interaction of the PE beam with the substrate. In addition to Pt–CH_3 bond cleavage, previous studies have shown that the surface intermediate generated by the DEA process is decomposed further by electron-stimulated C–H bond-breaking processes, ultimately creating platinum atoms encased in a carbonaceous matrix [9]. The overall FEBID process is shown schematically in Fig. 5.19.

5.4.4 The Case of $Pt(PF_3)_4$

Tetrakis(trifluorophosphine)platinum is a carbon-free precursor that exists as a high-vapour-pressure liquid (68 Torr at 20°C). $Pt(PF_3)_4$ has been used in CVD to create pure platinum films on

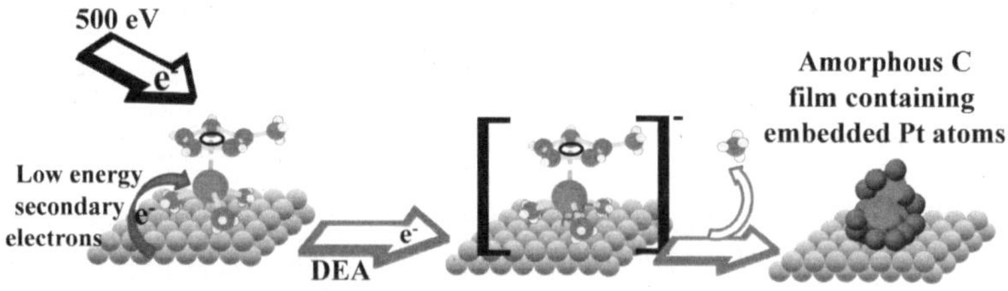

Figure 5.19 Electron-stimulated decomposition of $MeCpPtMe_3$. The incident primary beam causes low-energy SEs to be created from the substrate that attach to adsorbed $MeCpPtMe_3$ molecules to form a negative ion. These unstable species falls apart via a dissociative electron attachment (DEA) process through the cleavage of one $Pt-CH_3$ bond. The remaining fragments remain trapped on the surface, creating a carbonaceous film containing embedded Pt atoms.

a variety of heated (200°C–300°C) substrates, in UV-laser-induced metallization, and in the selective area deposition of platinum silicide on Si(100) [123–125]. Although $Pt(PF_3)_4$ is not as widely used as $MeCpPtMe_3$ in FEBID, nanowires created from $Pt(PF_3)_4$ have been shown to have resistivity values orders of magnitude lower than those created with $MeCpPtMe_3$. Indeed, the lowest resistivity for a line deposit of $Pt(PF_3)_4$ is 26 μΩ·cm, a value only 2.5 times greater than that of pure platinum (10.42 μΩ·cm) [126].

Figure 5.20 shows the structure of an adsorbed $Pt(PF_3)_4$ molecule being electron-irradiated. For an electron spectroscopy, such as XPS, $Pt(PF_3)_4$ is uniquely well suited to a detailed analysis of the effect that electron irradiation has on composition and bonding environment, because of its lack of hydrogen. Electron-based spectroscopies are essentially "blind" to the presence of hydrogen due to its lack of core-level electrons. Figure 5.21 shows XPS data that illustrates the effect of comparatively short irradiation times (electron doses $<1.5 \times 10^{15}$ e^-/cm^2) on nanometer-thick films of $Pt(PF_3)_4$ adsorbed on amorphous carbon substrates. The Pt(4f) spectral envelope broadens to lower binding energy as the platinum atoms are reduced during the initial deposition step [11].

A determination of the dissociation mechanism, however, relies on analyzing how the relative atom concentration and speciation of the elements contained within the ligands change under the influence of electron irradiation. During the initial period of

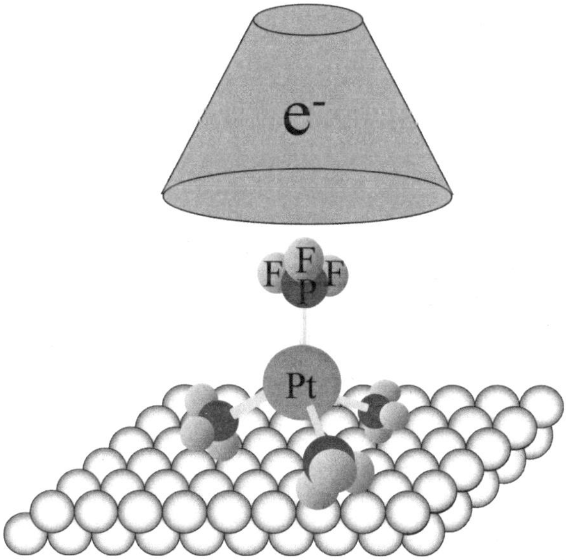

Figure 5.20 Adsorbed $Pt(PF_3)_4$ precursor irradiated with a broad electron beam.

irradiation, Fig. 5.22 shows that phosphorus and fluorine atoms are lost from the adsorbate layer. By integrating the spectral intensity within the P(2p) and F(1s) regions for multiple $Pt(PF_3)_4$ films exposed to electron doses $<1.5 \times 10^{15}$ e^-/cm^2, Fig. 5.22 reveals that during this initial period of irradiation the concentrations of adsorbed phosphorus and fluorine atoms both decrease to 75% of their initial values [11]. Moreover, the rate of electron-stimulated phosphorus and fluorine atom loss is the same, causing the ratio of phosphorus to fluorine atoms remaining in the film to remain constant (bottom panel in Fig. 5.22). Collectively, these observations indicate that the initial step in the decomposition of adsorbed $Pt(PF_3)_4$ molecules proceeds via the loss of one of the PF_3 ligands [11], analogous to the loss of one CH_3 group from $MeCpPtMe_3$ [9].

Gas-phase studies by May et al. have shown that the overwhelming product produced during the interactions of low-energy (<10 eV) electrons with $Pt(PF_3)_4$ is $Pt(PF_3)_3^-$ (see Table 5.2) [10]. Indeed, the relative intensity of $Pt(PF_3)_3^-$ is approximately

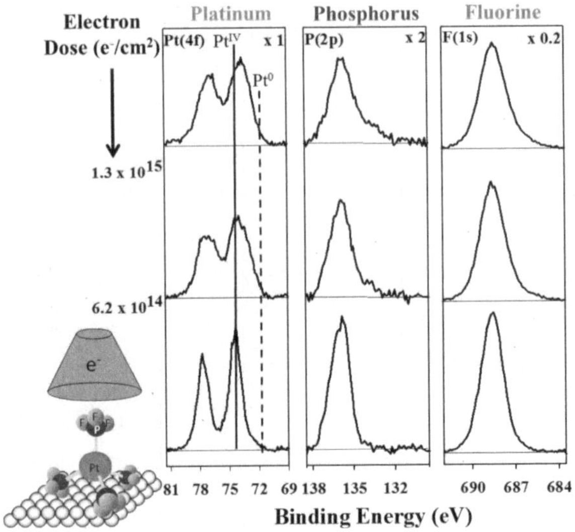

Figure 5.21 Effect of comparatively small electron doses ($<4 \times 10^{15}$ e^-/cm^2) on the Pt(4f), P (2p), and F(1s) XPS regions of a nanometer-scale $Pt(PF_3)_4$ film. Adapted with permission from Ref. [11]. Copyright (2011) by the American Chemical Society.

200 times greater than the next most likely product, $Pt(PF_3)F^-$. In contrast, results from broad-band electron ionization Fourier-transform ion cyclotron resonance (FT-ICR) using 20 eV electrons reveal that $Pt(PF_3)^+$ and $Pt(PF_3)_2^+$ are the two most prominent ions produced from $Pt(PF_3)_4$ via DI [10]. Thus, a comparison of gas-phase and surface science studies leads to the same conclusion as for $MeCpPtMe_3$, notably that in FEBID, dissociation of adsorbed $Pt(PF_3)_4$ precursor molecules occurs via a DEA process, with the low-energy electrons being produced by the interaction of the PE beam with the substrate [10, 11]. The overall deposition process can be represented by the following equation, shown schematically in Fig. 5.23:

$$Pt(PF_3)_{4(ads)} + e^- \rightarrow [Pt(PF_3)_4]^-_{(ads)}$$
$$\rightarrow Pt(PF_3)_{3(ads)} + PF_{3(g)} \uparrow \quad (5.23)$$

The dominance of DEA as opposed to DI for $Pt(PF_3)_4$ is also consistent with the cross-section measurements obtained by May et al. in their gas-phase studies [10]. The cross section for the loss of

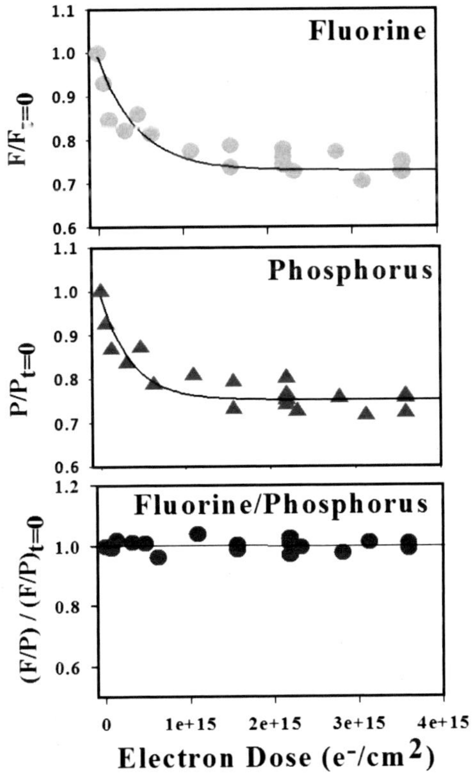

Figure 5.22 Changes in the fractional coverages of phosphorus ($P/P_{t=0}$) and fluorine ($F/F_{t=0}$) and the ratio of fluorine/phosphorus (F/P) atoms in a $Pt(PF_3)_4$ film during comparatively small electron doses ($<4 \times 10^{15}$ e$^-$/cm^2). Adapted with permission from Ref. [11]. Copyright (2011) American Chemical Society.

one PF_3 ligand by DEA was found to be approximately 20,000 pm^2, a very large value compared to more typical values of 14 pm^2 and 860 pm^2 for the production of O from CO_2 and N_2O by DEA, respectively [10]. A value of 20,000 pm^2 is comparable to the large cross sections observed at low energies for multiply halogenated hydrocarbons [127]. The combination of UHV surface science and gas-phase data therefore suggest that for $Pt(PF_3)_4$ the DEA cross sections are much larger than the equivalent DI values and as a result the DEA process dominates the dissociation.

Table 5.2 Fragments detected when gas-phase $Pt(PF_3)_4$ is irradiated by electrons with ≤12 eV energy

Fragment	Relative cross section
$Pt(PF_3)_3^-$	1
$Pt(PF_3)F^-$	6.1E-3
$Pt(PF_3)^-$	5.0E-3
$Pt(PF_3)_2^-$	4.5E-3
$Pt(PF_3)F^-$	1.3E-3
F^-	4.40E-4
Pt^-	9.09E-5

Source: Adapted from Ref. [10] with permission from the PCCP Owner Societies.

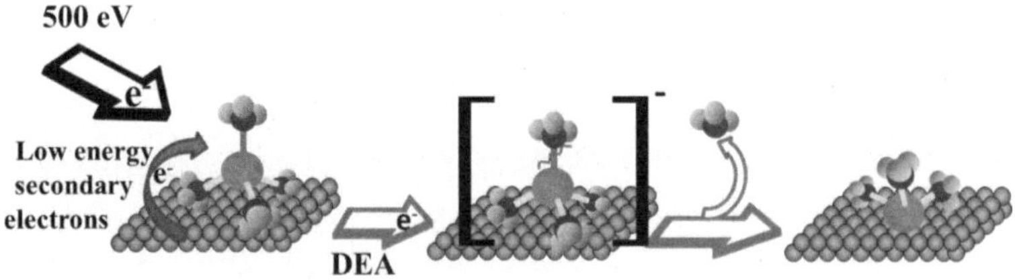

Figure 5.23 Summary of the decomposition step for $Pt(PF_3)_4$. The incident primary beam causes low-energy SEs to be created from the substrate that attach to adsorbed $Pt(PF_3)_4$ molecules to form a negative ion. These unstable species fall apart via a dissociative electron attachment (DEA) process through the cleavage of one Pt—PF_3 bond and the release of a single PF_3 ligand.

Analysis of XPS data obtained on $Pt(PF_3)_4$ films exposed to electron doses in excess of approximately 1.5×10^{15} e^-/cm^2, shown in Fig. 5.24, allow us to follow the fate of the adsorbed $Pt(PF_3)_3$ species generated by the initial deposition step [11]. Studying the fate of these intermediates is also important in FEBID because the extremely high electron fluxes used in electron microscopes mean that any surface-bound species will be exposed to the effects of continuous electron bombardment. At this stage of the reaction it should be noted that UHV surface science data can no longer be meaningfully referenced to gas-phase studies, as the latter only

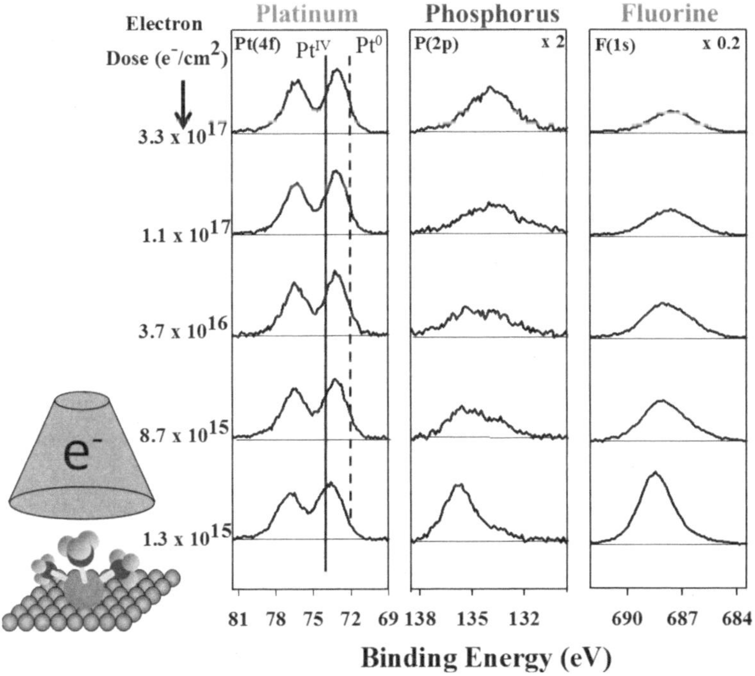

Figure 5.24 Effect of larger electron doses ($>1.3 \times 10^{15}$ e^-/cm^2) on the Pt(4f), P(2p), and F(1s) XPS regions of a nanometer-scale $Pt(PF_3)_4$ film. Adapted with permission from Ref. [11]. Copyright (2011) American Chemical Society.

provides a point of comparison for the initial step. Figure 5.24 shows that the Pt(4f) region remains unchanged for electron doses in excess of approximately 1.5×10^{15} e^-/cm^2, indicating that the local electronic environment around the Pt atoms remains constant [11]. In contrast, as shown in Figs. 5.24 and 5.25, the larger electron doses cause a continued decrease in the concentration of adsorbed fluorine atoms, although the phosphorus content in the film remains constant at 75% of its initial value (see Fig. 5.25). These observations indicate that electron irradiation of adsorbed $Pt(PF_3)_3$ species does not cause any further PF_3 ligand ejection but instead decomposes the remaining PF_3 ligands by P–F bond cleavage and ejection of fluoride ions into the gas phase; thus,

$$Pt(PF_3)_{3(ads)} + xe^- \rightarrow PtP_3F_{9-x(ads)} + xF^-_{(g)}\uparrow \rightarrow\rightarrow PtP_{3(ads)}. \quad (5.24)$$

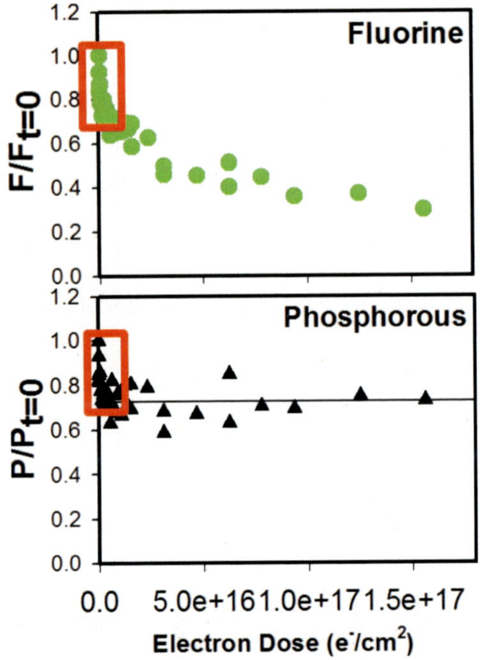

Figure 5.25 Changes as a function of electron dose in the fractional coverage of phosphorus ($P/P_{t=0}$) and fluorine ($F/F_{t=0}$) atoms exposed to electron irradiation. For each plot the red box indicates the electron dose regime where the $Pt(PF_3)_4$ molecules are undergoing the (initial) decomposition step. Adapted with permission from Ref. [11]. Copyright (2011) American Chemical Society.

For $Pt(PF_3)_4$ precursor molecules the sequence of electron-stimulated reactions in FEBID can therefore be summarized as a PF_3 ligand ejection step followed by P—F bond cleavage and loss of fluoride ions (see Fig. 5.26) [11]. In contrast to the initial deposition step, the origin of the electrons responsible for P—F bond cleavage (primary, secondary, etc.) in the second step cannot be unambiguously determined due to the inability to compare to gas-phase studies and the lack of theoretical understanding of the elementary steps involved in FEBID. However, it should be noted that the size limit of any deposit created by FEBID only depends on the details of the initial deposition step. Our analysis points towards a deposition step initiated by low-energy electrons generated by the

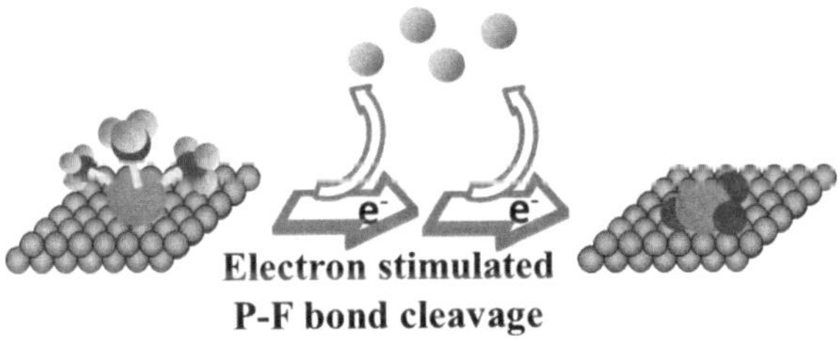

Figure 5.26 Electron-stimulated decomposition of the adsorbed $Pt(PF_3)_3$ species produced in the initial electron-stimulated decomposition of adsorbed $Pt(PF_3)_4$ molecules. Further electron irradiation causes P—F bond cleavage and the release of fluorine from the film. Adapted with permission from Ref. [11]. Copyright (2011) American Chemical Society.

interactions of the PE beam with the substrate. Since these low-energy SEs are ejected from the substrate across an area larger than the PE beam, we predict that there will be a finite limit to the smallest-size deposits that can be created by FEBID from $Pt(PF_3)_4$ and that the dimension of any deposit will always be greater than that of the incident electron beam.

In a broader sense, the reaction sequence identified for $Pt(PF_3)_4$ reflects a general trend that our UHV surface science studies have uncovered in terms of the effect of electron irradiation on adsorbed precursor molecules. Notably, that *while the initial deposition step leads to the ejection of ligands into the gas phase (CH_3 from $MeCpPtMe_3$ and PF_3 from $Pt(PF_3)_4$), subsequent electron interactions with the remaining ligands tend to proceed by electron-stimulated decomposition rather than desorption* [9, 11, 13, 106]. For example, in the case of metal carbonyls, another popular type of organometallic precursor used in FEBID, the initial deposition step proceeds via CO desorption [13, 106]. However, subsequent electron-stimulated reactions of the partially decarbonylated fragments left behind result in decomposition of the CO ligands, producing graphitic carbon and reactive oxygen species that often react with the central metal atom to form a metal oxide. Thus, the overall process can be represented as follows:

Deposition step (CO desorption):

$$M(CO)_{x(ads)} + e^- \rightarrow M(CO)_{x-y(ads)} + yCO_{(g)} \uparrow \qquad (5.25)$$

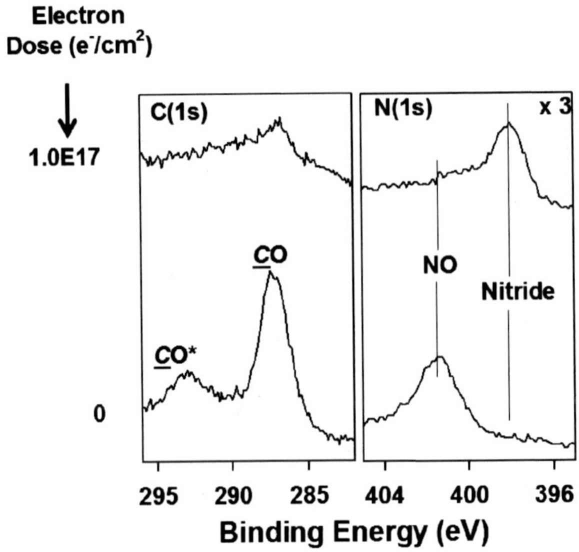

Figure 5.28 Evolution of the C(1s) and N(1s) XPS regions for a film of $Co(CO)_3NO$ molecules exposed to an electron dose of 1.00×10^{17} e^-/cm^2. Adapted with permission from Ref. [13]. Copyright (2011) American Chemical Society.

loss of only a single carbonyl atom, with a maximum cross section at <1eV. In sharp contrast, DI results mainly in the formation of Co^+, with a peak at approximately 70 eV [12, 114]. Thus, the comparison between the UHV surface science and gas-phase studies in this case do not point clearly to the predominance of DEA or DI as the mechanism responsible for the precursor's fragmentation. Indeed, by convoluting the absolute cross-section values determined from the gas-phase measurements with the SE yield measured when a metal surface (Ni(111)) is exposed to 400 eV incident electrons, it is possible to derive a qualitative sense of the damage yield that adsorbed $Co(CO)_3NO$ molecules will experience (Fig. 5.29) during FEBID [12, 135]. The benefit of such a plot of damage yield is that the integral of the DEA (red) curve at low energies (<10 eV) and the DI (blue) curve at higher energies (>20 eV) provides a direct measure of the relative contribution that these two processes are expected to play in the precursor's initial fragmentation step [14].

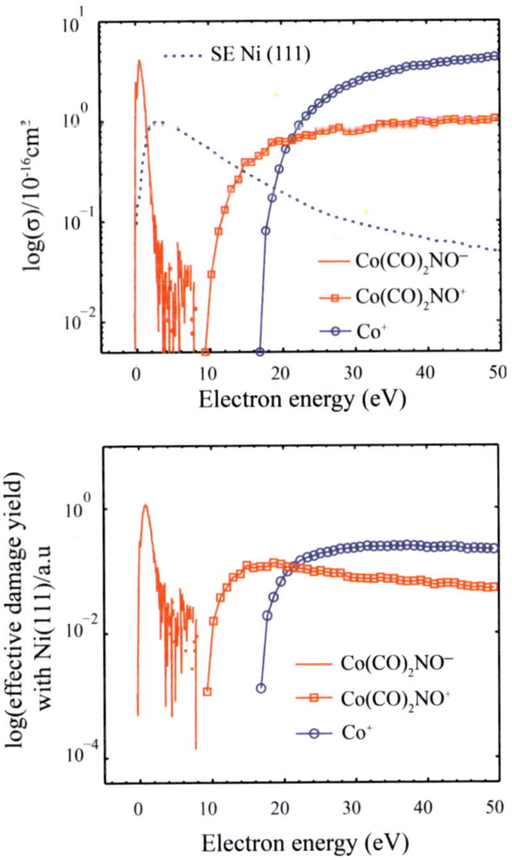

Figure 5.29 (Top) Combined plot of Ni (111) SE yield as well as (i) the partial cross section for single CO loss through DEA (red solid line) and DI (red open square) and (ii) Co^+ formation through DI of $Co(CO)_3NO$. (Bottom) Effective damage yield (i.e., product of respective partial cross section and normalized SE yield) derived from the data shows in the top panel [14] (https://creativecommons.org/licenses/by/2.0/).

Analysis of Fig. 5.29, obtained from the absolute reaction cross sections determined from the gas-phase measurements, suggests that both DI and DEA processes are likely to play a significant role in the fragmentation of this precursor, qualitatively consistent with the UHV surface science data [12]. Analysis of the gas-phase data indicates that this is largely a consequence of the higher than usual cross sections measured for the DI process. Thus, it is apparent that DEA is not necessarily the dominant fragmentation pathway for every EBID precursor [12].

Comparisons between gas-phase and surface science studies can be further complicated by the greater variety of reaction pathways availablefor molecules in a condensed phase (surface) environment, rather than an isolated (gas-phase) medium. For example, the NO group dissociates when adsorbed $Co(CO)_3NO$ molecules are electron-irradiated (see Fig. 5.28), although there is no evidence for such a process in the gas phase. Another type of complication occurs for organometallic precursors that contain multidentate ligands, such as $Cu^{II}(hfac)_2$, where UHV surface science studies are unable to identify the initial deposition step because the relative bulk of the ligands precludes their ejection into the gas phase, even if they were to be released from the central metal atom. This can even lead to secondary reactions of the precursors with the fragments resulting from an adjacent initial deposition event [17]. These considerations underline the importance of using a combination of as many techniques as possible when studying the fundamental chemistry underlying FEBID.

5.5 Outlook

It is clear from the contents of this chapter that advances in FEBID technology require not only fundamental insight in the electron-induced chemistry of the applied precursors. Also, a highly interdisciplinary approach to FEBID research is needed to cover all aspects of the process. This also includes synthetic chemists capable of making new precursors containing ligands that can be removed with ease in the FEBID process and that are devised according to fundamental insights into its electron-induced chemistry.

The need for the different fields to work together and thus solve the same problems inherent in FEBID from different angles is obvious. Therefore, research aiming at improved FEBID precursors and processes requires a concerted approach linking fundamental knowledge on electron-driven chemistry in FEBID with rational design and chemical synthesis of novel, superior precursor molecules. Such an approach has been realized in the form of a multidisciplinary coordinated research network within Europe (COST Action CELINA, [136]). CELINA has in fact networked the relevant research communities, including both experiment and theory, and, most importantly, synthetic chemists to design new precursors that could, once studied by all sides, establish FEBID as a powerful industrial technology on a large scale.

The combined effort of researchers from the fields of chemical synthesis, electron-induced chemistry, and the actual FEBID process has, for instance, brought forward novel bimetallic precursors [137, 138]. These types of precursors enable to precisely control the composition of alloy nanostructures produced by FEBID. The network has also led to higher purities of metallic deposits, as exemplified by deposits with high Ag content for plasmonic applications achieved through the use of a novel precursor with carboxylate ligands [139]. A systematic study of precursors with different ligands has started to reveal trends that show how a precursor needs to be designed to perform well in FEBID [140]. And, finally, surface science studies revealing the mechanisms by which deposit purification procedures [30, 32] act are ongoing [141, 142] and will hopefully lead to the increased emergence of simple and robust new FEBID processes. All these examples show that FEBID is a field that rapidly evolves and fundamental knowledge on electron-induced chemistry plays an important role in this progress.

References

1. Utke, I., Hoffmann, P., and Melngailis, J. (2008). Gas-assisted focused electron beam and ion beam processing and fabrication. *J. Vac. Sci. Technol., B*, **26**, pp. 1197–1276.
2. Van Dorp, W. F., van Someren, B., Hagen, C. W., and Kruit, P. (2005). Approaching the resolution limit of nanometer-scale electron beam-induced deposition. *Nano Lett.*, **5**, pp. 1303–1307.
3. Huth, M., Porrati, F., Schwalb, C., Winhold, M., Sachser, R., Dukic, M., Adams, J., and Fantner, G. (2012). Focused electron beam induced deposition: a perspective. *Beilstein J. Nanotechnol.*, **3**, pp. 597–619.
4. van Dorp, W. F., Beyer, A., Mainka, M., Gölzhäuser, A., Hansen, T. W., Wagner, J. B., Hagen, C. W., and De Hosson, J. Th. M. (2013). Focused

electron beam induced processing and the effect of substrate thickness revisited. *Nanotechnology*, **24**, p. 345301.

5. Botman, A., Mulders, J. J. L., and Hagen, C. W. (2009). Creating pure nanostructures from electron-beam-induced deposition using purification techniques: a technology perspective. *Nanotechnology*, **20**, p. 372001.
6. Bernau, L., Gabureac, M., Erni, R., and Utke, I. (2010). Tunable nanosynthesis of composite materials by electron-impact reaction. *Angew. Chem. Int. Ed.*, **49**, pp. 8880–8884.
7. Lukasczyk, T., Schirmer, M., Steinrück, H.-P., and Marbach, H. (2008). Electron-beam-induced deposition in ultrahigh vacuum: lithographic fabrication of clean iron nanostructures. *Small*, **4**, pp. 841–846.
8. Engmann, S., Stano, M., Matejčík, Š., and Ingólfsson, O. (2012). Gas phase low energy electron induced decomposition of the focused electron beam induced deposition (FEBID) precursor trimethyl (methylcyclopentadienyl) platinum(IV) ($MeCpPtMe_3$). *Phys. Chem. Chem. Phys.*, **14**, pp. 14611–14618.
9. Wnuk, J. D., Gorham, J. M., Rosenberg, S., van Dorp, W. F., Madey, T. E., Hagen, C. W., and Fairbrother, D. H. (2009). Electron induced surface reactions of the organometallic precursor trimethyl (methylcyclopentadienyl) platinum(IV). *J. Phys. Chem. C*, **113**, pp. 2487–2496.
10. May, O., Kubala, D., and Allan, M. (2012). Dissociative electron attachment to $Pt(PF_3)_4$ - a precursor for focused electron beam induced processing (FEBIP). *Phys. Chem. Chem. Phys.*, **14**, pp. 2979–2982.
11. Landheer, K., Rosenberg, S. G., Bernau, L., Swiderek, P., Utke, I., Hagen, C. W., and Fairbrother, D. H. (2011). Low energy electron induced decomposition and reactions of adsorbed tetrakis (trifluorophosphine) platinum ($Pt(PF_3)_4$). *J. Phys. Chem. C*, **115**, pp. 17452–17463.
12. Engmann, S., Stano, M., Papp, P., Brunger, M. J., Matejčík, Š., and Ingólfsson, O. (2013). Absolute cross sections for dissociative electron attachment and dissociative ionization of cobalt tricarbonyl nitrosyl in the energy range from 0 eV to 140 eV. *J. Chem. Phys.*, **138**, p. 044305.
13. Rosenberg, S. G., Barclay, M., and Fairbrother, D. H. (2013). Electron beam induced reactions of adsorbed cobalt tricarbonyl nitrosyl ($Co(CO)_3NO$) molecules. *J. Phys. Chem. C*, **117**, pp. 16053–16064.
14. Thorman, R. M., Ragesh Kumar, T. P., Fairbrother, D. H., and Ingólfsson, O. (2015). The role of low-energy electrons in focused electron beam induced deposition: four case studies of representative precursors. *Beilstein J. Nanotechnol.*, **6**, pp. 1904–1926.

15. Wnuk, J. D., Rosenberg, S. G., Gorham, J. M., van Dorp, W. F., Hagen, C. W., and Fairbrother, D. H. (2011). Electron beam deposition for nanofabrication: insights from surface science. *Surf. Sci.*, **605**, pp. 257–266.

16. Spencer, J. A., Rosenberg, S. G., Barclay, M., Wu., Y.-C., McElwee-White, L., and Fairbrother, D. H. (2014). Understanding the electron-stimulated surface reactions of organometallic complexes to enable design of precursors for electron beam-induced deposition. *Appl. Phys. A: Mater. Sci. Process.*, **117**, pp. 1631–1644.

17. Warneke, J., van Dorp, W. F., Rudolf, P., Stano, M., Papp, P., Matejcik, S., Borrmann, T., and Swiderek, P. (2015). Acetone and the precursor ligand acetylacetone: distinctly different electron beam induced decomposition? *Phys. Chem. Chem. Phys.*, **17**, pp. 1204–1216.

18. Klein, C., Loeschner, H., and Platzgummer, E. (2013). A multibeam mask writer for nodes of 11 nm or less. *SPIE Newsroom*, doi: 10.1117/2.1201311.005210.

19. Utke, I. and Gölzhäuser, A. (2010). Small, minimally-invasive, direct: electrons induce local nano-sized reactions of adsorbed functional molecules. *Angew. Chem. Int. Ed.*, **49**, pp. 9328–9330.

20. (a) Walz, M.-M., Schirmer, M., Vollnhals, F., Lukasczyk, T., Steinrück, H.-P., and Marbach, H. (2010). Electrons as "invisible ink": fabrication of nanostructures by local electron beam induced activation of SiOx. *Angew. Chem. Int. Ed.*, **49**, pp. 4669–4673. (b) Fowlkes, J. D., Winkler, R., Lewis, B. B., Stanford, M. G., Plank, H., and Rack, P. D. (2016). Simulation-guided 3D nanomanufacturing via focused electron beam induced deposition. *ACS Nano*, **10**(6), pp. 6163–6172. (c) Winkler, R., Lewis, B. B., Fowlkes, J. D., Rack, P. D., and Plank, H. (2018). High-fidelity 3D-nanoprinting via focused electron beams: growth fundamentals. *ACS Appl. Nano Mater.*, **1**(3), pp. 1014–1027. (d) Huth, M., Porrati, F., and Dobrovolskiy, O. V. (2018). Focused electron beam induced deposition meets materials science. *Microelectron. Eng.*, **185–186**, pp. 9–28.

21. Bret, T., Jonckheere, R., Van den Heuvel, D., et al. (2012). Closing the gap for EUV mask repair. *Extreme Ultraviolet (EUV) Lithography III Book Series: Proceedings of SPIE*, **8322**, Article Number: 83220C.

22. Bret, T., Hofmann, T., and Edinger, K. (2014). Industrial perspective on focused electron beam-induced processes. *Appl. Phys. A: Mater. Sci. Process.*, **117**, pp. 1607–1614.

23. http://www.nanotools.com/lines/blue-line/index.html.

24. (a) Friedli, V., Wanzenboeck, H. D., and Utke, I. (2012). Gas injection systems for FEB and FIB processing: theory and experiment, in *Nanofabrication Using Focused Ion and Electron Beams: Principles and Applications*, Utke, I., Moshkalev, S., and Russell, P., eds., Oxford University Press, New York, pp. 126–183. (b) Fowlkes, J. D., Winkler, R., Lewis, B. B., Fernandez-Pacheco, A., Skoric, L., Sanz-Hernandez, D., Stanford, M. G., Mutunga, E., Rack, P. D., and Plank, H. (2018). High-fidelity 3D-nanoprinting via focused electron beams: computer-aided design (3BID). *ACS Appl. Nano Mater.*, **1**(3), pp. 1028–1041.
25. Friedli, V. and Utke, I. (2009). Optimized molecule supply from nozzle-based gas injection systems for focused electron- and ion-beam induced deposition and etching: simulation and experiment. *J. Phys. D: Appl. Phys.*, **42**, p. 125305.
26. Bret, T., Afra, B., Becker, R., Hofmann, Th., Edinger, K., Liang, T., and Hoffmann, P. (2008). Gas assisted focused electron beam induced etching of alumina. *J. Vac. Sci. Technol., B*, **27**, pp. 2727–2731.
27. Melngailis, J., Moshkalev, S., and Utke, I. (2012). Current challenges in focused ion and electron beam materials processing, in *Nanofabrication Using Focused Ion and Electron Beams: Principles and Applications*, Utke, I., Moshkalev, S., and Russell, P., eds., Oxford University Press, New York, pp. 755–781.
28. Bret, T. and Hoffmann, P. (2012). How to select compounds for focused charged particle beam assisted etching and deposition, in *Nanofabrication Using Focused Ion and Electron Beams: Principles and Applications*, Utke, I., Moshkalev, S., and Russell, P., eds., Oxford University Press, New York, pp. 108–125.
29. Perentes, A. and Hoffmann, P. (2007). Focused electron beam induced deposition of Si-based materials from SiO_xCy to stoichiometric SiO_2: chemical compositions, chemical-etch rates, and deep ultraviolet optical transmissions. *Chem. Vap. Deposition*, **13**, pp. 176–184.
30. Shawrav, M. M., Taus, P., Wanzenboeck, H. D., Schinnerl, M., Stöger-Pollach, M., Schwarz, S., Steiger-Thirsfeld, A., and Bertagnolli, E. (2016). Highly conductive and pure gold nanostructures grown by electron beam induced deposition. *Sci. Rep.*, **6**, 34003, doi: 10.1038/srep34003.
31. Plank, H., Noh, J. H., Fowlkes, J. D., Lester, K., Lewis, B. B., and Rack, P. D. (2014). Electron-beam-assisted oxygen purification at low temperatures for electron-beam-induced Pt deposits: towards pure and high-fidelity nanostructures. *ACS Appl. Mater. Interfaces*, **6**, pp. 1018–1024.

32. Geier, B., Gspan, C., Winkler, R., Schmied, R., Fowlkes, J. D., Fitzek, H., Rauch, S., Rattenberger, J., Rack, P. D., and Plank, H. (2014). Rapid and highly compact purification for focused electron beam induced deposits: a low temperature approach using electron stimulated H_2O reactions. *J. Phys. Chem. C*, **118**, pp. 14009–14016.

33. Begun, E., Dobrovolskiy, O. V., Kompaniiets, M., Sachser, R., Gspan, C., Plank, H., and Huth, M. (2015). Post-growth purification of Co nanostructures prepared by focused electron beam induced deposition. *Nanotechnology*, **26**, 075301 (11pp), doi: 10.1088/0957-4484/26/7/075301.

34. Porrati, F., Sachser, R., Schwalb, C. H., Frangakis, A. S., and Huth, M. (2011). Tuning the electrical conductivity of Pt-containing granular metals by postgrowth electron irradiation. *J. Appl. Phys.*, **109**, p. 063715.

35. Sachser, R., Reith, H., Huzel, D., Winhold, M., and Huth, M. (2014). Catalytic purification of directly written nanostructured Pt microelectrodes. *ACS Appl. Mater. Interfaces*, **6**, pp. 15868–15874.

36. Puydinger dos Santos, M. V., Velo, M. F., Domingos, R. D., Zhang, Y., Maeder, X., Guerra-Nunez, C., Best, J. P., Béron, F., Pirota, K. R., Moshkalev, S., Diniz, J. A., and Utke, I. (2016). Annealing-based electrical tuning of cobalt–carbon deposits grown by focused-electron-beam-induced deposition. *ACS Appl. Mater. Interfaces*, **8**, pp. 32496–32503, doi: 10.1021/acsami.6b12192.

37. Brintlinger, T., Fuhrer, M. S., Melngailis, J., Utke, I., Bret, T., Perentes, A., Hoffmann, P., Abourida, M., and Doppelt, P. (2005). Electrodes for carbon nanotube devices by focused electron beam induced deposition of gold. *J. Vac. Sci. Technol., B*, **23**, pp. 3174–3177.

38. Fernández-Pacheco, A., De Teresa, J. M., Córdoba, R., and Ibarra, M. R. (2009). Magnetotransport properties of high-quality cobalt nanowires grown by focused-electron-beam-induced deposition. *J. Phys. D: Appl. Phys.*, **42**, p. 055005.

39. Serrano-Ramón, L., Córdoba, R., Rodríguez, L. A., Magen, C., Snoeck, E., Gatel, C., Serrano, I., Ibarra, M. R., and De Teresa, J. M. (2011). Ultrasmall functional ferromagnetic nanostructures grown by focused electron-beam induced deposition. *ACS Nano*, **5**, pp. 7781–7787.

40. Gavagnin, M., Wanzenboeck, H. D., Belic, D., and Bertagnolli, E. (2012). Synthesis of individually tuned nanomagnets for nanomagnet logic by direct write focused electron beam induced deposition, *ACS Nano*, **7**, pp. 777–784.

41. Costanzi, B. N., Riazanova, A. V., Dahlberg, E. D., and Belova, L. M. (2014). In situ manufacture of magnetic tunnel junctions by a direct write process. *Appl. Phys. Lett.*, **104**, p. 222401.
42. Höflich, K., Yang, R. B., Berger, A., Leuchs, G., and Christiansen, S. (2011). The direct writing of plasmonic gold nanostructures via electron beam induced deposition. *Adv. Mater.*, **23**, pp. 2657–2661.
43. Pohlit, M., Begun, E., Porrati, F., Huth, M., and Mueller, J. (2014). FEBID-prepared nanocluster building blocks of artificial square spin ice: stray-field studies of thermal dynamics, 5th workshop on focused electron beam induced processing (FEBIP2014), pp. 115–116.
44. Porrati, F., Sachser, R., and Huth, M. (2014). Magnetoresistance of granular Pt-C nanostrucures close to the metal-insulator transition. *J. Phys.: Condens. Matter.*, **26**, p. 085302.
45. http://www.zeiss.com/semiconductor-manufacturing-technology/en_de/products-solutions/photomask-systems/mask-repair.html#highlights.
46. http://www.nanotools.com/
47. Kim, S. K., Kulkarni, D. D., Rykaczewski, K., Henry, M., Tsukruk, V. V., and Fedorov, A. G. (2012). Fabrication of an ultra-low-resistance, ohmic contact to MWCNT-metal interconnect using graphitic carbon by electron beam induced deposition (EBID). *IEEE Trans. Nanotechnol.*, **11**, pp. 1223–1230.
48. Fedorov, A. G., Kim, S., Henry, M., Kulkarni, D., and Tsukruk, V. V. (2014). Focused electron beam induced processing (FEBIP) for emerging applications in carbon nanoelectronics. *Appl. Phys. A: Mater. Sci. Process.*, **117**, pp. 1659–1674.
49. Huth, M. (2010). From electronic correlations to strain-sensing applications. *J. Appl. Phys.*, **107**, p. 113709.
50. Huth, M., Kolb, F., and Plank, H. (2014). Dielectric sensing by charging energy modulation in a nano-granular metal. *Appl. Phys. A*, **117**, pp. 1689–1696.
51. Gabureac, M., Bernau, L., Utke, I., and Boero, G. (2010). Granular Co–C nano-Hall sensors by focused-beam induced deposition. *Nanotechnology*, **21**, p. 115503.
52. Gabureac, M. S., Bernau, L., Boero, G., and Utke, I. (2013). Single superparamagnetic bead detection and direct tracing of bead position using novel nanocomposite nano-hall sensors. *IEEE Trans. Nanotechnol.*, **12**, pp. 668–673.

53. Schwalb, C. H., Grimm, Ch., Baranowski, M., Sachser, R., Porrati, F., Reith, H., Das, P., Müller, J., Völklein, F., Kaya, A., and Huth, M. (2010). A tunable strain sensor using nanogranular metals. *Sensors*, **10**, pp. 9847–9856.

54. Utke, I., Jenke, M. G., Röling, C., Thiesen, P. H., Iakovlev, V., Sirbu, A., Mereuta, A., Caliman, A., and Kapon, E. (2011). Polarization stabilization of vertical cavity surface emitting lasers by minimally-invasive focused electron beam triggered chemistry. *Nanoscale*, **3**, pp. 2718–2722.

55. Kolb, F., Schmoltner, K., Huth, M., Hohenau, A., Krenn, J., Klug, A., List, E. J. W., and Plank, H. (2013). Variable tunneling barriers in FEBID based PtC metal-matrix nanocomposites as a transducing element for humidity sensing. *Nanotechnology*, **24**, p. 305501.

56. Murakami, K., Wakaya, F., and Takai, M. (2007). Observation of fringe-like electron-emission pattern in field emission from Pt field emitter fabricated by electron-beam-induced deposition. *J. Vac. Sci. Technol., B*, **25**, pp. 1310–1314.

57. (a) Porrati, F., Begun, E., Winhold, M., Schwalb, Ch. H., Sachser, R., Frangakis, A. S., and Huth, M. (2012). Room temperature $L1_0$ phase transformation in binary CoPt nanostructures prepared by focused-electron-beam-induced deposition. *Nanotechnology*, **23**, 185702. (b) De Teresa, J. M., and Fernandez-Pacheco, A. (2014). Present and future applications of magnetic nanostructures grown by FEBID. *Appl. Phys. A*, **117**(4), pp. 1645–1658.

58. (a) Utke, I., Moshkalev, S., and Russell, P., eds. (2012). *Nanofabrication Using Focused Ion and Electron Beams: Principles and Applications*, Oxford University Press, New York. (b) Kim, S., Fröch, J. E., Christian, J., Straw, M., Bishop, J., Totonjian, D., Watanabe, K., Taniguchi, T., Toth, M., and Aharonovich, I. (2018). Photonic crystal cavities from hexagonal boron nitride. *Nature Communications*, **9**, Article no. 2623. (c) Bishop, J., Fronzi, M., Elbadawi, C., Nikam, V., Pritchard, J., Fröch, J. E., Duong, N. M. H., Ford, M. J., Aharonovich, I., Lobo, C. J., and Toth, M. (2018). Deterministic nanopatterning of diamond using electron beams. *ACS Nano*, **12**(3), pp. 2873–2882.

59. Joy, D. (1995). *Monte Carlo Modeling for Electron Microscopy and Microanalysis*, Oxford University Press, New York.

60. Silvis-Cividjian, N. and Hagen, C. W. (2006). *Electron-Beam-Induced Nanometer-Scale Deposition*, Advances in Imaging and Electron Physics, Vol. 143, Hawkes, P. W., ed., Elsevier, San Diego and London.

61. Reimer, L. (2010). *Scanning Electron Microscopy: Physics of Image Formation and Microanalysis*, Springer Series in Optical Sciences, Vol. 45, Hawkes, P., ed., Springer Verlag, Heidelberg.
62. Cailler, M. and Ganachaud, J.-P. (1990). Secondary electron emission from solids II. Theoretical descriptions, in *Scanning Microscopy Supplement 4*, Schou, J., Kruit, P., and Newbury, D. E., eds., Scanning Microscopy International, AMF O'Hare, Chicago, pp. 81–110.
63. Bishop, H. E. (1976). The history and development of Monte Carlo methods for use in X-ray microanalysis, in *Use of Monte Carlo Calculations in Electron Probe Microanalysis and Scanning Electron Microscopy, NBS Special Publication 460*, Heinrich, K. F. J., Newbury, D. E., and Yakowitz, H., eds., Dept. of Commerce/National Bureau of Standards, pp. 5–14.
64. Newbury, D. E. and Myklebust, R. L. (1981). A Monte Carlo electron trajectory simulation for AEM, in *Analytical Electron Microscopy 1981*, Geiss, R. H., ed., San Francisco Press, San Francisco, pp. 91–98.
65. Czyzewski, Z., O'Neil MacCallum, D., Romig, A., and Joy, D. C. (1990). Calculations of Mott scattering cross section. *J. Appl. Phys.*, **68**, pp. 3066–3072.
66. Reimer, L. and Lodding, B. (1984). Calculation and tabulation of Mott cross-sections for large-angle electron-scattering. *Scanning*, **6**, pp. 128–151.
67. Browning, R., Li, T. Z., Chui, B., Ye, J., Pease, R. F. W., Czyzewski, Z., and Joy, D. C. (1994). Empirical forms for the electron/atom elastic scattering cross sections from 0 to 30 keV. *J. Appl. Phys.*, **76**, pp. 2016–2022.
68. Gauvin, R. and Drouin, D. (1993). A formula to compute total elastic Mott cross sections. *Scanning*, **15**, pp. 140–150.
69. Ding, Z. J. and Shimizu, R. (1996). A Monte Carlo modeling of electron interaction with solids including cascade secondary electron production. *Scanning*, **18**, pp. 92–113.
70. Ding, Z. J. and Shimizu, R. (2000) Monte Carlo simulation study of secondary electron emission and surface excitation. *Inst. Phys. Conf. Ser. No. 165: Symposium 8, 2nd Conf. Int. Union Microbeam Analysis Societies*, Kailua-Kona, Hawaii, 9–14 July 2000, IOP Publishing Ltd., London, pp. 279–280.
71. Ding, Z. J., Tang, X. D., and Shimizu, R. (2001). Monte Carlo study of secondary electron emission. *J. Appl. Phys.*, **89**, 718–726.

72. Seah, M. P. and Dench, W. A. (1979). Quantitative electron spectroscopy of surfaces: a standard data base for electron inelastic mean free paths in solids. *Surf. Interface Anal.*, **1**, pp. 2–11.

73. Kanaya, K. and Okayama, S. (1972) Penetration and energy-loss theory of electrons in solid targets. *J. Phys. D: Appl. Phys.*, **5**, pp. 43–58.

74. Kieft, E. and Bosch, E. (2008). Refinement of Monte Carlo simulations of electron–specimen interaction in low-voltage SEM. *J. Phys. D: Appl. Phys.*, **41**, p. 215310.

75. Joy, D. C. (1995). A database on electron-solid interactions. *Scanning*, **17**, pp. 270–275. Available through http://web.utk.edu/~srcutk/htm/interact.htm. A web-based database based on this collection is being maintained by E Napchan: http://www.mc-set.com/bse6/lbload.php.

76. Corbridge, J., Dennison, J. R. and Nickles, N. (2003). Effects of bandgap on secondary electron emission for graphitic carbon semiconductors. *Bull. Am. Phys. Soc.*, **48**, p. 1146.

77. Shimizu, R. and Ding, Z. J. (1992). Monte Carlo modelling of electron-solid interactions. *Rep. Prog. Phys.*, **55**, pp. 487–531.

78. Agostinelli, S., et al. (for the Geant4 consortium) (2003). GEANT4 – a simulation toolkit. *Nucl. Instrum. Methods Phys. Res., Sect. A*, **506**, pp. 250–303.

79. Fitting, H. J., Schreiber, E., Kuhr, J. C., and von Czarnowski, A. (2001). Attenuation and escape depths of low-energy electron emission. *J. Electron Spectrosc. Relat. Phenom.*, **119**, pp. 35–47.

80. Schreiber, E. and Fitting, H. (2002). Monte Carlo simulation of secondary electron emission from the insulator SiO_2. *J. Electron Spectrosc. Relat. Phenom.*, **124**, pp. 25–37.

81. Palik, E. D., ed. (1985). *Handbook of Optical Constants of Solids*, Academic, New York.

82. Palik, E. D., ed. (1991). *Handbook of Optical Constants of Solids II*, Academic, New York.

83. Palik, E. D., ed. (1998). *Handbook of Optical Constants of Solids III*, Academic, New York.

84. Henke, B. L., Gullikson, E. M., and Davis, J. C. (1993). X-ray interactions: photoabsorption, scattering, transmission, and reflection at E = 50–30000 eV, Z = 1–92. *At. Data Nucl. Data Tables*, **54**, pp. 181–342. Database available at http://henke.lbl.gov/optical constants/getdb2.html.

85. Cullen, D., Perkins, S., and Seltzer, S. (1991). Tables and graphs of electron interaction cross 10 eV to 100 GeV derived from the LLNL evaluated electron data library (EEDL), Z = 1-100. *Lawrence Livermore National Laboratory Evaluated Electron Data Library*, UCRL-50400, **31**.
86. Verduin, T. (2017). Quantum noise effects in e-beam lithography and metrology. Thesis, Delft University of Technology, http://repository.tudelft.nl/.
87. Kunz, R. R., Allen, T. E., and Mayer, T. M. (1987). Selective area deposition of metals using low-energy electron beams. *J. Vac. Sci. Technol., B*, **5**, pp. 1427–1431.
88. Smith, D. A., Fowlkes, J. D., and Rack, P. D. (2007). A nanoscale three-dimensional Monte Carlo simulation of electron-beam-induced deposition with gas dynamics. *Nanotechnology*, **18**, p. 265308.
89. Alman, D. A., Ruzic, D. N., and Brooks, J. N. (2000). A hydrocarbon reaction model for low temperature hydrogen plasma and an application to the Joint European Torus. *Phys. Plasmas*, **7**, pp. 1421–1432.
90. Yoshimura, N., Hirano, H., and Etoh, T. (1983). Mechanism of contamination build-up induced by fine electron probe irradiation. *Vacuum*, **33**, pp. 391–395.
91. Uesugi, K. and Yao, T. (1992). Nanometer-scale fabrication on graphite surface by scanning tunneling microscopy. *Ultramicroscopy*, **42–44**, pp. 1443–1445.
92. van Dorp, W. F., Hagen, C. W., Crozier, P. A., and Kruit, P. (2008). Growth behavior near the ultimate resolution of nanometer-scale focused electron beam-induced deposition. *Nanotechnology*, **19**, p. 225305.
93. Van Oven, J. C., Berwald, F., Berggren, K. K., Kruit, P., and Hagen, C. W. (2011). Electron-beam-induced deposition of 3-nm-half-pitch patterns on bulk Si. *J. Vac. Sci. Technol., B*, **29**, p. 06F305.
94. van Dorp, W. F. and Hagen, C. W. (2008). A critical literature review of focused electron beam induced deposition. *J. Appl. Phys.*, **104**, p. 081301.
95. Rack, P. D., Randolph, S., Deng, Y., Fowlkes, J., Choi, Y., and Joy, D. C. (2003). Nanoscale electron-beam-stimulated processing. *Appl. Phys. Lett.*, **82**, pp. 2326–2328.
96. Vickerman, J. C. and Gilmore, I. S. (2009). *Surface Analysis: The Principal Techniques*, 2nd ed., John Wiley & Sons, USA.
97. Randolph, S. J., Fowlkes, J. D., and Rack, P. D. (2006). Focused, nanoscale electron-beam-induced deposition and etching. *Crit. Rev. Solid State Mater. Sci.*, **31**, 55–89.

98. Wnuk, J. D., Gorham, J. M., Rosenberg, S. G., van Dorp, W. F., Madey, T. E., Hagen, C. W., and Fairbrother, D. H. (2010). Electron beam irradiation of dimethyl-(acetylacetonate) gold(III) adsorbed onto solid substrates. *J. Appl. Phys.*, **107**, p. 054301.

99. van Dorp, W. F., Wnuk, J. D., Gorham, J. M., Fairbrother, D. H., Madey, T. E., and Hagen, C. W. (2009). Electron induced dissociation of trimethyl(methylcyclopentadienyl)platinumIV: total cross section as a function of incident electron energy. *J. Appl. Phys.*, **106**, p. 074903.

100. Igaki, J. Y., Kanda, K., Haruyama, Y., Ishida, M., Ochiai, Y., Fujita, J. I., Kaito, T., and Matsui, S. (2006). Comparison of FIB-CVD and EB-CVD growth characteristics. *Microelectron. Eng.*, **83**, pp. 1225–1228.

101. Fujii, T., Iwasaki, K., Munekane, M., Takeuchi, T., Hasuda, M., Asahata, T., Kiyohara, M., Kogure, T., Kijima, Y., and Kaito, T. (2005). A nanofactory by focused ion beam. *J. Micromech. Microeng.*, **15**, pp. S286–S291.

102. Frey, L., Lehrer, C., and Ryssel, H. (2003). Nanoscale effects in focused ion beam processing. *Appl. Phys. A: Mater. Sci. Process.*, **76**, pp. 1017–1023.

103. McElwee-White, L. (2009). Mechanism-based design of pecursors for MOCVD. *ECS Trans.*, **25**, pp. 161–171.

104. McElwee-White, L. (2006). Design of precursors for the CVD of inorganic thin films. *Dalton Trans.*, pp. 5327–5333.

105. Perentes, A., Sinicco, G., Boero, G., Dwir, B., and Hoffmann, P. (2007). Focused electron beam induced deposition of nickel. *J. Vac. Sci. Technol., B*, **25**, pp. 2228–2232.

106. Rosenberg, S. G., Barclay, M., and Fairbrother, D. H. (2013). Electron induced reactions of surface adsorbed tungsten hexacarbonyl ($W(CO)_6$). *Phys. Chem. Chem. Phys.*, **15**, pp. 4002–4015.

107. Takeguchi, M., Shimojo, M., and Furuya, K. (2005). Fabrication of alpha-iron and iron carbide nanostructures by electron-beam induced chemical vapor deposition and postdeposition heat treatment. *Jpn. J. Appl. Phys., Part 1*, **44**, pp. 5631–5634.

108. Rosenberg, S. G., Landheer, K., and Hagen, C. W. (2012). Substrate temperature and electron fluence effects on metallic films created by electron beam induced deposition. *J. Vac. Sci. Technol., B*, **30**, p. 051805.

109. Botman, A., Mulders, J. J. L., Weemaes, R., and Mentink, S. (2006). Purification of platinum and gold structures after electron-beam-induced deposition. *Nanotechnology*, **17**, pp. 3779–3785.

110. Botman, A., Hagen, C. W., Li, J., Thiel, B. L., Dunn, K. A., Mulders, J. J. L., Randolph, S. J., and Toth, M. (2009). Electron postgrowth irradiation of

platinum-containing nanostructures grown by electron-beam-induced deposition from $Pt(PF_3)_4$. *J. Vac. Sci. Technol., B*, **27**, pp. 2759–2763.

111. Arumainayagam, C. R., Lee, H.-L., Nelson, R. B., and Haines, D. R. (2010). Low-energy electron-induced reaction in condensed matter. *Surf. Sci. Rep.*, **65**, pp. 1–44.

112. Wnorowski, K., Stano, M., Matias, C., Denifl, S., Barszczewska, W., and Matejčík, Š. (2012). Low-energy electron interactions with tungsten hexacarbonyl - $W(CO)_6$. *Rapid Commun. Mass Spectrom.*, **26**, pp. 2093–2098.

113. Wnorowski, K., Stano, M., Barszczewska, W., Jówko, A., and Matejčík, Š. (2012). Electron ionization of $W(CO)_6$: appearance energies. *Int. J. Mass Spectrom.*, **314**, pp. 42–48.

114. Engmann, S., Stano, M., Matejcik, S., and Ingólfsson, O. (2011). The role of dissociative electron attachment in focused electron beam induced processing: a case study on cobalt tricarbonyl nitrosyl. *Angew. Chem. Int. Ed.*, **50**, pp. 9475–9477.

115. van Dorp, W. F., van Someren, B., Hagen, C. W., Kruit, P., and Crozier, P. A. (2006). Statistical variation analysis of sub-5-nm-sized electron-beam-induced deposits. *J. Vac. Sci. Technol., B*, **24**, pp. 618–622.

116. Sychugov, I., Nakayama, Y., and Mitsuishi, K. (2009). Composition control of electron beam induced nanodeposits by surface pretreatment and beam focusing. *J. Phys. Chem. C*, **113**, pp. 21516–21519.

117. Koops, H. W. P., Kaya, A., and Weber, M. (1995). Fabrication and characterization of platinum nanocrystalline material grown by electron-beam induced deposition. *J. Vac. Sci. Technol., B*, **13**, pp. 2400–2403.

118. Frabboni, S., Gazzadi, G. C., and Spessot, A. (2007). TEM study of annealed Pt nanostructures grown by electron beam-induced deposition. *Physica E*, **37**, pp. 265–269.

119. Frabboni, S., Gazzadi, G. C., Felisari, L., and Spessot, A. (2006). Fabrication by electron beam induced deposition and transmission electron microscopic characterization of sub-10-nm freestanding Pt nanowires. *Appl. Phys. Lett.*, **88**, p. 213116.

120. Botman, A., Hesselberth, M., and Mulders, J. J. L. (2008). Improving the conductivity of platinum-containing nano-structures created by electron-beam-induced deposition. *Microelectron. Eng.*, **85**, pp. 1139–1142.

121. Botman, A., Hesselberth, M., and Mulders, J. J. L. (2008). Investigation of morphological changes in platinum-containing nanostructures created by electron-beam-induced deposition. *J. Vac. Sci. Technol., B*, **26**, pp. 2464–2467.

122. Botman, A., de Winter, D. A. M., and Mulders, J. J. L. (2008). Electron-beam-induced deposition of platinum at low landing energies. *J. Vac. Sci. Technol., B*, **26**, pp. 2460–2463.

123. Rand, M. J. (1973). Chemical vapor-deposition of thin-film platinum. *J. Electrochem. Soc.*, **120**, pp. 686–693.

124. Schröder, H., Kompa, K. L., Masci, D., and Gianinoni, I. (1985). Investigation of UV-laser induced metallization - platinum from $Pt(PF_3)_4$. *Appl. Phys. A*, **38**, pp. 227–233.

125. Hsu, D. S. Y., Troilo, L. M., Turner, N. H., and Pierson, K. W. (1995). Selective area platinum silicide film deposition using a molecular precursor chemical beam source. *Thin Solid Films*, **269**, pp. 21–28.

126. Barry, J., Ervin, M. H., Molstad, J., Wickenden, A., Brintlinger, T., Hoffman, P., and Meingailis, J. (2006). Electron beam induced deposition of low resistivity platinum from $Pt(PF_3)_4$. *J. Vac. Sci. Technol., B*, **24**, pp. 3165–3168.

127. Aflatooni, K. and Burrow, P. D. (2000). Total cross sections for dissociative electron attachment in dichloroalkanes and selected polychloroalkanes: the correlation with vertical attachment energies. *J. Chem. Phys.*, **113**, pp. 1455–1464.

128. Utke, I., Hoffmann, P., Berger, R., and Scandella, L. (2002). High-resolution magnetic Co supertips grown by a focused electron beam. *Appl. Phys. Lett.*, **80**, pp. 4792–4794.

129. Lau, Y. M., Chee, P. C., Thong, J. T. L., and Ng, V. (2002). Properties and applications of cobalt-based Material produced by electron-beam-induced deposition. *J. Vac. Sci. Technol., A*, **20**, pp. 1295–1302.

130. Gazzadi, G. C., Mulders, H., Trompenaars, P., Ghirri, A., Affronte, M., Grillo, V., and Frabboni, S. (2011). Focused electron beam deposition of nanowires from cobalt tricarbonyl nitrosyl ($Co(CO)_3NO$) precursor. *J. Phys. Chem. C*, **115**, pp. 19606–19611.

131. Gazzadi, G. C., Mulders, J. J. L., Trompenaars, P., Ghirri, A., Rota, A., Affronte, M., and Frabboni, S. (2011). Characterization of a new cobalt precursor for focused beam deposition of magnetic nanostructures. *Microelectron. Eng.*, **88**, pp. 1955–1958.

132. Ervin, M. H. and Nichols, B. M. (2009). Electron beam induced deposition of cobalt for use as single- and multiwalled carbon nanotube growth catalyst. *J. Vac. Sci. Technol., B*, **27**, pp. 2982–2985.

133. Boero, G., Utke, I., Bret, T., Quack, N., Todorova, M., Mouaziz, S., Kejik, P., Brugger, J., Popovic, R. S., and Hoffmann, P. (2005). Submicrometer Hall devices fabricated by focused electron-beam-induced deposition. *Appl. Phys. Lett.*, **86**, p. 042503.

beings are constantly exposed not only to natural radioactivity and cosmic rays but also to artificial sources of radiation during medical treatments (X-ray and computed tomography [CT] scans). On the other hand, high-energy radiation is used to treat tumours with the aim to kill cancer cells in a highly selective way. The importance of radiation treatment during cancer therapy is constantly increasing. It can be estimated that worldwide about half of the patients diagnosed with cancer are treated at least once with radiation therapy [117]. To increase the sensitivity of tumour tissue toward radiation, radiosensitizers are administered concomitant to the radiation treatment. DNA—the carrier of genetic information—is generally regarded as the most important target, whose damage has to be studied carefully. DNA is localized in the cell nucleus and the mitochondria, in which adenosine triphosphate is generated. DNA is composed of the four nucleobases guanine (G), adenine (A), cytosine (C), and thymine (T) and a backbone of alternating phosphate and 2′-deoxyribose units. Two DNA strands are coupled by hydrogen bonds between the complementary nucleobases G-C and A-T to form the well-known DNA double helix. In the cell nucleus the double helix is wrapped around eight histone proteins to form a nucleosome, and the nucleosomes are further organized into 10 nm and 30 nm fibers, which finally make up the chromatin [14, 31]. Within the cell the DNA is rather localized and before a primary radiation quantum hits the DNA it must traverse the cell membrane and the cytoplasm, leaving a track of highly reactive secondary particles that can attack the surrounding medium.

The sequence of events leading to the final biological effects is very complex, and it is useful to order the different physical, chemical, and biological processes by their timescale (Fig. 6.1). About 80% of the cell consists of water; thus the direct interaction of the primary radiation with H_2O is the most important initial reaction step. The first interaction is excitation or ionization of the target, which represents the physical stage. The excitation and ionization events take place within femtoseconds (10^{-15} s, fs) or even shorter. Direct electronic excitation of DNA can initiate photophysical and photochemical processes that are generally referred to as *direct damage*. In time-resolved femtosecond laser excitation experiments the excited-state dynamics of DNA nucleobases and model

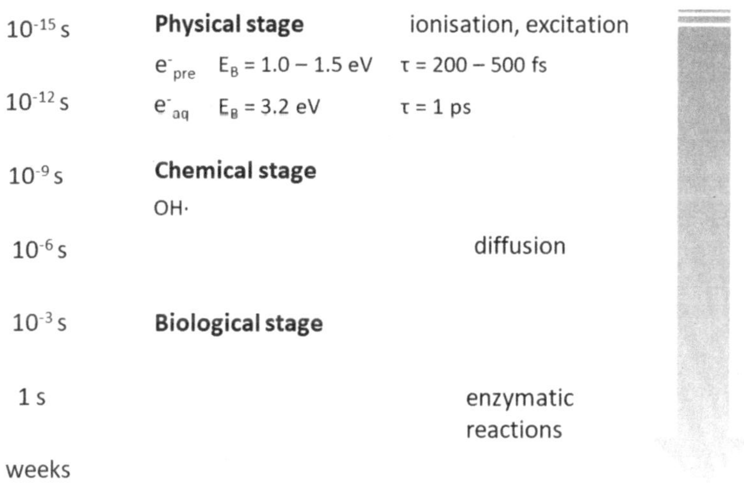

Figure 6.1 Scheme of the processes relevant for DNA radiation damage ordered by their timescale. e^-_{pre}, prehydrated electron; e^-_{aq}, solvated electron; E_B, binding energy; τ, lifetime.

compounds was investigated in quite a detail and it was shown that the excited-state lifetimes of DNA nucleobases are generally very short (i.e., in the subpicosecond regime) [21]. This is due to strong vibronic coupling of the excited and the ground state, giving rise to conical intersections resulting in fast recovery of the ground state [95]. The ultrafast photophysical processes protect the DNA against UV damage. Nevertheless, direct electronic excitation of DNA nucleobases can initiate photochemical processes such as the formation of cyclobutane pyrimidine dimers, which represent pivotal intermediates for mutagenesis and cancerogenesis, especially skin cancer.

More important than the damage caused by direct excitation of DNA nucleobases is the damage due to secondary reactive species that are formed by ionization of molecules in the vicinity of DNA (*indirect damage*). Ionization of water molecules results in formation of secondary electrons and water cations, which can further react or decompose to form hydroxyl radicals and protons (water radiolysis):

$$H_2O + h\nu \rightarrow H_2O^+ + e^- \rightarrow H^+ + OH\cdot + e^-$$

All products of the water radiolysis are highly reactive, and a reasonable estimation of the relative reactivity of hydroxyl radicals and secondary electrons with respect to DNA strand breaks, and thus their importance for radiation damage, succeeded only very recently. Accordingly, the DNA strand break yield caused by so-called prehydrated electrons is twice as high as that caused by OH radicals [73].

Secondary electrons are dissolved by water molecules within picoseconds (10^{-12} s, ps) to form the long-known solvated electrons [89]. The reactivity of solvated electrons is rather low compared to the preceding transition states when the electrons are gradually slowed down from their initial energy to form at first a prehydrated state before complete solvation. According to recent photoionization experiments the binding energies and lifetimes of the prehydrated and hydrated electrons are 1.0–1.5 eV and 200–500 fs, and 3.2 eV and 1.0 ps, respectively [66].

The physical stage encompasses the events between the initial excitation/ionization to the complete solvation of secondary electrons, whereas the chemical stage is defined by the diffusion-controlled reactions that take place within nanoseconds to milliseconds. This concerns mainly the reactions induced by the hydroxyl radicals that have been studied for decades. Both secondary electrons and OH radicals can induce severe damage to DNA such as DNA single- and double-strand breaks and base loss.

The modification of the DNA structure triggers a biological response (thus defining the biological stage, which takes place on long timescales from seconds to days or weeks), that is, the damage is recognized by enzymes, which trigger biochemical reaction cascades with the aim to repair the DNA damage. A missing or faulty repair then leads to mutations or cell death. Single-strand breaks (SSBs) and base losses can be repaired rather easily by using the complementary DNA strand as a template. However, particularly severe are DNA double-strand breaks (DSBs) that lead in most cases to cell death.

After defining the basic series of events we now focus on the importance of secondary electrons for DNA radiation damage, which was recognized within the past 15 years. In the present chapter we will show how secondary electrons can damage DNA and its

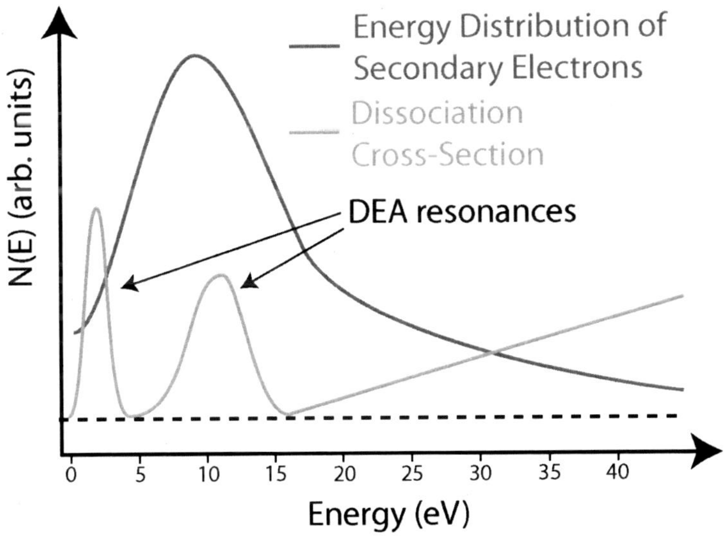

Figure 6.2 Scheme showing the energy distribution of secondary electrons due to the primary ionization of water by high-energy radiation. Depending on the type and energy of primary radiation, the most probable energy for secondary electrons is at 9–10 eV [75]. The dissociation cross section at low energy shows resonant structures due to the initial ionization mechanism, dissociative electron attachment (DEA). The nonresonant processes at higher energy are less relevant due to the small amount of secondary electrons.

constituents. To begin with, we have to consider how many electrons are actually generated by high-energy radiation and what energy they do possess. Answers are provided by Monte Carlo simulations of liquid water irradiated with protons in the megaelectron volt range, which are used in hadron therapy [75]. It was demonstrated that about 80% of the energy of the initial projectile is distributed to secondary electrons by ionization of water molecules, with the energy distribution of electrons peaking around 10 eV [75]. Thus, the energy of secondary electrons is rather small; nevertheless they can efficiently damage DNA (Fig. 6.2). This was demonstrated in 2000 by irradiation of plasmid DNA with low-energy electrons of well-defined energy [16].

Plasmid DNA is circular double-stranded DNA with a few thousand base pairs and originates from bacteria. Due to the circular

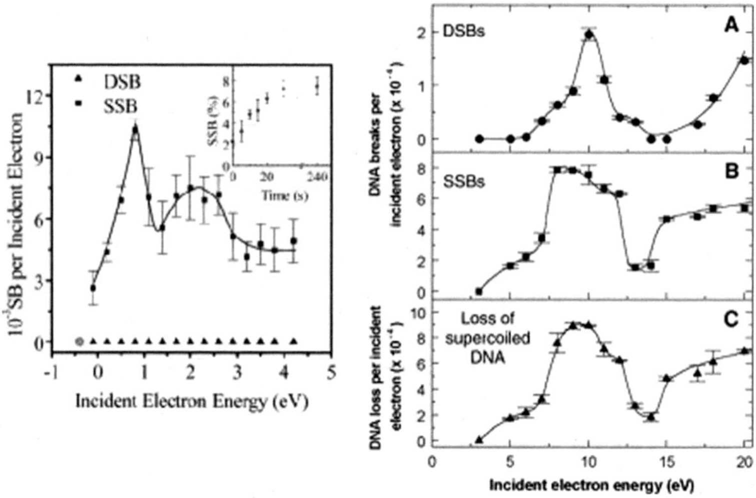

Figure 6.3 DNA single- and double-strand break yields determined by irradiation of plasmid DNA with low-energy electrons (left: 0–4 eV; right: 3–20 eV) and subsequent analysis by agarose gel electrophoresis. (Left) Reprinted (figure) with permission from Ref. [68]. Copyright (2004) by the American Physical Society. (Right) Reprinted from Ref. [16]. Copyright 2000, The American Association for the Advancement of Science.

structure plasmids have an intrinsic strain leading to a supercoiled state. Plasmids are often used in DNA damage research since strand breaks lead to a strong change in morphology that can be easily detected and quantified by agarose gel electrophoresis (AGE). An SSB leads to a loss of the strain resulting in a circular form, and a DSB (two SSBs in opposite strands in close proximity) cuts the plasmid DNA completely, resulting in linear fragments. In AGE the supercoiled, circular, and linear forms of DNA can be easily separated and very small amounts of damaged material can be detected.

Boudaiffa et al. used this technique to determine SSBs and DSBs as a function of electron energy in the regime of 3–20 eV [16]. As shown in Fig. 6.3, it was found that the strand break yields exhibit distinct maxima in the energy regime below 15 eV [46], with the highest DSB yield at 10 eV. That is, the highest DSB yield coincides with the most probable energy of secondary electrons in water radiolysis according to the Monte Carlo simulations mentioned before (Fig. 6.2.), which is an extremely important result. In a study published in 2005 the energy scale was extended to even lower values (0–4 eV), and remarkably the SSB yield exhibits a global maximum right below 1 eV (Fig. 6.3.), that is, the absolute value of the SSB yield at 1 eV is even higher than at energies >20 eV [68].

To put the electron energy values in some context it is advantageous at this point to recall some energy values for related physicochemical processes in molecules. For the ionization of DNA nucleobases, an energy of 8–9 eV is required (lowest vertical ionization potentials) [47], and the electronic excitation of nucleobases requires around 4–5 eV (the lowest π–π^* transitions, which lie in the UV-B range) [21]. The bond dissociation energy of a carbon–carbon bond is approximately 3–4 eV (3.9 eV in ethane) [119], whereas the energy for vibrational excitation is in the range of 0.1–0.5 eV. Thus, the electron energies involved in DNA strand breakage are in the range of vibrational or electronic excitation.

In summary, we can conclude that the damage caused by *high-energy* radiation is to a large extent due to reactions induced by *low-energy* particles. The pivotal mechanism is dissociative electron attachment (DEA):

$$e^- (\varepsilon) + AB \rightarrow AB^- \rightarrow A^- + B$$

The incoming electron with specific energy ε occupies a formerly empty molecular orbital (MO) to form a transient negative ion (TNI), and the negative ion state might be unstable with respect to dissociation. In the next section it will be explained in detail how low-energy electrons react with the nucleobases thymine and uracil.

6.2 Electron Attachment to Thymine and Uracil: A Test Case Study for the Description of Molecular Damage by Low-Energy Electrons

After defining the fundamental problem in Section 6.1 the question arises, How should we do an experiment that could be useful to describe radiation damage of cells by low-energy electrons on the molecular level? For the scientist asking this question one fundamental question arises at this stage. Should experiments be done in the gas phase, with thin films on surfaces, or in solution? As discussed in Section 6.1 experiments with thin films (as carried out by the group of Leon Sanche in Canada) are a very beneficial approach that allows also the study of complex molecules like DNA.

However, a large number of research groups preferred the study of molecules in the gas phase. The latter probably provides the best possibility to create a well-defined collision system (in our case a low-energy electron and a biomolecule in the gas phase). For gas-phase experiments a beam of biomolecules is crossed with an electron beam in the simplest configuration and the charged reaction products are analyzed by mass spectrometry (MS). Thin-film studies are faced with the problem of possible film charging by the irradiation of films by electrons, which substantially changes the electric field situation on the surface. In such cases the kinetic energy, which is a crucial experimental parameter, becomes less defined and the effective fragmentation yield may be altered. MS with gas-phase samples on the other hand avoids this problem except such high electron currents are used that space charge effects becoming severe. An additional problem of surface experiments, where formed anions need to desorb from the surface in order to be detected by MS, is the required kinetic energy in order to desorb. Since heavy mass fragments as well as parent anions have in general low kinetic energy and the film becomes polarized by the presence of the negative charge, often strong discrimination of heavy mass fragments exists. Such discrimination effects are substantially reduced for mass spectrometric studies in the gas phase (though also there hardly avoidable).

In mass spectrometric investigations usually two different strategies of measurements are followed, either the top-down approach [41] or the bottom-up approach. A typical application of the former is protein or DNA sequencing, that is, biomacromolecules are brought into the gas phase (e.g., by matrix-assisted laser desorption/ionization [MALDI] or electrospray ionization [ESI]), and resulting ions are monitored by MS or by various collision reactions investigated by tandem mass spectrometry (MS/MS) techniques. Such techniques like ESI could be also used in the investigation of low-energy electron interaction with macrobiomolecules but, however, might imply challenging problems in the interpretation of results (in addition to experimental challenges). To give an example, the experimental spectra may reveal a large number of fragmentation peaks whose formation can be explained then by a multitude of different reaction pathways. In such case the support by theory would be crucial; however, the size of the system under study exceeds by far the current capability of electron collision theory. Further it should be noted that experimentally not every type of mass spectrometer can be used for the investigation of macromolecules. For example, quadrupole mass spectrometers widespread among groups studying electron attachment have a limited mass range up to few kilodaltons and can be applied therefore only for low-mass macromolecules.

Hence the bottom-up approach was thought to be more fruitful. This implies to start with the exploration of small molecules or building blocks of the macromolecule, collect data, and then stepwise increase the complexity of the molecules in order to see which reactions remain and which modifications of the electron attachment process may occur in terms of resonance energies and fragmentation pathways. The size of building blocks of DNA and proteins are in the range of current computational efforts, that is, joint experimental/theoretical efforts allow for a reasonable precise understanding of electron attachment processes to simple biomolecules. Next, a prime example of such combined studies on electron attachment to the simple biomolecules thymine and uracil is presented.

Thymine and uracil are nucleobases of DNA and RNA, respectively. Figure 6.4 shows the molecular structure of the uracil molecule. Thymine has an identical structure, except a methyl group at the C5 position replaces the hydrogen atom present in uracil. The molecules have a six-membered ring structure with nitrogen atoms at ring positions 1 and 3. Thus they also belong to the class of pyrimidines. Thymine can be combined with the furanose ring of the monosaccharide deoxyribose at the 1N nitrogen, which leads to the formation of the nucleoside thymidine. Moreover thymine can be phosphorylated with up to three phosphate groups, which leads—as for the other nucleotides—to the formation of thymidine-

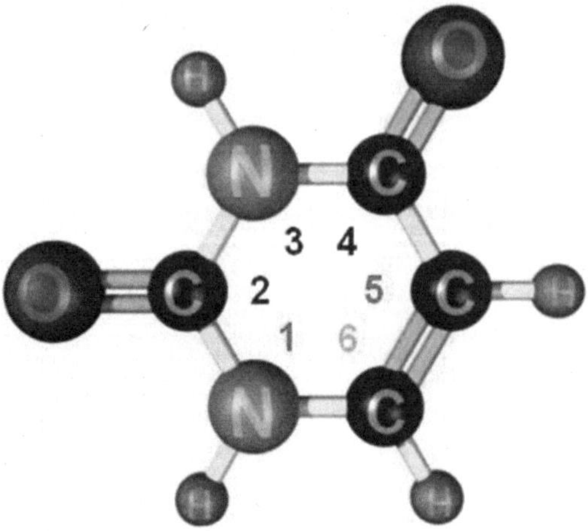

Figure 6.4 Schematic molecular structure of uracil. Reprinted with permission from Springer Customer Service Centre GmbH: Springer Nature, *The European Physical Journal D - Atomic, Molecular, Optical and Plasma Physics*, Ref. [20], Copyright (2002).

mono-, di-, or tri-phosphate (TMP, TDP, or TTP), respectively. The molecular mass of thymine is 126 u (uracil: 111 u) and thus suitable for all common types of mass spectrometers used to study electron attachment. The molecules are powders under ambient conditions with too low vapour pressure in order to form a molecular beam (e.g., the vapour pressure of thymine at 391 K was determined to be only 0.26 Pa [108]). Thus the powders need to be heated in order to achieve sufficient sublimation rates for a molecular beam.

When starting an electron attachment experiment with a molecule, one of the basic questions is, Do we observe a molecular anion? As described in the previous chapters, electron attachment to a molecule means formation of a TNI:

$$e^- + M \rightarrow TNI \rightarrow \cdots \tag{6.1}$$

Hence the question is whether the lifetime of the TNIs is sufficient to be detected in crossed-beam experiments. The latter have a typical timescale between anion formation and detection in the microsecond regime, that is, if the lifetime is shorter, the TNI cannot be detected by MS (though resonant TNI formation might have occurred). In this case the decay mechanism is spontaneous electron emission, or if the electron stays on the nuclear framework DEA, the dissociation may lead to the formation of a stable anion (through the product of a dissociation reaction). The stability of the anion can be also expressed by the electron affinity (EA), that is, the binding energy of the excess electron to the neutral species. For systems with a (as per definition) negative EA, that is, the anionic system lies in energy above the neutral system, no finally stable anion can be formed, because the state is unstable against autodetachment. For systems with positive EA the state might be stable when the kinetic energy of the captured electron and the released EA can be accommodated in the anion (e.g., by vibrational excitation) or transferred to the surrounding. While for isolated systems the latter is only possible by photon emission (which is, however, a rather slow process compared to autodetachment and dissociation), transfer of vibrational energy is an important process for molecules in an environment. In this case a TNI may be stabilized substantially, which leads to formation of molecular anions observable also in crossed-beam experiments using clusters instead of isolated molecules. This technique is introduced in Section 6.4.1. This leads then to the general question of whether results for building blocks in the gas phase correspond to the situation in which a free ballistic electron is formed in the surrounding of DNA in the cell and then resonantly captured by a pyrimidine base. This topic will be discussed further in Section 6.3 in detail.

Due to the possibility of autodetachment the TNI formation itself is not investigated only by MS experiments but also by electron transmission spectroscopy (ETS). In this type of spectroscopy an electron beam is directed through a cell containing the vapour of the molecule under study. The transmitted electron current through the cell is measured as a function the initial electron energy, that is, the total scattering cross section is determined. A sharp dip therein indicates the resonant formation of a TNI [5]. The peak energy can be associated with the vertical attachment energy, since electronic transitions occur within the Franck–Condon principle. The group of P. D. Burrow studied ETS with the DNA bases and uracil, and (supported by calculations of MO energies) they assigned three dips observed for each compound to occupation of the lowest antibonding orbitals of π symmetry (see Fig. 6.5a–c). For example,

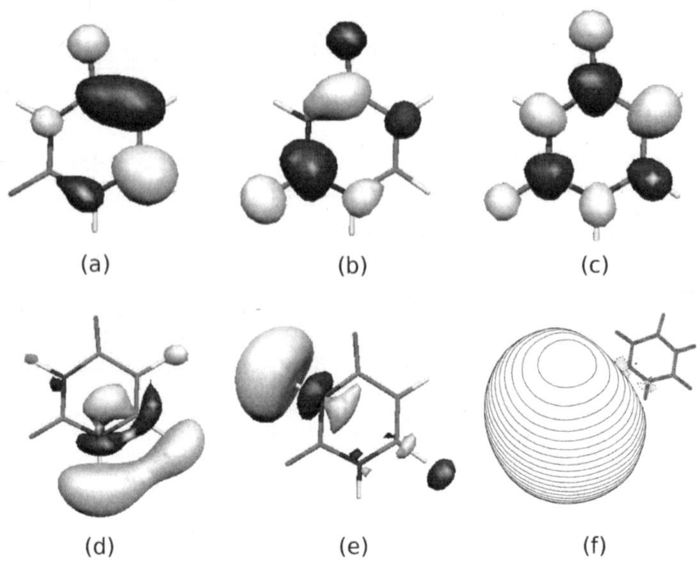

Figure 6.5 Sketches of the (a–c) three lowest π^* molecular orbitals with the energies of the corresponding TNI states, (d, e) the two lowest σ^* molecular orbitals with the energies of the corresponding TNI states, and (f) contour representations of the dipole-bound anion state of uracil. (a–e) Reprinted from Ref. [17], with the permission of AIP Publishing. (f) Reprinted from Ref. [29], Copyright (1999), with permission from Elsevier.

in case of thymine the $\pi_{1,2,3}{}^*$ resonances were experimentally observed at 0.29, 1.71, and 4.05 eV, respectively [5].

In contrast, resonances ascribed to the lowest empty valence orbitals of σ symmetry (see Fig. 6.5d,e) are hardly observable in ETS. This can be explained by the shorter lifetime of σ^* resonances, and they are repulsive along some nuclear coordinates, which leads to broad peaks [17]. In addition for the nucleobase the three pronounced π^* resonances tend to mask the σ^* resonance in the ETS spectrum. Thus orbital energies for the latter have been derived only by theory. All are listed in Table 6.1.

It should be noted that an excess electron is not necessarily captured in one of the lowest MOs (valence state); another way of binding is possible when the molecule has a sufficiently high dipole moment. If the latter succeeds the now accepted value of about ~2 D, a molecule should be able to form a dipole-bound

Table 6.1 (Virtual) orbital energies in electron volts of three lowest molecular orbitals of π^* and σ^* symmetry [17]

	Resonancc					
	π_1^*	π_2^*	π_3^*	σ_1^*	σ_2^*	σ_3^*
Thymine	0.29	1.71	4.05	2.40	3.42	3.75
Uracil	0.22	1.58	3.83	2.26	3.67	3.87

anion. Nucleobases have in general high dipole moments (e.g., 4.63 D for uracil [51]), which allow for the formation of dipole-bound states. The electron thereby resides in a Rydberg state far outside the molecular frame, as shown in Fig. 6.5f. It is weakly bound ($<$100 meV), which has been proven in various photo- and electric field-detachment experiments, for example, Ref. [93]. Stable dipole-bound parent anions of nucleobases were formed upon Rydberg electron transfer. In this method dipole-bound parent anions of isolated DNA bases were formed in charge exchange collisions between laser-excited (Rydberg) Xe atoms and bare neutral nucleobase molecules [28]. Dipole-bound states may also form by free-electron capture, but they decay quickly, for example, by electron emission that does not allow the direct observation of a stable dipole-bound parent anion upon free-electron capture. However, it was shown by theory that this Rydberg state may couple with a valence state before autodetachment [105], which leads subsequently to dissociation of the anion.

6.2.1 Electron Attachment to Thymine and Uracil: The $(M - H)^-$ Ion Yield and the Search for Bond and Site Selectivity

While ETS probes the energetics of TNI formation, the stability of the TNI upon dissociation into fragment anions and neutral(s) can be studied by crossed-beam experiments equipped with mass spectrometers. The first gas-phase DEA study with nucleobases carried out in 1998 reported the formation of stable parent anions of thymine and cytosine on mass spectrometric timescales [45]; however, later studies showed that this assignment obviously

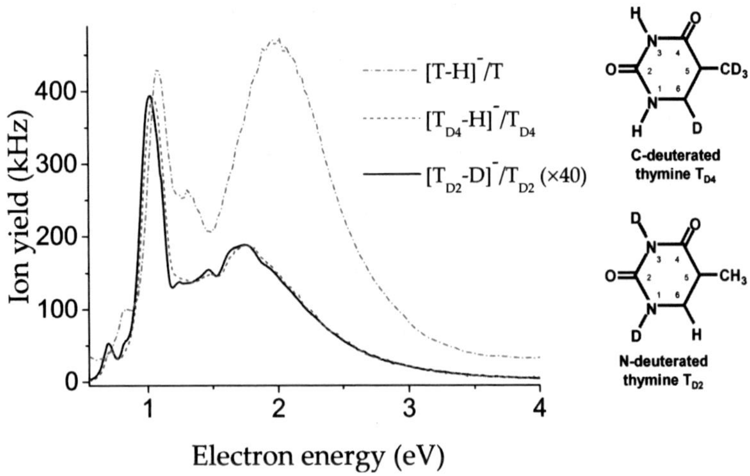

Figure 6.7 Measurement of the $(M - H)^-$ for native thymine T and thymine deuterated at carbon positions T_{D4} (see structures on the right). Also shown is the measurement of $(M - H)^-$ from thymine deuterated at nitrogen positions (T_{D4}). Adapted from Ref. [26], Copyright (2008), with permission from Elsevier.

water. Figure 6.7 shows the ion yield observed for $(M - H)^-$ from thymine deuterated at carbon positions. The $(M-D)^-$ loss from this labeled molecule was completely absent. The $(M - H)^-$ yield is virtually identical with that for the native molecule, which indicates high bond selectivity. The complementary experiment with labeled nitrogen positions delivers ion yield for $(T - D)^-$; however, as shown in Fig. 6.7, the resonant yield is modified in terms of peak positions and also peak heights. Such modifications can be explained by isotope effects that change the energetics as well as the dynamics of DEA. In the former mainly the lower zero-point energy for the deuterated species leads to (blue)-shifting of resonances, while the longer dissociation time of fragment (anions) involving deuterated species leads to red-shifting due to increased autodetachment and a lower cross section. For a detailed discussion on the resonance shifts we refer to Ref. [26]. Since the autodetachment rate may vary for the particular electronic states decaying and different dissociation mechanisms, the relative peak heights may change strongly, as seen in the Fig. 6.7, where the intensity of the sharp peak at lower energies decreases significantly more (by approximately factor 40) than the 1.7 eV resonance (~ factor 20). The former is among the largest isotopes effects in DEA ever measured and results from the particular dissociation mechanism involving quantum mechanical tunnel effects discussed later.

The results shown in Fig. 6.7 indicate already that the sharp peaks at low energies and the broad resonance have a different origin and this was proven by measurements with partially methylated thymine and uracil [80]. By the replacement of a hydrogen at one nitrogen position by a methyl group the energetics will change because, for example, for hydrogen loss from the substituted site, then a C–H bond needs to be cleaved compared to the (generally weaker) N–H bond. Hence one can speak of the blocking of a site for hydrogen loss. The results indicate remarkable site selectivity in the DEA process: For 3-methyluracil, when the N3 site is blocked, the sharp peaks at low electron energies are present only and the broad resonance at 1.8 eV vanished. In the case of 1-methylthymine, when the N1 site is blocked, the complementary results are observed. Hence, here one obtains the case of 100% selectivity, that is, by tuning the electron energy only one specific bond in the molecule is broken. This behavior is in strong contrast to electron ionization, where above the ionization threshold (e.g., at 70 eV, where the ionization cross section has its maximum) often strong fragmentation can be observed at one electron energy, that is, a number of bonds are broken in competitive fragmentation channels. It should be noted that electron-induced dissociation was shown to be controllable not only for gas-phase molecules in the interaction with free electrons but also for molecules deposited on a surface in scanning tunneling microscopy (e.g., [74]).

Concerning the assignment of the site-selective resonances, a complete picture of the TNI formation and its dissociation was drawn on the basis of the ETS data and the calculations of MOs shown in Fig. 6.5 [17]. The sharp peaks were assigned to an initial dipole-bound anion state (DBS), as shown in Fig. 6.5f. This state acts as a doorway state of DEA because it mixes with the strongly repulsive σ_2^* state, which leads to an avoided crossing in the potential energy of the DBS and the σ_2^* state along the N1–H reaction coordinate. This avoided crossing leads to a potential barrier. Excited vibrational states may then decay by tunneling of the hydrogen through the barrier as long they lie above the dissociation limit. Hence the first sharp resonance can be ascribed to tunnel effects and they were identified as vibrational Feshbach resonances. This model involving a DBS is consistent with the experimental data shown in Fig. 6.6; however, the data cannot directly prove that a DBS is involved. In contrast, studies on the electron attachment dynamics indicated that valence anions may immediately form in uracil without the assistance of a DBS [52]. As discussed earlier, the broad resonance at 1.8 eV is formed by another, less controversial mechanism; in this case the TNI is initially formed in a π_2^* resonance, which then mixes with the repulsive σ_2^* state. The TNI finally relaxes through hydrogen loss from the N3 position to form $[M-H]^-$. Such pronounced site selectivity is not a particular feature of the pyrimidine bases shown; instead it was shown for a number of other molecules like, for example, methanol, simple organic acids [76], and other biomolecules like sugars [11] and amino acids [33].

Comparing now the experimental peak positions/onset for the site-selective resonances with the computed threshold one could claim that the site selectivity is only controlled by the threshold because below 1.4 eV the $\pi_2{}^*/\sigma_2{}^*$ dissociation channel is not opened, while the DBS/$\sigma_2{}^*$ is. Such condition on thresholds is not always necessary to obtain site-selective bond cleavage as the example of H^- for thymine and uracil shows. The corresponding ion yield is shown in Fig. 6.8, which shows a series of partially overlapping resonances at about 5.5, 6.8, 8.5, and 10 eV. Atomic hydrogen has a much lower EA ($\sim$0.75 eV) than $(M - H)^-$, that is, the endothermicity of H^- formation is considerable higher than $(M - H)^-$ and all sites become just thermodynamically available above 4–5 eV [79]. By measurements with partially deuterated or methylated nucleobases it was shown that each of the four resonances is formed by hydrogen loss from a different position of the thymine molecule. Thus we have here the case of a true bond- and site-selective process that is not formed just by limitations due to differences in the threshold energy. Instead it is related to particular electronic structures of the associated TNIs formed at the different resonance energies [81]. In this case the mechanism

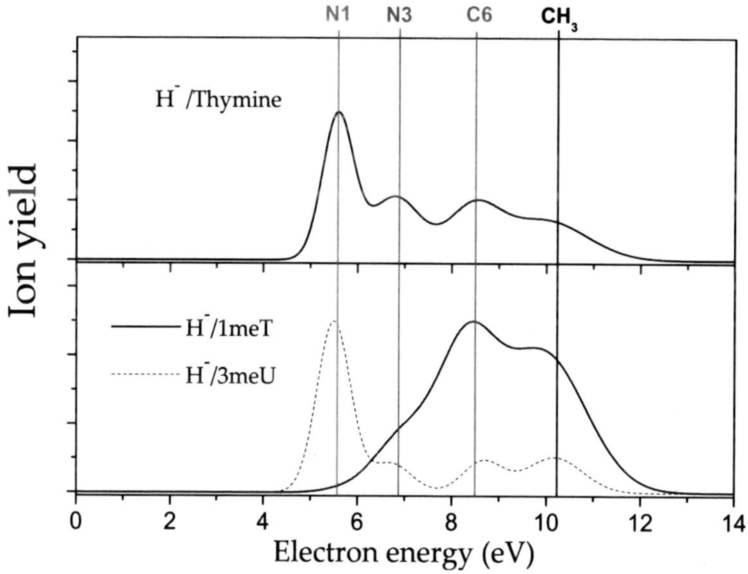

Figure 6.8 Ion yield of H^- upon DEA to thymine (upper panel). The high site selectivity in bond cleavage was shown by studies with partially deuterated/methylated pyrimidine bases. For example, the lower panel shows the H^- formation of thymine methlyated at the N1 position (1-meT) and uracil methylated at the 3N position (3-meU). Reprinted (figure) from Ref. [81]. Copyright (2005) by the American Physical Society.

of site selectivity may be viewed as electron capture in an MO in the vicinity of the hydrogen atom to be cleaved. Provided the electron occupies a highly-antibonding-repulsive σ^* state (initially occupied or by mixing with a doorway orbital), the dissociation process is very quick (on vibrational timescales) with insufficient vibrational redistribution in the molecule in this case (leading to single-bond cleavage). This is in contrast in terms of the quasi-equilibrium theory [116], where an energy equilibrium between the rovibrational degrees of freedom in the molecule exists. The latter assumption would provide less selective bond breaking for a specific electron energy but is the basis of many types of collision processes like electron and photon ionization.

In addition to the DEA arising from the cleavage of a single N–H or C–H bond, DEA to nucleobases show many more fragmentation channels that are also accompanied by dissociation of the ring structure. As shown in Refs. [6, 7] the total cross section for negative ion formation as the function of the electron energy has two ranges with considerable ion production: below 3 eV, where mainly formation of $(M - H)^-$ contributes, and between 4 and 10 eV, where other DEA channels sum up to similar cross-section height as at low energies. The main contribution at higher electron energies comes from NCO^- [32], which is formed by dissociation of the pyrimidine ring. As can be deduced from Fig. 6.4, three ways exist to detach NCO^-, all of them include breakage of N–H and C–N bonds, while one pathway also includes cleavage of a C–C bond. In such complex dissociation, detailed knowledge of the dissociation dynamics is highly valuable. MS/MS techniques were experimentally used, which indicated a sequential dissociation process instead of a simultaneous cleavage of multiple bonds. In the former case the metastable decay of $(M - H)^-$ into NCO^- was observed, that is, first the N–H bond is broken and subsequently the bonds in the ring are released within few microseconds. This metastable decay is also resonant and show two main peaks at 3.5 and 5.5 eV. Interestingly the site selectivity of hydrogen loss is also present for the metastable decay reaction. The first resonance was not observed for thymine methylated at N3 site (3-meT), while the second vanishes if 1-meT is used. Thus the site of the NCO^- excision in the molecule is strongly electron energy dependent and can be well localized since it is controlled by the site of initial hydrogen loss. In both cases the internal energy of $(M - H)^-$ is sufficient to induce further bond cleavage in the fragment anion. However, the comparison with classical dynamics simulations and results from MALDI experiments [34], where the metastable decays of the deprotonated parent anion into NCO^- was investigated, showed that the 5.5 eV peak must arise from an electronically excited state absent in MALDI and the simulations.

In summary, electron attachment to pyrimidine molecules in the gas phase leads to strong dissociation of the molecule and is highly site selective as a function of the electron energy. The formation of fragment anions may proceed by single-bond cleavages as well as complex dissociation reactions. The former reactions are quick on timescales of vibrational motion (occurring along purely repulsive energy surfaces), while the latter also proceed up to the microsecond timescale.

6.2.2 Attachment Cross Section of Nucleobases

By the crossed-beam experiments introduced before the ion yield resulting from an electron attachment reaction is measured, which can be also related to a reaction cross section, that is, a measure of the probability for a certain type of reaction. Thereby absolute cross sections are highly desired, for example, as input for various (Monte Carlo) codes [64] of the linear energy transfer of ionizing radiation in biological matter. However, experimentally it is challenging to obtain absolute cross-section values. The measured ion yield is correlated to the cross section σ_{ion} by the following formula:

$$i_{\text{ion}} = \sigma_{\text{ion}} \cdot i_{\text{e}} \cdot N_{\text{t}} \cdot l, \tag{6.2}$$

where i_{ion} is the measured ion current, i_{e} is the used electron current, N_{t} is the density of target molecules, and l is the interaction length between electrons and neutrals. Each of these experimental parameters may have substantial uncertainties, that is, for crossed-beam experiments often the unknown cross section is determined relative to a well-known electron attachment cross section, where the corresponding ion yields are measured under the same conditions. However, in early determinations of absolute electron attachment cross sections for thymine liquids or gases were used as relative gases [24]. In this case the target density of thymine molecules sublimated in an oven close to the interaction zone with the electron beam was substantially higher than measured with a pressure gauge connected to chamber (due to condensation on unheated surfaces in the vacuum chamber). Since the pressure of gases and liquids measured by the gauge more or less corresponds to this in the interaction region, the derived attachment cross sections may have considerable systematic errors. Early measurements report a DEA cross section of about 5×10^{-16} cm^2 for the broad 1.8 eV resonance of $(\text{T} - \text{H})^-$ [4], while later measurements with other techniques reported a substantially lower value. For example Burrow and coworkers [7] used a so-called total ionization tube, where no mass spectrometer is attached, but by the measurement of the ion current on the walls of the interaction volume the collection of all ions formed by electron attachment is ensured. By first measuring the ion yield for electron ionization where (semiclassical) formulas exist to calculate the total ionization cross section at the maximum and then reducing the electron energy and measuring negative ion yield, the electron attachment cross section can be determined. The advantage of this method is that no other calibrant gas needs to be used, that is, systematic errors due the pressure measurement can be avoided. One disadvantage of this method may be that no mass analysis is used, that is, absolute partial anion cross sections are not determined. However, it should be noted that the resonant nature of electron attachment delivers nevertheless the possibility to determine a partial cross section for anion formation, because fragment anions may form exclusively at specific electron energies. For example, in the present case of thymine the partial cross section of $(T - H)^-$ at 1.8 eV is equal to the total cross section. Burrow and coworkers derived a value of 3×10^{-19} cm^2, that is, almost 3 orders of magnitude lower than in the earlier MS experiments.

Later one experiment used a standard crossed-beam setup with a quadrupole mass spectrometer, but by heating of the whole setup (inclusive chamber walls) condensation of molecules on cold spots was avoided [56]. In that study a value of 4×10^{-17} cm^2 was derived for thymine, which they saw as an upper limit of the DEA cross section, while the cross section derived by Burrow and coworkers was seen as the lower limit. For comparison, theoretical calculations using R-matrix theory predicted for uracil, an $(M - H)^-$ cross section of about 4×10^{-18} cm^2 [35]. Experiments with the total ionization tube showed that the $(M - H)^-$ cross section is very similar for thymine and uracil (4.4×10^{-19} cm^2 for uracil vs. 3×10^{-19} cm^2 for thymine) [6], that is, indeed a cross section on the order of 10^{-18} cm^2 can be also expected for thymine.

6.3 From Model Compounds to Complex Systems

In the previous sections we have learned that low-energy electrons can induce strand breaks by DEA in plasmid DNA consisting of several thousand base pairs. To understand the DEA process in such complex systems a bottom-up approach was followed and DEA to the single nucleobases thymine and uracil in the gas phase has been extensively described. The investigations in the gas phase are a good indication that electron attachment is indeed the initial step in DNA strand breakage. However, it is still not understood, how a strand breakage of "real" DNA in a biologically relevant environment is induced by low-energy electrons. Here we will discuss some questions that need to be considered when the complexity of the system under study is gradually increased, starting from model compounds in the gas phase and approaching biologically relevant systems.

6.3.1 The Initial Step: Where Does Electron Attachment Take Place?

DNA strand breakage is due to the breaking of a chemical bond within the DNA backbone, which consists of the sugar 2′-deoxyribose and a phosphate group. Consequently, the electron attachment to one of the nucleobases must be followed by a charge transfer to the DNA backbone in conjunction with cleavage of a bond such as the phosphodiester bond. This process is difficult to observe experimentally but can be modeled by quantum-chemical methods. The formation of a TNI is modeled by adding an electron to the nucleobase or the phosphate group of a nucleotide in the geometry of the neutral system. The *vertical electron affinity* is a measure of the stability of the resulting anion and a first estimation of the transition energy when going from the neutral to the anionic system (Fig. 6.9). Other relevant thermodynamic data that can be obtained from quantum chemical calculations is the *adiabatic electron affinity* (representing the stability of the anion in a relaxed geometry) and the *vertical detachment energy* (giving a hint for the lifetime of the anion). Subsequent to the vertical electron attachment, a strand breakage is modeled by elongating a specific bond, such as the C–O bond between the sugar and the phosphate group (phosphodiester bond), and calculating the activation barrier for the breaking of the bond, resulting in a neutral and anionic fragment. After the bond breaking the negative charge is located on the phosphate moiety, but there are different possibilities where the initial charge is located. The general process was first described theoretically by the group of Jack Simons [15]. In various subsequent studies by different groups [37–39, 55, 60, 61] it was discussed under which conditions the extra electron is initially located on the nucleobase or the phosphate group and which specific bond in the DNA backbone is subsequently broken. In summary it was found that electron attachment at very low energies (down to 0 eV) is possible for all nucleotides (T, C, A, and G) with subsequent strand breakage. The calculations agree that a direct electron attachment to the nucleobases is possible at lower energies than direct attachment to the phosphate group and that the breaking of the C3′–O3′ bond should have the lowest-energy barrier. Another computational approach is the use of quantum scattering methods by treating the TNIs as genuine scattering states with finite lifetimes. In this way fragmentation pattern and possible competitions between the decay pathways of different metastable species can be explored. as was demonstrated with several model compounds [10, 36, 69].

From the experimental point of view it is challenging to prove these pathways, since nucleosides (nuclebase + sugar) and nucleotides (nucleoside + phosphate) are thermally labile. Both gas-phase and condensed-phase studies using thermal evaporation of these model substances must be carefully performed to prevent decomposition of the target species. The direct attachment of low-energy electrons to the DNA backbone, that is, either the phosphate group or the sugar unit was studied using model compounds such as phosphoric acid esters [54], the simplified sugar model tetrahydrofuran [107], and the sugars 2′-deoxyribose [78] and ribose [11]. The gas-phase studies indicate that electron attachment to the sugar unit is possible on the one hand close to 0 eV presumably via dipole-bound states and on the other hand at energies above 7 eV via core-excited resonances (i.e., involving electronic excitation of the TNI). A multitude of fragmentation reactions corresponding to potential strand breakage in the more complex DNA follows the electron attachment. Similar to DEA to thymine the DEA reactions in ribose are highly site selective, with the C5 site being preferably excised as a neutral fragment [11]. Also a direct attachment to the phosphate group is possible at a variety of electron energies, as was demonstrated in phosphoric acid esters such as dibutyl phosphate [54]. At an energy of about 2.5 eV DEA to dibutyl phosphate proceeds by cleavage of the C–OP bond, which corresponds to a strand break in DNA and represents the same bond cleavage predicted by the aforementioned quantum-chemical calculations.

In the meantime DEA to a series of nucleosides has been studied and the first experimental investigation on DEA to a complete nucleotide (deoxycytidine monophosphate) was published in 2012 [58]. In this study it was confirmed that an initial electron attachment to the nucleobase followed by phosphodiester bond breakage is very likely. In general it is likely that both fragmentation pathways contribute to DNA damage but are operative at different electron energies.

6.3.2 Which Resonances and Reactions Are Preserved in Larger Systems?

When considering larger systems consisting of individual subunits the question Which DEA resonances and reactions are preserved in the extended molecular systems? arises. This question was tackled by using a nucleotide surrogate, namely tetraacetylribose [12]. The study indicated that the most important and characteristic resonances of the individual subunits are still operative. As demonstrated in Section 6.2, substituents with double bonds or aromatic systems give rise to π^* shape resonances typically observable in the range of 0.5–3 eV. In tetraacetylribose π^* shape resonances of the acetyl C=O groups have been observed at 1–3 eV, that is, the same energy range of π^* shape resonances of the nucleobases in DNA nucleotides (in addition to a dipole-bound state at 1 eV). In tetraacetylribose additional signals arise close to 0 eV and around 9 eV that are ascribed to the central sugar ring. The low-energy signal is assigned to a dipole-bound state due to the sugar ring, which is still accessible although there are no free hydroxyl groups as in the free sugar. Investigations of the nucleoside thymidine show the same tendency, that is, DEA close to 0 eV that is most likely due to electron attachment to the sugar unit and DEA at 0.5–3 eV

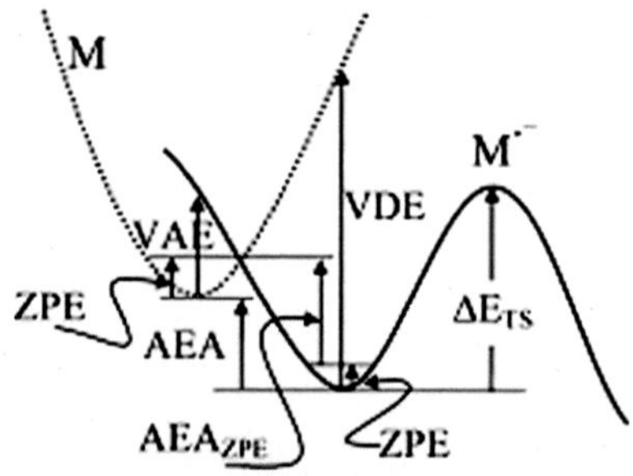

Figure 6.9 Simplified 2D potential energy diagram illustrating the electron attachment process $M + e^- \rightarrow M^{\cdot-}$ and important quantities that can be modeled by quantum chemical calculations. AEA, adiabatic electron affinity; VAE, vertical attachment energy; ZPE, zero-point energy; VDE, vertical detachment energy; TS, transition energy. Reprinted with permission from Ref. [40]. Copyright (2012) American Chemical Society.

that is most probably due to electron attachment to the nucleobase [13, 82].

Nevertheless, thorough and systematic studies of complex molecules consisting of several building blocks still need to be performed. Several investigations have been done in the condensed phase by deposition of oligonucleotides onto metal substrates and detection of desorbing anionic fragments during electron irradiation [1, 91]. However, the information is limited that can be gained from these experiments since only very small anions are desorbing from the thin films (as mentioned earlier, due to a polarization barrier that traps the anions in the films and the kinetic energy of large fragments is too small to escape from the surface).

6.3.3 How Are Reactions and Resonances Modified by an Environment?

Almost all the experiments described before have been performed in ultrahigh vacuum (UHV) using dry DNA or DNA model compounds. However, in the biological environment DNA is embedded in water (more precisely in a buffer with comparatively high salt concentrations), surrounded by proteins and other biomolecules and of course at atmospheric pressure.

Quantum-chemical studies on dissolved nucleotides have tackled the question of how the presence of water molecules influences the energetics of anions and reaction pathways after electron attachment [55]. The general picture is less conclusive than in the studies on bare nucleotides. By comparing different computational studies on dissolved nucleotides it can be concluded that upon solvation (i) the adiabatic electron affinities of anions increase and (ii) the activation barriers of phosphodiester cleavage increase, although the G nucleotide (3′-dGMPH) appears to be rather unaffected by the presence of water molecules [62, 97]. However, the conclusion that water serves as a protection of the nucleotides against electron-induced damage is a bit premature. When "realistic" conditions are considered, that is, a thermally fluctuating environment at room temperature, then the C3′–O3′ bond breaking is still possible with small activation barriers of 6 kcal/mol [101].

Experimental studies of dissolved DNA are scarce. Ptasinska et al. reported that thin films of short oligonucleotides covered by three monolayers of water show an anion desorption that is increased by a factor of 1.6, thus indicating more effective DNA damage in the presence of water [83]. Further experiments have been performed in solution using presolvated electrons produced by ultrafast photoionization [66, 111]. In these studies the reactivity of prehydrated electrons is emphasized; however, it cannot be derived how the aqueous environment modifies the electron-induced DNA damage. Very recently microhydrated thymine has been studied experimentally and theoretically, indicating that the presence of water molecules can suppress dissociation reactions in some cases (see also the discussion in Section 6.4.1) [53, 100].

6.3.4 Are There Additional Interactions in Larger Systems That Influence Electron-Induced Reactions?

The association of the two complementary DNA strands to form the DNA double helix is governed by noncovalent interactions, that is, by hydrogen bonding acting on a plane between the complementary nucleobases arranged in Watson–Crick configuration and by vertically oriented stacking interactions. In contrast to common belief the stability of the double helix cannot be attributed to the hydrogen bonding alone but is to a large extent due to the π–π stacking interactions between the aromatic systems of adjacent nucleobases [114]. These intermolecular interactions can lead to modified electronic properties of DNA strands compared to the individual building blocks, and in electron-induced processes new reaction pathways might open up.

Hydrogen bonding is due to electrostatic interactions and is strongly directional. In the Watson–Crick base pairs (A–T with two intermolecular hydrogen bonds and G–C with three hydrogen bonds) a proton transfer between the nucleobases can occur. This process can be triggered by electron attachment to one of the nucleobases, resulting in a hydrogenated nucleobase radical. Ab initio calculations by Dabkowska et al. demonstrated that a second electron attachment to the hydrogenated species results in a barrier-free cleavage of the phosphodiester bond, which represents an alternative pathway for electron-induced DNA strand breakage [22].

The stacking interactions between two nucleobases are based on dispersion interactions between the two aromatic systems, which is less directional than hydrogen bonding. The π–π interactions are largest between structurally similar molecules, that is, between the same nucleobase species, and are strongest for guanine bases [106]. These interactions have been explored for quite some time in molecular electronics, where DNA is discussed to be used as a conducting polymer [30]. Since stacking interactions directly modify the DNA's electronic properties, it can be assumed that also the interaction with low-energy electrons is modified in a way that strongly depends on the nucleobase sequence.

A sequence dependence of electron-induced DNA damage was demonstrated in different experiments, on the one hand, using high-performance liquid chromatography (HPLC) analysis of electron-irradiated short oligonucleotides [63] and, on the other hand, using self-assembled monolayers of DNA immobilized on gold substrates [87, 104]. The HPLC experiments have been conducted with the oligonucleotide trimers TXT (X = T, C, A, G), and it was shown that the TTT sequence is the most sensitive and the TGT sequence the least sensitive with respect to total damage upon irradiation with 10 eV electrons. Between 9% and 20% of the total damage is due to phosphodiester cleavage, that is, a strand break. The sensitivity of the TTT sequence is qualitatively ascribed to the fact that T has the highest EA of all nucleobases: EA(T) = 0.20 eV, EA(C) = 0.03 eV, EA(G) = −0.07 eV, EA(A) = −0.28 eV [89]. Nevertheless, it should be noted that the highest yield of internal strand breaks in the trimers was observed for the TGT sequence. A very clear result about the role of G in nucleotide sequences was obtained with low-energy photoelectron transmission spectroscopy (LEPET) of DNA self-assembled monolayers (SAMs) [87]. In LEPET photoelectrons are released from a gold surface using an excimer laser operated at 193 nm, and the electrons transmitted through the SAMs escaping into vacuum are detected. By using nucleotide sequences with different G content it was demonstrated that the amount of electrons trapped in the DNA SAM is proportional to the number of G in the DNA sequence. From this result it is not clear whether the electrons trapped in the DNA actually lead to any damage, but recently it was shown that also the SSB yield increases with the G content [104]. For the latter experiments a novel technique was developed that is based on damage detection using fluorescence spectroscopy of DNA microarrays [103]. In these investigations the sensitivity of G containing sequences is qualitatively explained by the low vertical ionization potential (VIP) of G: VIP(G) = 8.24 eV, VIP(A) = 8.44 eV, VIP(C) = 8.94 eV, VIP(T) = 9.14 eV [18]. Another experimental approach is the use of DNA origami platforms to expose specific DNA sequences to low-energy electrons. The strand breakage is quantified by atomic force microscopy, a detailed description is given below in Section 6.4. In preliminary experiments using this technique a particular sensitivity of A containing sequences was found [49, 84].

Interestingly, the composition of particular DNA sequences can lead to new properties that are not simply due to the properties of single nucleotides. An example is the human telomeric sequence TTAGGG. Telomeres are the single-stranded repeats at the chromosome ends, the length of which determines the lifetime of a cell. Electron transmission experiments revealed a cooperative effect in the electron trapping, that is, the telomere sequence appears to be more sensitive than other structurally similar sequences, even if they contain more G than the telomere [67]. Subsequent calculations of the ionization potentials of different sequences including the

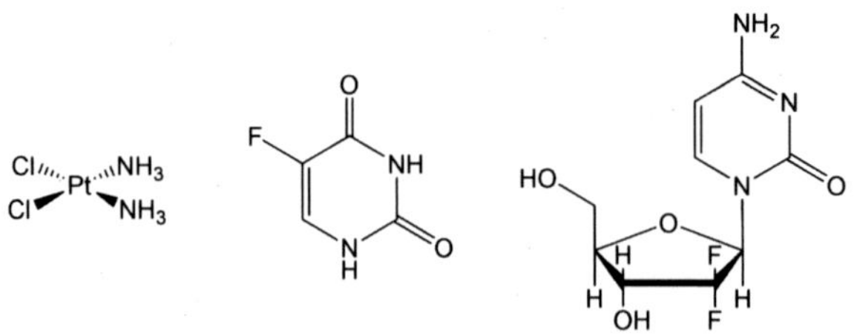

Figure 6.10 Molecular structures of the most widely used therapeutics in radiochemotherapy.

telomeric sequence revealed a lower ionization potential of TTAGGG than for TTGGGG, although the latter contains more G bases [19]. This cooperative effect might explain the biological importance of the telomeres, which might serve as a hole trap to protect the DNA.

6.3.5 What is the Physicochemical Mechanism of Radiosensitization?

Radiosensitizers are administered in radiotherapy to increase the sensitivity of tumour cells toward high-energy radiation. Cisplatin, 5-fluorouracil (5-FU), and gemcitabine (Fig. 6.10) belong to the most widely applied radiosensitizers in concomitant radiochemotherapy [98]. Apart from a pure biological function as antimetabolites [113], these radiosensitizers can also act on a physicochemical level, that is, by enhanced interaction with low-energy electrons [92].

Cisplatin ($PtCl_2(NH_3)_2$) binds to the N7 sites of two adjacent G bases in dsDNA and induces a kink in the double helical structure, which is assumed to be the relevant DNA modification resulting in cell death [88]. The initial step of DNA binding is a release of the Cl atoms that can be mediated by presolvated electrons [66]. Recently, it was shown that a single low-energy electron can react with cisplatin via DEA close to 0 eV, resulting in a simultaneous release of the two Cl atoms, which could activate cisplatin for DNA binding resulting in increased effectiveness of cisplatin upon irradiation [57]. The particular role of low-energy electrons for the radiosensitizing action of cisplatin was demonstrated by irradiating thin films of plasmid DNA with electrons of specific energy and determination of the SSB and DSB yield by gel electrophoresis. The SSB yield was found to increase by a factor of 3.7 upon irradiation of a cisplatin/DNA film with 10 eV electrons, whereas the DSB yield increases by a factor of 1.7 [91].

5-FU and gemcitabine are well-known therapeutics that are clinically used for several years or even decades. From the physicochemical point of view it is interesting to note that the presence of fluorine atoms (in the case of 5-FU at the base and in gemcitabine at the sugar unit) increases the electron attachment cross section, and hence it can be assumed that the incorporation of these compounds into DNA directly results in a higher reactivity with secondary electrons [59, 85]. DEA to a structurally similar compound (5-bromouracil [5-BrU]) has been studied extensively in different experimental setups, and all studies agree that 5-BrU acts as an electron scavenger [1, 2, 112]. The formation of Br^- close to 0 eV is a very effective process with a cross section of 4×10^{-14} cm^2 resulting also in formation of a neutral uracil radical, which can be regarded as a precursor for DNA strand breakage [3]. Very recently, also fluorinated and brominated adenine derivatives have been demonstrated to be efficient radiosensitizers [84, 96].

The detailed understanding of fundamental physicochemical mechanisms of radiosensitization is necessary with the aim to improve radiation therapy and treatment protocols and the design of novel optimized therapeutic agents.

Another promising strategy is the use of noble metal nanoparticles as radiosensitizing agents. Their effect is due to enhanced formation of low-energy electrons upon irradiation with high-energy radiation [92].

6.4 Examples of Experimental Techniques for Studies of Electron Interaction with Biomolecules

As was described in the preceding sections it is very challenging for the experimentalist to obtain relevant (i.e., interpretable) data on DNA damage for complex systems. Complex systems can be simply larger than single DNA building blocks, for instance, complete nucleotides, oligonucleotides of defined sequence, or even higher-order DNA structures. Another parameter (in addition to the size of the system) is the environment. For the study of larger target structures several experimental approaches have been developed in the recent past and many are still under development. These include photoelectron transmission experiments [87], the microarray approach combined with fluorescence spectroscopy [91, 104], the ultrafast electron transfer studies using femtosecond laser spectroscopy [66], the evaporation of thermally labile molecules using laser-induced acoustic desorption [13], and many others.

In this last section we will discuss two experimental techniques that allow for the study of electron induced processes in different environments, (i) the use of cluster beams and (ii) the novel use of DNA nanostructures for irradiation experiments.

6.4.1 Electron Attachment Experiments with He Droplets Doped with Biomolecules

One experimental technique to form more complex targets than small building blocks for low-energy electron interaction are cluster expansions. Clusters are defined as aggregates of atoms or molecules and have generally an intermediate size between individual atoms and aggregates large enough to be called bulk matter. Clusters still differ from bulk matter because a large fraction of their component particles are on the surface but they offer experiments with all benefits of MS under high vacuum.

Clusters can be formed in several ways; however, there are basically three important techniques for cluster formation [115]:

- Supersonic expansion: An atomic or molecular gas is expanded under high pressure through a small hole into vacuum. Thereby the relative energy between the expanding gas particles decreases strongly and enables condensation to clusters in the adiabatic expansion.
- Gas aggregation sources: In this technique atoms or molecules are injected into a stationary or streaming gas. Their energy is decreased, which leads to formation

of clusters (analogous to smoke and cloud formation in nature).

- Surface sources: Clusters are released from solid surfaces upon the interaction of photons or heavy particle impact.

All three techniques are capable to form also biomolecular clusters as targets for electron irradiation. It is important to mention that the supersonic technique itself offers two main possibilities to form clusters of biomolecules: (i) Only a sufficiently cold cluster matrix is formed in the expansion and biomolecules are picked up in a later stage, or (ii) similar to a gas aggregation source biomolecules are vaporized into carrier gas and expanded together through a nozzle (seeded beam expansion). Both of the techniques have been used already, which offer both advantages and disadvantages. Before discussing recent seeded beam expansion experiment in the following, we summarize the first experiments with clusters with nucleobases using the pick-up technique. Figure 6.11 shows a typical experimental setup. In this case helium droplets were used as an inert cluster matrix. Large helium clusters (or due to their size in the nanometer regime also called helium nanodroplets) can be formed by expansion of precooled helium gas with a low temperature ($T = \sim 10$ K) under high pressure ($\sim$2 MPa) through a small nozzle into high vacuum [109]. Due to limited vacuum systems of experiments and high gas flux in most experiment, only nozzles with a diameter of 5 μm can be used. The expansion conditions decide on the final mean droplet size, where a lower temperature and higher pressure significantly increase the mean size. For example, for the expansion temperature of 9 K and expansion pressure of 2 MPa, the droplet mean size is about $\sim 10^5$ He atoms, which corresponds to a droplet diameter of about 20 nm. The He droplets formed are superfluid. Superfluidity implies a very low viscosity, which allows a frictionless stream of helium through capillaries and thinnest channels and, moreover, an infinitely high heat conductivity. Additionally, a particle in the superfluid liquid moves without any drag, as long as the relative velocity is below the Landau speed of 58 m/s. Helium droplets are also characterized by a very low temperature of 0.37 K. The droplet represents therefore a heat bath, where any species embedded is cooled down to temperature of the droplet by evaporative cooling. As a rule of thumb about 1600 He atoms are released from a droplet where 1 eV energy was deposited. Due to their large size, droplets have high cross sections for collisions with other atoms or molecules. This property is exploited for the pick-up technique, that is, after the expansion chamber a differentially pumped pick-up chamber is placed (separated by a skimmer in order to avoid interference of shock waves with the droplet beam), where the vapour of the biomolecule under study is introduced. Since biomolecules like nucleobases have a low vapour pressure, a crucible is placed in the vacuum chamber, which is connected to a pick-up cell placed in the chamber. The helium droplets pass through the small volume of the pick-up cell and embed biomolecules in a statistical process. The number of molecules embedded follows a modified Poisson distribution and can be controlled by the density of vapour in the pick-up cell, that is, experimentally controlled by the temperature of the crucible. Once a biomolecule is picked up, it is quickly cooled down to the temperature of the droplet. Closed-shell species like biomolecules are heliophilic and are present close to the center of the droplet.

When more than one molecule is embedded in a droplet easily clusters of biomolecules are formed due to the low temperature of the helium droplet. Quantum chemical calculations on thymine–thymine, adenine–adenine and thymine–adenine base pairs showed that similar to Watson–Crick base pairs, the most stable configurations are thereby planar and stabilized by hydrogen bonds. Their stabilization energies are 0.5–0.7 eV [77]. For Watson–Crick base pairs values up to 0.5 eV have been reported in the literature [44]. After formation of doped droplets, they are irradiated by low-energy electrons and resulting anions are monitored by MS. Thereby should be mentioned that helium droplets (with all helium atoms in the electronic ground state) are strongly electrophobic, that is, the captured electron does not localize at a helium atom but instead is located in an electron bubble, that is, a cavity with a radius of 17 Å. Further an electron needs about 1.5–2 eV to enter the conduction band of a droplet, that is, resonances are blue-shifted by this amount. The bubble migrating in the droplet may undergo emission into vacuum or finally localize on the biomolecule, which leads to anion formation. It is important to note that no

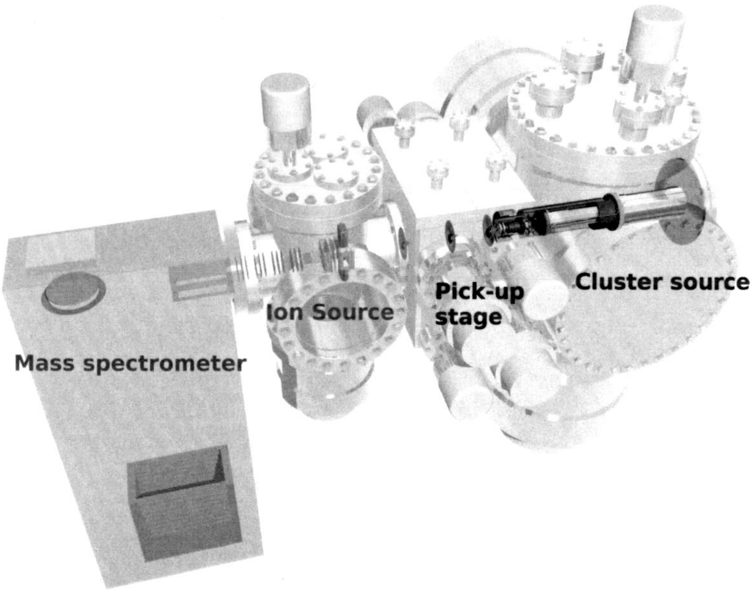

Figure 6.11 Schematic view of supersonic expansion experiment for production of helium droplets doped with biomolecular species combined with electron ionization/attachment mass spectrometry. Adapted from Ref. [94] with kind permission of the author.

indication of a stable large He/biomolecule droplet anion can be observed (also because of the finite range of mass spectrometers) and instead only bare small cluster, molecular, or fragment anions of biomolecules can be observed with sparsely attachment of few helium atoms. This indicates that the anionic species is expelled out of the droplet once formed; however, as discussed next the time elapsed till the expulsion from the droplet is still sufficient for significant interaction between the biomolecular anion and the helium matrix.

This interaction leads then to a substantial change of the product anion yields observed by MS compared to the isolated situation. For example, as mentioned before, no long-lived valence nucleobase anion can be observed for the bare molecule, which can be attributed to the short lifetime against autodetachment and dissociation. In helium droplets a stabilization of the temporary negative ion state occurs where the He/biomolecule environment apparently allows to dissipate the kinetic energy of the captured electron as well as the EA released [25]. As mentioned before, nucleobases may form also dipole-bound anions, which for bare anions undergo autodetachment or they act as doorway states of dissociation. It was shown by photodetachment/photoelectron spectroscopy experiments that solvation of a nucleobase (e.g., in water clusters or in rare gases) changes, however, the characteristic of the excess electron binding (from dipole to valence bonding with a strong increase of the vertical detachment energy/electron affinity [93]). Thus it can be expected that the molecular anions observed from electron attachment to the doped helium droplets are valence anions instead dipole-bound anions. Those are formed in a resonance close to 2 eV, as shown in Fig. 6.12, where the peak position of the molecular (cluster) anions is artificial due to the initial repulsive interaction of the electron with the helium droplet. In contrast, for an electrophilic matrix (with positive EA) or bare nucleobase clusters, a 0 eV peak may be expected. In the attachment measurements the dimer, that is, base pair anions, showed the highest yields [25]. Measurements were also carried out with vaporizing thymine and adenine in two crucibles, that is, mixed nucleobase clusters were formed. Thereby it should be noted that according to quantum chemical calculations, the resulting structure of the gas-phase base pair has not the Watson–Crick base pair configuration like in DNA (though planar and hydrogen bonded). However, the latter can be achieved by the use of 1-methylthymine and 9-methyladenine, that is, methylated species that mimic the gylcosidic bonds of the nuclobases to the sugar molecules of DNA. For a base pair the question on the localization of excess electron is an important question because it will also determine the mobility of an excess charge between the two strands of DNA. This issue was studied by means of anion photoelectron spectroscopy combined with theory [86]. From the measured and calculated vertical detachment energy it was concluded that the captured electron is initially located at the thymine moiety of the pair and a subsequent barrier-free proton transfer from adenine to thymine results into a complex of a hydrogenated neutral thymine and deprotonated adenine. Hence $(AT)^-$ cannot be characterized as T^- solvated by the adenine [86].

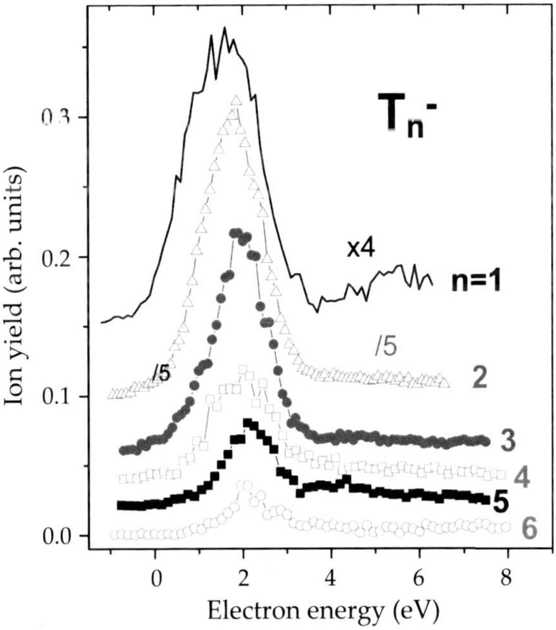

Figure 6.12 (Cluster) parent anions of thymine formed by electron attachment to thymine clusters formed in helium nanodroplets. Reprinted (figure) from Ref. [25]. Copyright (2006) by the American Physical Society.

The experiments showed, however, that dissociation of a nucleobase cannot be avoided in spite of the presence of a matrix. However, the fragmentation pattern is strongly modified in terms of exclusive presence of the dehydrogenated parent anion as fragment, while other fragment anions observed for the bare molecule like NCO^- are not detected for thymine embedded in helium droplets [27]. This is a direct consequence of the different dissociation mechanisms responsible for the occurrence of fragment anions. While fast direct dissociation processes along purely repulsive energy surfaces cannot be prevented by a surrounding matrix, slow dissociation processes with multiple bond cleavage, which require considerable redistribution of vibrational energy in the target anion, can be frozen due to fast dissipation of the excess energy to the surrounding. As discussed before nucleobases show a hybrid nature in this respect, and, for example, NCO^- was shown to form by

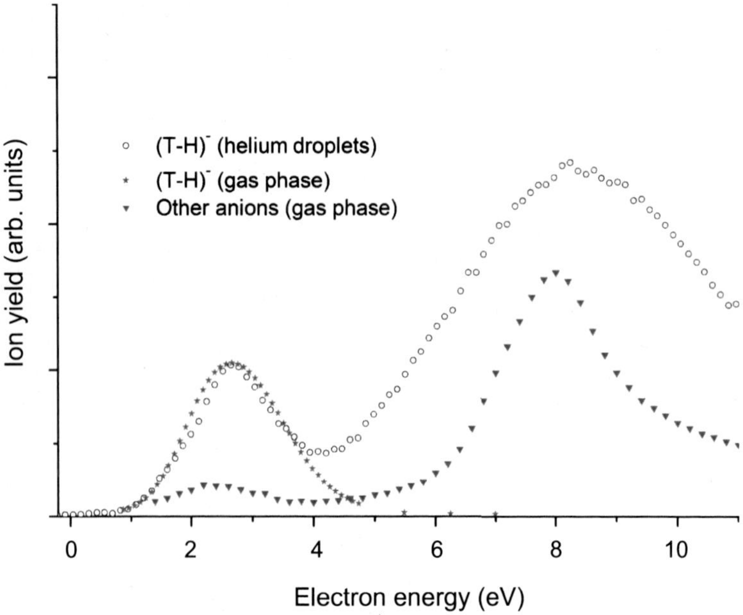

Figure 6.13 Ion yield of $(T-H)^-$ of thymine measured in helium droplets (open circles), $(T-H)^-$ of isolated thymine (full stars), and sum of all other product anions (with the main contribution from NCO^- observed upon DEA to the corresponding bare molecule (full triangles). The gas-phase data is shifted upward by 1.5 eV to account for the energy required for the electrons to penetrate the droplet. For better clarity the ion yield of the other anion in the gas phase above 3 eV is multiplied by a factor of 10. Note that the electron energy resolution of these measurements was ~1 eV, that is, the sharp peaks in the gas-phase data visible in Fig. 6.6. are not resolved. Reproduced from Ref. [27] with permission. Copyright (2008) by John Wiley and Sons.

initial fast hydrogen loss with subsequent bond cleavages within the aromatic ring [32]. This ring cleavage can be prevented by the helium matrix, which leads to the frozen intermediate product $(M-H)^-$ observed in MS (see Fig. 6.13). Thus for $(M-H)^-$ formed in the droplets same resonance energies can be observed like for NCO^- ion yield observed in DEA to the bare nucleobase molecule.

Thus these measurements gave a first impressive view of the action of a matrix in electron attachment to a biomolecule, but it is clear that helium droplets do not provide a fully realistic scenario for water solvated biomolecules present in a cell. Most noticeable is the extreme cooling capability of helium droplets, which is the strongest of all matrices, and in addition, helium is an inert matrix, while for water surrounding significant polarization effects due to the polar nature of water may be expected. The presence of water may then significantly increase the electron capture abilities of biomolecules [40]. Therefore, the studies on bare hydrated biomolecular clusters have been intensified recent years. Using still the pick-up technique, electron ionization of pure water–amino acid molecules was investigated [70]. Very recently also an electron ionization study with (hydrated) tetrahydrofuran (THF) clusters was carried out [72]. Also in these cases fragmentation of the biomolecule was still present, and the protection of the biomolecule was by far not complete. For THF clusters it was shown that the fragment ion $(THF–OH)^+$ is exclusively a product from ionization of clusters. This result indicated that few electron ionization mass spectra reported in the literature for gas-phase THF were determined at experimental conditions, where the neutral beam also contained clusters [72].

In the case of electron capture, studies for nonhydrated and microhydrated pyrimidine clusters showed striking differences to the isolated pyrimidine molecule as well as pronounced hydration effects [71]. The monomer anion was detected in case of pure pyrimidine clusters. However, the ion yield showed only two broad features between $\sim$4 and 15 eV and no resonance close to 0 eV, which is in agreement with the negative EA of the molecule. The ion yield at higher electron energy was ascribed to form via core-excited resonances, which are metastable states at considerably larger intermolecular distances compared to the neutral. These states can be only stabilized in a sufficiently large cluster environment (at least a tetramer) with complete evaporation of the residuals [71].

The ion yield of the bare dimer anion shows, in contrast to the monomer, a strong peak close to 0 eV indicating a positive EA (see Fig. 6.14). The maximum of the core-excited resonance signal is about 10 times lower than the low-energy feature; however, in the case of electron attachment to hydrated pyrimidine clusters this ion yield of core-excited resonances becomes suppressed, as shown in Fig. 6.14, that is, the captured electron becomes emitted

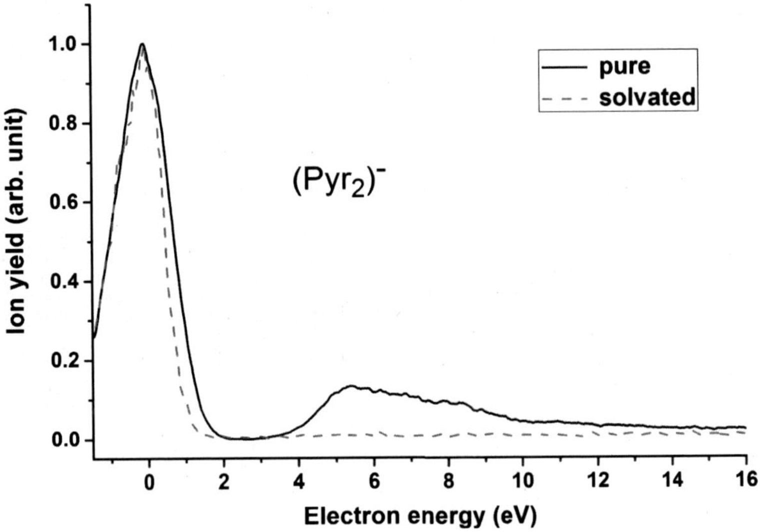

Figure 6.14 Ion yield of the bare dimer anion of pyrimidine (Pyr) formed upon electron attachment to nonhydrated and hydrated pyrimidine clusters. Ion yields accompanied by electronic excitation of the neutral precursor cluster are suppressed in the case of hydration (see text). Reproduced from Ref. [71] with permission. Copyright (2008) by John Wiley and Sons.

after the electronic excitation of the hydrated pyrimidine species. This result supports a hypothesis for the reaction sequence to a DSB by electron attachment of a single electron, as described in Ref. [8]: after electron capture accompanied by electronic excitation, autodetachment and formation of an OH radical by dissociation of the complex occur. This OH radical attacks the DNA and causes an SSB, while the released (slow) electron is responsible for the break of the opposite strand by well-known mechanisms.

In 2016, the first experimental electron attachment study of microhydrated thymine and uracil was carried out [53]; 20 years later than the first experiment on electron ionization [50]. In contrast to the helium droplet experiments presented earlier, no signals of $(M - H)^-$ were observed in the cumulative negative ion mass spectra in the energy range from 0 to 3.5 eV. In Ref. [53] two possible explanations for the favoring of the nondissociative channel were suggested: either a change of the electronic structure of the nucleobase due to the hydration occurs, or a fast transfer of excess energy to the surrounding solvent (caging) leads to the stabilization of the nucleobase anion. Finally, the former hypothesis was discarded, since electronic structure calculations indicated only a solvation energy of ~0.3 eV [102]. In other respects, scattering theory presented in Ref. [102] was in strong disagreement with the experiment, since an enhancement of the DEA cross section was predicted for a nucleobase in a water cluster environment.

At the end of this section, we should make two general remarks. First, it should be noted that the effect of water surrounding may strongly depend on the type of collision process. A previous study on collisions of neutral atoms (neon and sodium) with adenosine 5′-monophophate (AMP^-) at 50 keV collision energy showed a dramatic effect on fragmentation of AMP^- upon water solvation [65]. While for the bare anion significant fragmentation was observable, the fragmentation of AMP^- was decreased upon nanosolvation with few water molecules and was completely quenched upon solvation with 13 water molecules. In this case only the loss of water molecules from the evaporative ensemble was observed. Hence in collision of neutrals with AMP^- the fragmentation process can be understood as a vibrational predissociation process, which can be quenched by vibrational redistribution of excess energy in the water surrounding. In comparison, dissociation induced by low-energy electron attachment may also base on this mechanism; however, direct bond cleavage was also shown to be a important mechanism in electron attachment to biomolecules.

Another important point to mention in the context of experiments with mixed biomolecule-water clusters is the structure of the clusters formed. For example, a computational study of AMP^- (with up to 20 water molecules) [65] showed that water molecules agglomerate close to phosphate group, as shown in Fig. 6.15, that is, one cannot speak strictly of solvation of the molecule (beside the presence of several isomeric configurations). The same steric behavior turned out to be the case for pick-up experiments where biomolecules were picked up by a preformed water cluster formed in supersonic expansion [70]. Thus for water–biomolecule clusters in the gas phase the buffering of fragmentation can be rather

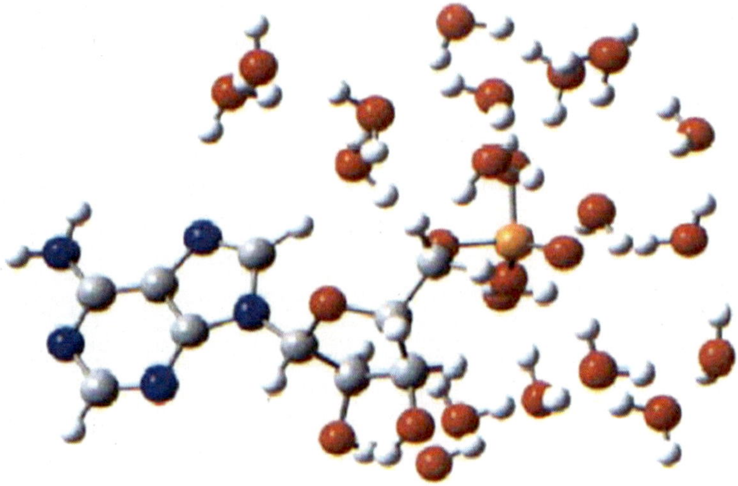

Figure 6.15 Possible structure of $AMP^{-}(H_2O)_{20}$ with agglomerated water molecules around the phosphate group. Color code: grey, carbon atom; blue, nitrogen; red, oxygen; orange, phosphor; small white, hydrogen. Reprinted (figure) with permission from Ref. [65]. Copyright (2006) by the American Physical Society.

ascribed to the water heat path taking over excess energy than a complete sterically shielding.

6.4.2 Electron-Induced Reactions in DNA Nanostructures

A central challenge in the investigation of irradiated oligonucleotides is the small penetration depth of low-energy electrons and the resulting small amount of damaged material. When using highly sensitive detection such as mass spectrometric analysis of desorbing anions (electron-stimulated desorption [ESD]), only small fragments can be detected and an assignment of observed reactions to relevant processes such as DNA strand breaks is difficult. On the other hand, traditional chemical analytical methods such as HPLC and agarose gel electrophoresis (AGE) have detection limits that do not allow for an analysis of medium-size well-defined oligonucleotides. HPLC analysis of material damaged by low-energy electrons is a tedious process limited to molecules consisting of not more than three or four nucleotides since the number of possible fragments increases with the size of the system and the concentration of each fragment decreases below the detection limit. An analysis by AGE is only possible with plasmid DNA [63], since a strand break leads to a strong change of the molecular shape that can be easily detected. However, plasmid DNA consists of several thousand base pairs, and no information can be obtained on the specific location of the damaged site.

Consequently, a highly sensitive detection of DNA strand breakage in well-defined oligonucleotides is desired. Additionally, the detection scheme should be versatile enough to easily modify the oligonucleotide structure (sequence, hybridization state, higher-order structures, inclusion of radiosensitizers) and the environment.

For this purpose, DNA nanoarrays can be constructed, in which defined oligonucleotide target sequences are placed on predefined positions, and subsequent to irradiation with low-energy electrons the damaged oligonucleotides are analyzed on a single-molecule level by atomic force microscopy [48]. For the construction of the DNA nanoarrays so-called DNA origami structures are used [110]. In the DNA origami technique a long single-stranded DNA scaffold (in most cases M13mp18 with 7249 nucleotides) is folded into specific 2D and 3D nanostructures simply by hybridization with a suitable set of short oligonucleotides [90]. In this way it is possible to form complex structures such as 2D rectangles and triangles [90] but also 3D boxes [9], spheres [42], and many others [99]. The DNA origami structures can be easily functionalized with well-defined oligonucleotides, and the DNA origami structures themselves serve only as a template. The preparation of the DNA origami structures is simple, since it is based on self-assembly. Several different target sequences can be placed on one template and the single-molecule detection allows for the analysis of electron-induced damage to multiple target sequences in one irradiation experiment [48, 49, 84, 85, 96].

The detection scheme for low-energy-electron-induced strand breaks is schematically shown in Fig. 6.16. The target sequences are prolonged from the staple strands and are thus located at predetermined positions on a triangular DNA origami structure. The 5′-end of the protruding sequences is modified with biotin

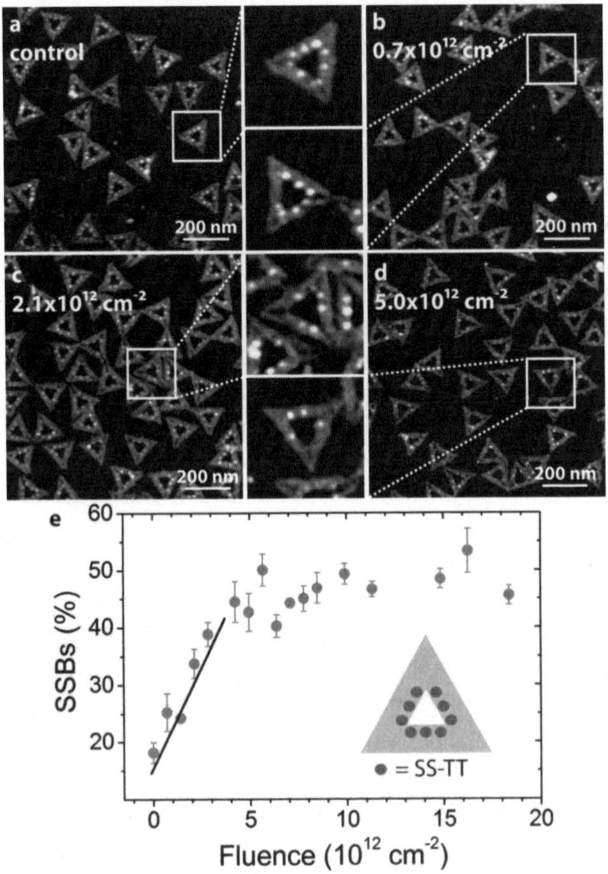

Figure 6.16 Irradiation of DNA origami templates with nine protruding target strands. (a) AFM image of a nonirradiated control sample. (b–d) Samples irradiated with different fluences of 18 eV electrons. The number of specifically bound streptavidin decreases with electron fluence, indicating low-energy-electron-induced strand breaks. (e) Number of single-strand breaks (in %) vs. electron fluence. Reprinted with permission from Ref. [48]. Copyright (2012) American Chemical Society.

to enable a binding to the protein streptavidin. The triangular DNA origami substrates are adsorbed on oxidized Si surfaces and subsequently irradiated with low-energy electrons under defined conditions (energy and fluence). After irradiation, the samples are rinsed to remove fragmentation products, and incubated with

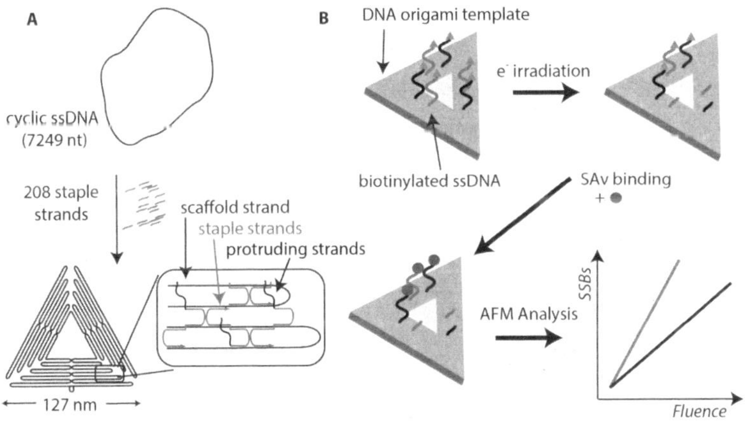

Figure 6.17 (A) Illustration of the folding of a single-stranded cyclic DNA strand into a triangular shape by hybridization with a suitable set of short artificial staple strands. (B) Detection scheme for the determination of DNA strand break yields after irradiation with low-energy electrons. Reprinted with permission from Ref. [48]. Copyright (2012) American Chemical Society.

streptavidin that binds to the biotin markers of the remaining intact strands. The addition of streptavidin is simply to visualize the position of intact protruding strands for atomic force microscope (AFM) analysis. The AFM analysis of a number of DNA origami structures thus allows for a statistical analysis of strand break events for each modified position separately [49, 85]. In Fig. 6.17 an example of a test system is shown to demonstrate the technique [48]. A DNA origami design with nine protruding target sequences is shown. Figure 6.16a–d shows AFM images of a control sample and samples irradiated with different electron fluences at an electron energy of 18 eV. The magnified images show clearly how the number of specifically bound streptavidin molecules (bright spots on the triangles) decreases with the fluence. This is due to electron-induced strand breaks in the protruding sequences leading to a loss of the biotin marker and thus the binding position for streptavidin. A detailed fluence dependence of the relative number of strand breaks is shown in Fig. 6.16e. The strand break yield is obtained from the slope of the linear fit to the low-fluence regime.

With the DNA origami technique a fast and parallel determination of DNA strand break yields with unprecedented control over the DNA's primary and secondary structure is possible [84, 96]. In this way, the investigation of DNA radiation damage in its most natural environment, for example, DNA nucleosomes constituting the chromatin, now becomes feasible.

Acknowledgments

I.B. acknowledges support from the University of Potsdam, the Deutsche Forschungsgemeinschaft (DFG), and the Marie-Curie program of the European Union. S.D. would like to thank Prof. Paul Burrow, Prof. Paul Scheier, and Prof. Tilmann Märk. This work was partially supported by FWF (P30332, P22665, P24443, I-1015, and M-14450).

References

1. Abdoul-Carime, H., Dugal, P. C., and Sanche, L. (2000). Damage induced by 1-30 eV electrons on thymine- and bromouracil-substituted oligonucleotides. *Radiat. Res.*, **153**, pp. 23–28.
2. Abdoul-Carime, H., Huels, M. A., Illenberger, E., and Sanche, L. (2001). Sensitizing DNA to secondary electron damage: resonant formation of oxidative radicals from 5-halouracils. *J. Am. Chem. Soc.*, **123**, pp. 5354–5355.
3. Abdoul-Carime, H., Huels, M. A., Illenberger, E., and Sanche, L. (2003). Formation of negative ions from gas phase halo-uracils by low-energy (0–18 eV) electron impact. *Int. J. Mass Spectrom.*, **228**, pp. 703–716.
4. Abdoul-Carime H., Gohlke, S., and Illenberger, E. (2004). Site-specific dissociation of DNA bases by slow electrons at early stage of irradiation. *Phys. Rev. Lett.*, **92**, p. 168103-1-4.
5. Aflatooni, K., Gallup, G. A., and Burrow, P. D. (1998). Electron attachment energies of the DNA bases. *J. Phys. Chem. A*, **102**, pp. 6205–6207.
6. Aflatooni, K., Scheer, A. M., and Burrow, P. D. (2005). Dissociative electron attachment in uracil: total anion yield. *Chem. Phys. Lett.*, **408**, pp. 426–428.

7. Aflatooni, K., Scheer, A. M., and Burrow, P. D. (2006). Total dissociative electron attachment cross sections for molecular constituents of DNA. *J. Chem. Phys.*, **125**, p. 054301-1-5.

8. Alizadeh, E., Orlando, T. M., and Sanche, L. (2015). Biomolecular damage induced by ionizing radiation: the direct and indirect effects of low-energy electrons on DNA. *Annu. Rev. Phys. Chem.*, **66**, pp. 379–398.

9. Andersen, E. S., Dong, M., Nielsen, M. M., Jahn, K., Subramani, R., Mamdouh, W., Golas, M. M., Sander, B., Stark, H., Oliveira, C. L. P., Pedersen, J. S., Birkedal, V., Besenbacher, F., Gothelf, K. V., and Kjems, J., (2009). Self-assembly of a nanoscale DNA box with a controllable lid. *Nature*, **459**, p. 73-U5.

10. Baccarelli, I., Bald, I., Gianturco, F. A., Illenberger, E., and Kopyra, J. (2011). Electron-induced damage of DNA and of its components, experiments and theoretical modelings. *Phys. Rep.*, **508**, pp. 1–44.

11. Bald, I., Kopyra, J., and Illenberger, E. (2006). Selective excision of C5 from D-ribose in the gas phase by low-energy electrons (0-1 eV): implications for the mechanism of DNA damage. *Angew. Chem. Int. Ed.*, **45**, pp. 4851–4855.

12. Bald., I, Kopyra, J., Dabkowska, I., Antonsson, E., and Illenberger, E. (2007). Low energy electron-induced reactions in gas phase 1,2,3,5-tetra-O-acetyl-b-DD-ribofuranose: a model system for the behavior of sugar in DNA. *J. Chem. Phys.*, **126**, p. 074308.

13. Bald, I., Dabkowska, I., and Illenberger, E. (2008). Probing biomolecules by laser-induced acoustic desorption: electrons at near zero electron volts trigger sugar-phosphate cleavage. *Angew. Chem. Int. Ed.*, **47**, pp. 8518–8520.

14. Ball, P. (2003). Portrait of a molecule. *Nature*, **421**, pp. 421–422.

15. Berdys, J., Anusiewicz, I., Skurski, P., and Simons, J. (2004). Damage to model DNA fragments from very low-energy (<1 eV) electrons. *J. Am. Chem. Soc.*, **126**, pp. 6441–6447.

16. Boudaiffa, B., Cloutier, P., Hunting, D., Huels, M. A., and Sanche, L. (2000). Resonant formation of DNA strand breaks by low-energy (3 to 20 eV) electrons. *Science*, **287**, pp. 1658–1660.

17. Burrow, P. D., Gallup, G. A., Scheer, A. M., Denifl, S., Ptasinska, S., Märk, T. D., and Scheier, P. (2006). Vibrational Feshbach resonances in uracil and thymine. *J. Chem. Phys.*, **124**, p. 124310-1-7.

18. Cauet, E., Dehareng, D., and Lievin, J. (2006). Ab initio study of the ionization of the DNA bases: ionization potentials and excited states of the cations. *J. Phys. Chem. A.*, **110**, pp. 9200–9211.

19. Cauet, E. (2011). Unique hole-trapping property of the human telomere sequence. *J. Biomol. Struct. Dyn.*, **29**, pp. 557–561.
20. Coupier, B., Farizon, B., Farizon, M., Gaillard, M. J., Gobet, F., de Castro Faria, N. V., Jalbert, G., Ouaskit, S., Carre, M., Gstir, B., Hanel, G., Denifl, S., Feketeova, L., Scheier, P., and Märk, T. D. (2002). Inelastic interactions of protons and electrons with biologically relevant molecules. *Eur. Phys. J. D*, **20**, pp. 459–468.
21. Crespo-Hernandez, C. E., Cohen, B., Hare, P. M., and Kohler, B. (2004). Ultrafast excited state dynamics of in nucleic acids. *Chem. Rev.*, **104**, pp. 1977–2019.
22. Dabkowska, I., Rak, J., and Gutowski, M. (2005). DNA strand breaks induced by concerted interaction of H radicals and low-energy electrons: a computational study on the nucleotide of cytosine. *Eur. Phys. J. D*, **35**, pp. 429–435.
23. Denifl, S., Ptasinska, S., Cingel, M., Matejcik, S., Scheier, P., and Märk, T. D. (2003). Electron attachment to the DNA bases thymine and cytosine. *Chem. Phys. Lett.*, **377**, pp. 74–80.
24. Denifl, S., Ptasinska, S., Probst, M., Hrusak, J., Scheier, P., and Märk, T. D. (2004). Electron attachment to the gas-phase DNA bases cytosine and thymine. *J. Phys. Chem. A*, **108**, pp. 6562–6569.
25. Denifl, S., Zappa, F., Mähr, I., Lecointre, J., Probst, M., Märk, T. D., and Scheier, P. (2006). Mass spectrometric investigation of anions formed upon free electron attachment to nucleobase molecules and clusters embedded in superfluid helium droplets. *Phys. Rev. Lett.*, **97**, p. 043201-4.
26. Denifl, S., Sulzer, P., Zappa, F., Moser, S., Kräutler, B., Echt, O., Bohme, D. K., Märk, T. D., and Scheier, P. (2008). Isotope effects in dissociative electron attachment to the DNA base thymine. *Int. J. Mass. Spectrom.*, **277**, pp. 296–299.
27. Denifl, S., Zappa, F., Mauracher, A., Ferreira da Silva, F., Bacher, A., Echt, O., Märk, T. D., Bohme, D. K., and Scheier, P. (2008). Dissociative electron attachment to DNA bases near absolute zero temperature. freezing dissociation intermediates. *Chem. Phys. Chem.*, **9**, pp. 1387–1389.
28. Desfrancois, C., Abdoul Carime, H., and Schermann, J. P. (1996). Electron attachment to isolated nucleic acid bases. *J. Chem. Phys.*, **104**, pp. 7792–7794.
29. Dolgounitcheva, O., Zakrzewski, V. G., and Ortiz, J. V. (1999). Structure and electron detachment energies of uracil anions. *Chem. Phys. Lett.*, **307**, pp. 220–226.

30. Endres, R. G., Cox, D. L., and Singh, R. R. P. (2004). Colloquium: the quest for high-conductance DNA. *Rev. Mod. Phys.*, **76**, pp. 195–214.

31. Felsenfeld, G., and Groudine, M. (2003). Controlling the double helix. *Nature*, **421**, pp. 448–453.

32. Ferreira da Silva, F., Matias, C., Almeida, D., García, G., Ingólfsson, O., Flosadottir, H. D., Ómarsson, B., Ptasinska, S., Puschnigg, B., Scheier, P., Limão-Vieira P., and Denifl, S. (2013). NCO−, a key fragment upon dissociative electron attachment and electron transfer to pyrimidine bases: site selectivity for a slow decay process. *J. Am. Soc. Mass Spectrom.*, **24**, pp. 1787–1797.

33. Flosadóttir, H. D., Denifl, S., Zappa, F., Wendt, N., Mauracher, A., Bacher, A., Jonsson, H., Märk, T. D., Scheier, P., and Ingolfsson, O. (2007). Combined experimental and theoretical study on the nature and the metastable decay pathways of the amino acid ion fragment $[M\text{-}H]^-$. *Angew. Chem. Int. Ed.*, **46**, pp. 8057–8059.

34. Flosadóttir, H. D., Ómarsson, B., Bald, I., and Ingolfsson, O. (2012). Metastable decay of DNA components and their compositions - a perspective on the role of reactive electron scattering in radiation damage. *Eur. Phys. J. D*, **66**, pp. 13–32.

35. Gallup, G. A., and Fabrikant, I. I. (2011). Vibrational Feshbach resonances in dissociative electron attachment to uracil. *Phys. Rev. A*, **83**, p. 012706-1-7.

36. Gorfinkiel, J. D. and Ptasinska, S. (2017). Electron scattering from molecules and molecular aggregates of biological relevance. *J. Phys. B: At. Mol. Opt. Phys.*, **50**, p. 182001.

37. Gu, J. D., Wang, J., and Leszczynski, J. (2006). Electron attachment-induced DNA single strand breaks: C-3′-O-3′ sigma-bond breaking of pyrimidine nucleotides predominates. *J. Am. Chem. Soc.*, **128**, pp. 9322–9323.

38. Gu, J. D., Xie, Y. M., and Schaefer, H. F. (2006). Near 0 eV electrons attach to nucleotides. *J. Am. Chem. Soc.*, **128**, pp. 1250–1252.

39. Gu, J. D., Wang, J., and Leszczynski, J. (2011). Low energy electron attachment to the adenosine site of DNA. *J. Phys. Chem. B*, **115**, pp. 14831–14837.

40. Gu, J., Leszczynski, J., and Schaefer III, H. F. (2012). Interactions of electrons with bare and hydrated biomolecules: from nucleic acid bases to DNA segments. *Chem. Rev.*, **112**, pp. 5603–5640.

41. Han, X., Jin, M., Breuker, K., and McLafferty, F. W. (2006). Extending top-down mass spectrometry to proteins with masses greater than 200 kilodaltons. *Science*, **314**, pp. 109–112.

42. Han, D., Pal, S., Yang, Y., Jiang, S., Nangreave, J., Liu, Y., and Yan, H. (2013). DNA gridiron nanostructures based on four-arm junctions. *Science*, **339**, pp. 1412–1415.
43. Hanel, G., Gstir, B., Denifl, S., Scheier, P., Probst, M., Farizon, B., Farizon, M., Illenberger, E., and Märk, T. D. (2003). Electron attachment to uracil. Effective destruction at subexcitation energies. *Phys. Rev. Lett.*, **90**, p. 188104-4.
44. Hobza, P., and Sandorfy, C. (1987). Nonempirical calculations on all the 29 possible DNA base pairs. *J. Am. Chem. Soc.*, **109**, pp. 1302–1307.
45. Huels, M. A., Hahndorf, I., Illenberger, E., and Sanche, L. (1998). Resonant dissociation of DNA bases by subionization electrons. *J. Chem. Phys.*, **108**, pp. 1309–1313.
46. Huels, M. A., Boudaiffa, B., Cloutier, P., Hunting, D., and Sanche, L. (2003). Single, double, and multiple double strand breaks induced in DNA by 3-100 eV electrons. *J. Am. Chem. Soc.*, **125**, pp. 4467–4477.
47. Hush, N. S. and Cheung, A. S. (1975). Ionization potentials and donor properties of nucleic acid bases and related compounds. *Chem. Phys. Lett.*, **34**, p. 11.
48. Keller, A., Bald, I., Rotaru, A., Cauet, E., Gothelf, K. V., and Besenbacher, F. (2012). Probing electron-induced bond cleavage at the single-molecule level using DNA origami templates. *ACS Nano*, **6**, pp. 4392–4399.
49. Keller, A., Rackwitz, J., Cauët, E., Liévin, J., Körzdörfer, T., Rotaru, A., Gothelf, K. V., Besenbacher, F., and Bald, I. (2014). Sequence-dependence of electron-induced DNA strand breakage revealed by DNA nanoarrays. *Sci. Rep.*, **4**, p. 7391.
50. Kim, S. K., Lee, W., and Herschach, D. R. (1996). Cluster beam chemistry: hydration of nucleic acid bases; ionization potentials of hydrated adenine and thymine. *J. Phys. Chem.*, **100**, pp. 7933–7937.
51. Kim, S. and Schaefer III, H. F. (2006). Effects of microsolvation on uracil and its radical anion: uracil(H2O)n(n=1-5). *J. Chem. Phys.*, **125**, p. 144305-1-9.
52. King, S. B., Yandell, M. A., Stephansen, A. B., and Neumark, D. M. (2013). Time-resolved radiation chemistry: dynamics of electron attachment to uracil following UV excitation of iodide-uracil complexes. *J. Chem. Phys.*, **141**, p. 224310.
53. Kocisek, J., Pysanenko, A., Farnik, M., and Fedor, J., (2016). Microhydration prevents fragmentation of uracil and thymine by low-energy electrons. *J. Phys. Chem. Lett.*, **7**, pp. 3401–3405.

54. Koenig, C., Kopyra, J., Bald, I., and Illenberger, E. (2006). Dissociative electron attachment to phosphoric acid esters: the direct mechanism for single strand breaks in DNA. *Phys. Rev. Lett.*, **97**, p. 018105.

55. Kohanoff, J., McAllister, M., Tribello, G. A., and Gu, B. (2017). Interactions between low-energy electrons and DNA: a perspective from first-principles simulations. *J. Phys.: Condens. Matter*, **29**, p. 383001.

56. Kopyra, J., Koenig-Lehmann, C., and Illenberger, E. (2009). On the absolute value for the cross-section of dissociative electron attachment (DEA) to the DNA base thymine. *Int. J. Mass Spectrom.*, **281**, pp. 89–91.

57. Kopyra, J., Koenig-Lehmann, C., Bald, I., and Illenberger, E. (2009). A single slow electron triggers the loss of both chlorine atoms from the anticancer drug cisplatin: implications for chemoradiation therapy. *Angew. Chem. Int. Ed.*, **48**, pp. 7904–7907.

58. Kopyra, J. (2012). Low energy electron attachment to the nucleotide deoxycytidine monophosphate: direct evidence for the molecular mechanisms of electron-induced DNA strand breaks. *Phys. Chem. Chem. Phys.*, **14**, pp. 8287–8289.

59. Kopyra, J., Keller, A., and Bald, I. (2014). On the role of fluoro-substituted nucleosides in DNA radiosensitization for tumor radiation therapy. *RSC Adv.*, **4**, pp. 6825–6829.

60. Kumar, A. and Sevilla, M. D. (2007). Low-energy electron attachment to 5′-thymidine monophosphate: modeling single strand breaks through dissociative electron attachment. *J. Phys. Chem. B*, **111**, pp. 5464–5474.

61. Kumar, A. and Sevilla, M. D. (2008). The role of pi sigma* excited states in electron-induced DNA strand break formation: a time-dependent density functional theory study. *J. Am. Chem. Soc.*, **130**, pp. 2130–2131.

62. Kumar, A. and Sevilla, M. D. (2009). Role of excited states in low-energy electron (LEE) induced strand breaks in DNA model systems: influence of aqueous environment. *Chem. Phys. Chem.*, **10**, pp. 1426–1430.

63. Li, Z. J., Cloutier, P., Sanche, L., and Wagner, J. R. (2010). Low-energy electron-induced DNA damage: effect of base sequence in oligonucleotide trimers. *J. Am. Chem. Soc.*, **132**, pp. 5422–5427.

64. Li, J.-L., Li, C.-Y., Qiu, R., Yan, C.-C., Xie, W.-Z., Zeng, Z., and Tung, C.-J. (2013). Comparison of direct DNA strand breaks induced by low energy electrons with different inelastic cross sections. *Nucl. Instrum. Methods Phys. Res., Sect. B*, **311**, pp. 27–36.

65. Liu, B., Bronsted, Nielsen, S., Hvelplund, P., Zettergren, H., Cederquist, H., Manil, B., and Huber, B. A. (2006). Collision-induced dissociation of

[3]. Bombardment of Earth's upper atmosphere by sunlight (vacuum ultraviolet [VUV], 6.2–12.4 eV, but not visible light), solar wind particles (energetic electrons and protons), magnetospheric particles, and cosmic ray particles (high-energy protons and atomic nuclei) causes ionization that results in the production of copious numbers of low-energy secondary electrons [4]. While secondary products such as excited species and ions produced from the high-energy radiation interacting with molecules cause some radiation chemistry, it is the low-energy secondary electron–molecule inelastic collisions that are thought to be the primary driving forces in a wide variety of radiation-induced chemical reactions [4].

Low-energy-electron-induced processes (excitation, dissociation, and ionization) in the *terrestrial atmosphere* may influence a wide range of complex phenomena, including chemical reactions. Recent results of radio- and gamma-emission measurements conducted during thunderstorms suggest that lightning initiation in Earth's atmosphere may be mediated by low-energy electrons produced by cosmic rays [5]. The role of cosmic rays in initiating lightning showers has now become an area of active research [6]. According to a recent hypothesis [7–10], the subject of significant current controversy [11, 12], low-energy electrons produced by cosmic rays, in addition to ultraviolet-visible (UV-Vis) photons from the sun, interact with CFCs to produce chlorine atoms that subsequently destroy ozone in the Antarctic. Dissociative electron attachment (DEA) to $NaHCO_3$ and $NaCO_3$ has been identified as a possible source of free Na atoms in the sporadic (sudden) sodium layers (SSL) found in Earth's upper atmosphere [13]. Low-energy electrons produced by the interaction of cosmic rays with molecules present in Earth's atmosphere may contribute to enhanced aerosol nucleation, the first step in cloud formation. This cosmic-ray-induced cloud formation mechanism is a subject of active research because this process may influence global warming [14].

A number of important *extraterrestrial atmospheric* processes are also thought to be mediated by low-energy electrons. DEA to HCN has been identified as the first step in the production of heavy (m/z ≈10,000) anions detected by the Cassini Plasma Spectrometer in the upper atmosphere of Titan [15]. Low-energy-electron-induced enhanced production of H_2O_2 is postulated to have implications for life and habitability on Mars [16].

Following a brief introduction to the vertical structure of Earth's atmosphere and the ionosphere, we present in this chapter an overview of low-energy (<20 eV) electron-induced processes relevant to terrestrial and extraterrestrial atmospheric processes.

7.2 Vertical Structure of Earth's Atmosphere

Earth's atmosphere is generally divided into four layers on the basis of how the temperature changes with height (Fig. 7.1). The troposphere, the bottom layer extending up to $\sim$10 km in altitude, is characterized by a temperature decrease with height increase. Because this first layer contains the preponderance of atmospheric mass, particularly the majority of atmospheric water vapour, the troposphere is the region largely responsible for Earth's weather. The stratosphere (10–50 km altitude), the second layer, not only still contains appreciable concentrations of many of the major tropospheric gases but also includes the ozone layer ($\sim$25 km altitude) that absorbs ultraviolet radiation from the sun. The temperature in the stratosphere generally increases with increasing height. Except for the occasional polar stratospheric clouds (PSCs), also known as nacreous clouds, the stratosphere is generally devoid of clouds. The mesosphere (50–80 km altitude) is characterized by temperature decrease with increasing height. This third layer is arguably the least studied of the layers due to its altitude being too high for most planes and too low for satellites [17]. The temperature in the fourth layer, the thermosphere, increases with height to values that exceed 1000°C.

The ionosphere (60–300 km altitude), the electrically conducting region of the mesosphere and thermosphere, plays an important role in the propagation of radio waves and the formation of auroras. On the basis of the ion density, which is equal to the electron density, the ionosphere is further subdivided into three regions. The D-region at an altitude of 60–85 km, the lowest ionospheric region, is situated in the mesosphere. The D-region disappears during nighttime because the relatively high pressures allow for

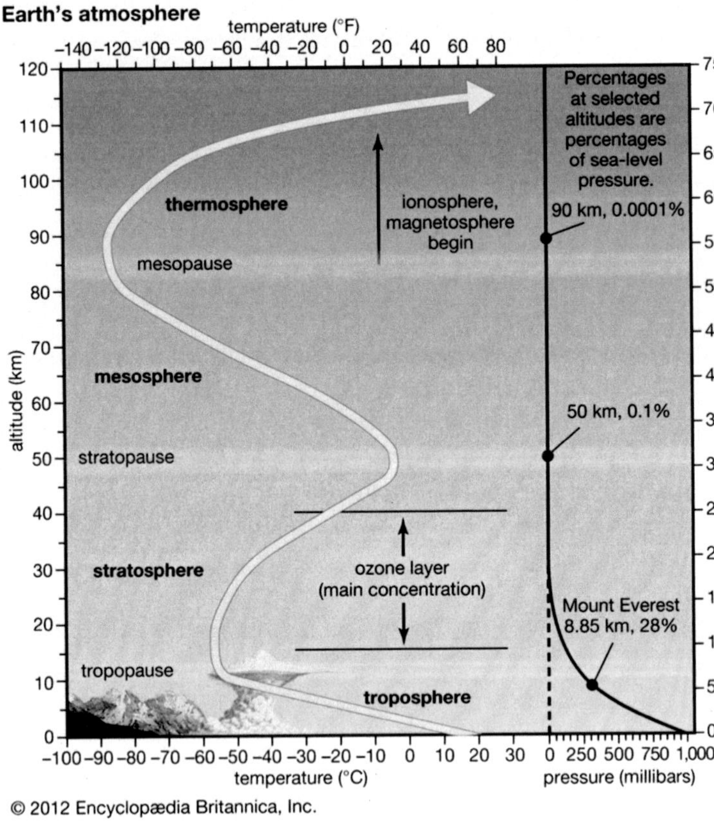

Figure 7.1 Vertical structure of Earth's atmosphere. Reproduced from Ref. [18] with permission from Encyclopædia Britannica, Inc.

the attachment of electrons to molecules to form anions such as CO_3^- and NO_3^- [3]. Photodetachment effectively competes with this process during daytime. The D-region is characterized by an electron density of 10^2–10^3 cm^{-3} [19]. The E-region, with an electron density of ~10^5 cm^{-3}, spans altitudes between 85 and 140 km. The formation of this region is mediated by soft X-rays and extreme ultraviolet (EUV) ($\lambda < 100$ nm) from the sun. The E-region vanishes at night because of the fast reaction between electrons and the positive ions of oxygen and nitrogen. The dominant molecular ions in this region are NO^+ and O_2^+ [3]. The F-region, the uppermost portion of the ionosphere observed at altitudes above 140 km, is characterized by an electron density that can exceed 10^6 cm^{-3} at some locations [17]. While the lower part of the F-region disappears at night, the higher region persists throughout the night [3].

7.3 Auroral Emissions Driven by Low-Energy-Electron-Induced Vibrational Excitation

The aurora borealis and aurora australis emissions, displays of shimmering bands and filaments of light in the night sky, result from collisions between atoms/molecules in Earth's upper atmosphere and energetic magnetospheric/solar wind particles mostly consisting of electrons [20]. For example, the most distinctive green color at 557.7 nm in the auroral emissions results from excited oxygen atoms at an altitude of 90–150 km undergoing a forbidden transition from ^1S to ^1D. Calculations suggest that ~30% of this auroral feature is due to direct electron impact [21]. Auroral emissions in the UV region have been attributed to radiative decay following electron impact electronic excitation of molecular nitrogen [22].

Low-energy (1–3 eV) electrons are hypothesized to play an important role in infrared auroral emissions [23]. Previously such emissions were attributed exclusively to chemiluminescence from vibrationally excited NO produced by the reaction of excited O and N atoms. The newly proposed mechanism involves electron capture by NO to form a temporary negative ion (TNI), which subsequently undergoes, within 10–100 fs, autodetachment to form a vibrationally excited NO molecule [23]. Because the formation of the TNI is a resonance process, the vibrational excitation cross section at energies below 5 eV for NO is enhanced, as shown in Fig. 7.2. Statistical equilibrium calculations suggest that over 30% of the infrared auroral emissions can be attributed to photon emission near 5 μm from the low-energy-electron-induced vibrationally excited NO [23]. Complete agreement between calculations and rocket measurements, however, has not yet been achieved [24].

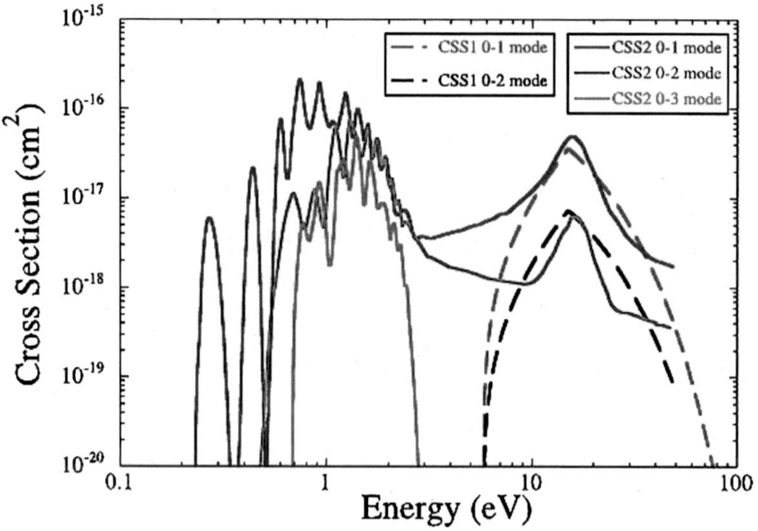

Figure 7.2 Vibrational excitation cross section as a function of incident electron energy. Results that take into account the formation of the TNI are labeled CSS2 and are shown as solid lines. Reproduced from Ref. [23] with permission. Copyright (2004) by John Wiley and Sons.

7.4 Role of Low-Energy Electrons in Triggering Lightning

Electrical breakdown in the atmosphere is a prerequisite for a lightning flash, an electrical discharge between a thundercloud and Earth [25]. It has been long known that the measured electric fields in thunderclouds are about 10 times smaller than those required for electrical breakdown [26]. Although statistical analysis of cosmic ray showers do not support the "relativistic runaway electron avalanche" mechanism initiating lightning [6], results of radio- and gamma-emission measurements conducted during thunderstorms suggest that lightning initiation in Earth's atmosphere is triggered by a "runaway breakdown," which is mediated by an avalanche of low-energy secondary electrons produced by cosmic rays [5]. In addition to cosmic rays, studies indicate that high-speed solar wind particles, consisting mostly of energetic electrons and protons, increase terrestrial lightning rates, which is estimated to be about

100 times per second [27]. This enhanced triggering of lightning by solar wind particles is also likely due to the cascade of low-energy secondary electrons producing runaway breakdowns similar to those produced by cosmic rays. A recent publication, however, notes that "processes leading to the initiation of lightning flashes are still not fully understood" [28].

7.5 Low-Energy-Electron-Induced Production of Negative Ions in the Ionosphere

DEA to ozone (O_3) present in the D-region of the ionosphere yields the anions O^- and $O_2{}^-$ [29].

$$O_3 + e^- \rightarrow [O_3{}^-] \rightarrow O^- + O_2$$

$$O_3 + e^- \rightarrow [O_3{}^-] \rightarrow O_2{}^- + O$$

These oxygen anions react with ozone to yield ozonide, the ozone anion ($O_3{}^-$), which subsequently reacts with CO_2 and NO to produce the three most important anions in the atmosphere: $CO_3{}^-$, NO^-, and $NO_3{}^-$ [30].

7.6 Enhanced Aerosol Nucleation Induced by Low-Energy Electrons?

A preponderance of evidence indicates that global warming is driven by the release into the atmosphere of greenhouse gases such as carbon dioxide during the burning of fossil fuels. Although experiments to simulate aerosol nucleation, conducted in the Cosmics Leaving Outdoor Droplets (CLOUD) chamber at CERN, the European Organization for Nuclear Research, suggest that cosmic rays enhance cloud formation [31], the extent to which this phenomenon may influence global warming remains uncertain [32]. Up to 50% of the global cloud condensation nuclei (CCN) may derive from nucleation of aerosol particles [33]. While changes in the sun's brightness are not thought to contribute more than 10% to global warming in the 20th century [34], the changes in solar activity influence the flux of cosmic rays reaching Earth's surface. A controversial hypothesis, put forward at the turn of the 21st century, suggests a correlation between global temperatures and solar activity [14]. According to this proposition, increased solar activity in the past several decades contributed to a cosmic ray flux reduction, which, in turn, caused an increase in global temperatures because of a decrease in cloud formation [14]. The attachment of cosmic-ray-induced low-energy secondary electrons to molecular oxygen is thought to produce O_2^-, which attracts water molecules (W_n) to yield the $O_2^- \cdot W_n$ complex. This step is posited to enhance the production of sulphuric acid aerosol particles via the following mechanism [35]:

$$O_2^- \cdot W_n + O_3 \rightarrow O_3^- \cdot W_{n-1} + O_2 + W_1$$

$$O_3^- \cdot W_n + SO_2 \rightarrow SO_3^- \cdot W_n + O_2$$

$$O_3^- + SO_2 \rightarrow SO_3^- + O_2$$

$$SO_3^- \cdot W_n \rightarrow H_2SO_4^- \cdot W_{n-1}$$

Although ionizing radiation has been recently shown to produce large (>50 nm) CCN [36], a multiple regression analysis of temperature and galactic cosmic ray (GCR) flux suggests that "there is little empirical evidence that links GCR to the recent global warming" [37]. A recent study found no significant correlation between global albedo, the fraction of the incident light that is reflected by a surface, and cosmic rays [38]. More recent experiments conducted in CLOUD at CERN also suggest that cosmic rays have little influence on the formation of sulphuric acid–dimethylamine clusters thought to be important for cloud formation [39]. Nevertheless, the role of cosmic-ray-derived low-energy-electron-induced aerosol nucleation remains an active area of research [40].

7.7 Chlorofluorocarbon Dissociation by Low-Energy Electrons: Implications for the Ozone Hole?

According to a hypothesis proposed at the turn of the 21st century, low-energy electrons produced by cosmic rays, in addition to UV-Vis photons from the sun, interact with CFCs to produce chlorine atoms that subsequently destroy ozone in the Antarctic. CFCs are man-made chemicals that were commonly used in refrigerants and aerosol sprays, starting in the late 1920s [41]. Most common among the CFCs are CFC-11 ($CFCl_3$) and CFC-12 (CF_2Cl_2). Commercial and domestic use of CFCs led to their release into the atmosphere, where they are a major culprit in the formation of the ozone hole in the stratosphere. The weak carbon–halogen bonding in CFCs allows for facile release of halogen atoms, which subsequently destroy ozone [1]. Due to CFCs' strong impact on stratospheric ozone, the use of CFCs was targeted for elimination by the Montreal Protocol, which called for signing nations to halt the use of several CFCs (including CFC-11 and CFC-12) by 1996 [42]. CFCs, initially replaced by hydrochlorofluorocarbons (HCFCs), have more recently been replaced by hydrofluorocarbons (HFCs), which have much lower ozone depletion potentials than CFCs and HCFCs and contain no chlorine, which dominates ozone processing [43]. However, despite reduction in their use and release to the atmosphere, CFCs remain the leading cause of ozone depletion because of their long lifetimes in the atmosphere (of the order of 50–100 years) [43].

7.7.1 Depletion of Ozone: Photochemical Mechanism

7.7.1.1 Stratospheric reactions

The traditionally accepted explanation for ozone depletion is that solar-produced UV photons (predominantly for wavelengths in the range of 175–220 nm [44]) lead to the dissociation of CFC molecules into a halogen and a free radical. This is the first step in a self-sustaining chain reaction that involves the processing and resulting destruction of O_3 initiated by the released halogen atom.

In 1974, Molina and Rowland proposed the following mechanism for O_3 depletion in the stratosphere, starting from CF_2Cl_2 and $CFCl_3$ molecules [45]:

Production of a free halogen atom:

$$CF_2Cl_2 + h\nu \rightarrow \cdot CF_2Cl + \cdot Cl$$

and

$$CFCl_3 + h\nu \rightarrow \cdot CFCl_2 + \cdot Cl$$

Depletion of ozone:

$$\cdot Cl + O_3 \rightarrow ClO\cdot + O_2$$

Formation of a free halogen atom:

$$ClO\cdot + O \rightarrow \cdot Cl + O_2$$

The final step in this reaction releases the Cl atom to react with another O_3 molecule, which allows for a continuation of the ozone depletion cycle.

While the reactions delineated above show the fundamental pathway for ozone depletion, other reactions also play an important role in determining the concentration of stratospheric ozone. Other molecules present in the stratosphere, such as methane (CH_4) and nitrogen dioxide (NO_2), can, at least temporarily, halt the depletion cycle by incorporating Cl into less reactive molecules such as HCl and $ClONO_2$ (chlorine nitrate) through the following gas-phase reactions [45, 46]:

$$\cdot Cl + CH_4 \rightarrow HCl + \cdot CH_3$$

and

$$ClO\cdot + NO_2 \rightarrow ClONO_2$$

In the absence of additional factors, these termination reactions alone should ensure that CFCs do not lead to the observed dramatic effect on ozone as the reactive •Cl is removed from the ozone depletion chain and incorporated into fairly inert molecules.

7.7.1.2 Polar stratospheric clouds

Ozone depletion undergoes a yearly cycle during which depletion is most pronounced in polar springs (e.g., Antarctica during the September through early November months) [47]. Researchers have tied the cyclical changes to a combination of processes, which start with the Antarctic winter.

With the onset of winter, temperatures within the stratosphere drop, leading to the formation of PSCs. These clouds contain both water ices as well as nitric acid trihydrate (NAT) particles, which consist of water and nitric acid in a molar ratio of 3:1 [48]. The formation of $HNO_3{\cdot}3H_2O$ ices is particularly important as these clouds remove nitrogen from the atmosphere, which can otherwise be used to remove Cl from the ozone depletion chain through the creation of inert compounds.

Within these clouds, heterogeneous reactions, mediated by an additional molecule/particle (M), lead to the release of reactive Cl_2 from the inert HCl and $ClONO_2$ compounds [48–50]:

$$ClONO_2(g) + HCl(s) \xrightarrow{M} HNO_3(s) + Cl_2(g)$$

Laboratory experiments emulating PSC conditions have successfully reproduced this heterogeneous reaction. The experiments also indicate that purely gas-phase reactions without mediating particles (e.g., particles comprising PSCs) are insignificant pathways for Cl_2 formation [51–53].

In addition to the formation of PSCs at the beginning of winter, a polar vortex also forms. A polar vortex is a cyclone that forms due to diminished polar sunlight, leading to stratospheric radiative cooling. High-velocity winds of the order 100 m/s are produced by this polar cooling because of the resulting pressure gradient between the cool polar region and the warmer midlatitude [48]. The Coriolis effect causes the winds to form a vortex around the pole. This polar vortex acts to isolate the atmospheric content inside the vortex from that outside, trapping PSCs, reactants, and products within the vortex core. The Cl_2 molecules released through heterogeneous reactions are contained within the vortex core near the pole.

As the polar spring arrives (beginning of September for Antarctica) and PSCs vanish, solar UV radiation returns to the poles and photodissociates Cl_2, releasing single Cl atoms, which can once again start the ozone depletion cycle for air trapped within the polar vortex. By late spring (around November for Antarctica), the polar vortex begins to disappear as the polar stratosphere begins to warm and the atmospheric air previously contained within the polar vortex is dispersed through the atmosphere. While the ozone is depleted most notably by Earth's poles during winter months, the overall depletion is felt extensively as the initially localized ozone depletion results in extended-latitude ozone thinning with atmospheric mixing after polar vortexes disappear.

Cosmic rays interacting with matter create large numbers of low-energy electrons. This phenomenon occurs within Earth's atmosphere, where it is estimated that cosmic rays create $\sim 4 \times 10^4$ electrons/MeV [54]. After Rowland and Molina's 1974 study pointing to CFC-led depletion of the ozone, the majority of research focused on photodissociation of CFCs. However, studies also began on electron interactions with CFCs.

Gas-phase experiments show that low-energy electrons can release chlorine from CFC-11 and CFC-12. Thus, the first step needed in initiating the ozone depletion cycle can be started by UV photons or low-energy electrons. Specifically, in irradiating $CFCl_3$ and CF_2Cl_2 molecules with low-energy electrons (<10 eV), Illenberger et al. found DEA resonances associated with Cl^- formation at 0 eV for $CFCl_3$ and 0.55 eV for CF_2Cl_2 [55]. The associated DEA cross sections for Cl^- formation were determined to be 9.5×10^{-15} cm^2 and 1.1×10^{-16} cm^2 for $CFCl_3$ and CF_2Cl_2, respectively [55]. These cross-sections are impressively ~2 to 6 orders of magnitude larger than UV absorption cross sections of 10^{-18} cm^2 at 175 nm to 10^{-20} cm^2 at 220 nm for both $CFCl_3$ and CF_2Cl_2 compounds at room temperature [56]. These initial gas-phase studies show that low-energy electrons will readily dissociate CFCs and release Cl^-, which can then shed the excess electron and initiate the ozone depletion cycle. Furthermore, the studies show that CFCs can potentially be dissociated more readily by electrons than by UV photons. However, while cosmic rays produce large numbers of secondary electrons in the stratosphere, the actual density of free electrons ($\sim 10^3$ per cm^3 near 85 km and under 10 per cm^3 below 60 km [57, 58]) is small since neutral molecules such as O_2 readily absorb free electrons to become anions [59].

7.7.1.3 The initial CRE mechanism

In 1999, Lu and Madey studied dissociation yields of Cl^- from low-energy-electron-induced reactions of CF_2Cl_2 coadsorbed with condensed polar and nonpolar gases on a Ru(0001) substrate at 25 K using electron-stimulated desorption ion angular distribution (ESDIAD) measurements [8]. The low-energy electrons in these experiments were produced by bombarding the Ru(0001) substrate with 250 eV electrons. Lu and Madey found that when CF_2Cl_2 is coadsorbed with polar gases (such as H_2O and NH_3), the Cl^- yield is enhanced by factors of $\sim 10^2$ (for H_2O) to $\sim 10^4$ (for NH_3) as compared to when CF_2Cl_2 is coadsorbed with nonpolar gases such Xe, Kr, and Ar [8].

The difference in Cl^- yields between the polar and nonpolar media was initially attributed to solvation of the secondary electrons in the polar media. The solvation process involves the reorienting of polar molecules in response to the presence of a free electron, ultimately leading to the electron being trapped within a Coulomb potential created by surrounding polar molecules. In polar liquids, this process may appear obvious as molecules can easily move and respond to local forces. However, even in polar ices, solvation can also occur [60] where the relatively weak intermolecular dipole–dipole forces will still allow for molecular reorientation in the presence of the electron. Such solvation is not expected for condensed nonpolar media.

Lu and Madey proposed that the solvated electron (e_S^-) in a condensed polar medium, such as water, could then tunnel through the polar media onto CF_2Cl_2, creating a vibrationally excited anion $CF_2Cl_2^{*-}$, which would subsequently dissociate [8]:

Electron solvation:

$$e^- + nH_2O \rightarrow e_S^-(H_2O)_n$$

Dissociation of CF_2Cl_2:

$$e_S^- + CF_2Cl_2 \rightarrow CF_2Cl_2^{*-} \rightarrow Cl^- + \cdot CF_2Cl$$

This last step releases Cl^- and so indicates a potential pathway for electron-initiated CFC dissociation to start the ozone depletion chain. Furthermore, Lu and Madey estimated dissociation cross sections for CF_2Cl_2 co-adsorbed with H_2O to be $\sim 1 \times 10^{-14}$ cm^2 and for NH_3 to be $\sim 6 \times 10^{-12}$ cm^2 [7]. These cross sections are 10^2–10^4 times larger than the gas-phase DEA cross section for CF_2Cl_2.

On the basis of these experiments (showing very large dissociation cross sections for CF_2Cl_2 co-adsorbed with polar media), Lu and Madey suggested a new possible pathway for ozone depletion, which begins with cosmic rays from outer space. Cosmic rays interact with the atmosphere, creating large numbers of low-energy electrons. During the polar winter, these low-energy electrons become solvated within the polar media comprising PSCs. The

proposed mechanism involving solvated electrons but is updated for presolvated electrons (e_p^-):

Electron presolvation:

$$e^- + nH_2O \rightarrow e_p^-(H_2O)_n$$

Dissociation of CF_2Cl_2:

$$e_p^- + CF_2Cl_2 \rightarrow CF_2Cl_2^{*-} \rightarrow Cl^- + CF_2Cl$$

Given the large number of electrons produced by cosmic rays, the large DEA cross sections for CFCs adsorbed on polar media compared to UV gas-phase absorption cross sections, and the link to PSCs that helps to explain the yearly cyclical ozone depletion, the CRE mechanism illustrates a possible additional pathway to ozone depletion besides the traditional photochemical explanation.

7.7.2 Additional Evidence for the CRE Mechanism and Questions

While the fundamental experimental work examining electron-induced reactions of CFCs adsorbed on ices are not of debate, the relative importance of these reactions in ozone depletion compared to the photochemical mechanism are.

7.7.2.1 Additional evidence for the CRE mechanism

If the CRE mechanism is to be a significant contributor to ozone depletion, there should be an observed correlation between atmospheric cosmic ray intensity and ozone levels [54, 59]. Lu and Sanche showed evidence of a latitudinal correlation between cosmic ray intensity and ozone content, indicating that cosmic rays are strongly present at latitudes with high total ozone (Fig. 7.5a). Hence, a potential source for ozone processing, cosmic rays, is present where there are significant levels of ozone that can be subsequently processed. In addition, they showed that, as a function of altitude, ozone loss is maximum near an altitude of 15 km where the cosmic ray ionization rate is largest (Fig. 7.5b).

Finally, they cited UARS satellite data, which tracks CFC concentrations as a function of latitude and time. These measurements indicate CFC concentrations are larger during the polar fall

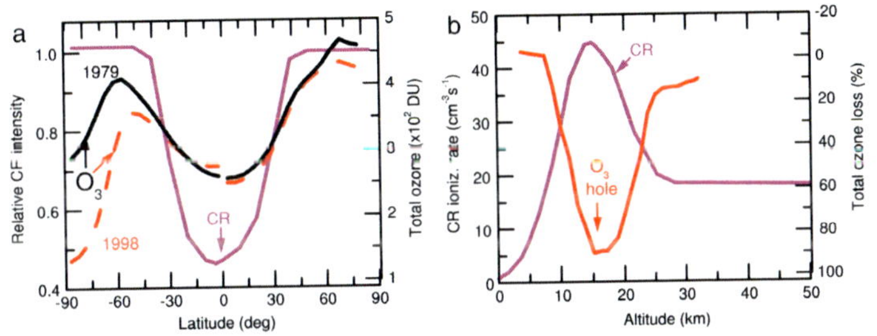

Figure 7.5 (a) Overlay of cosmic ray intensity (violet curve) as a function of latitude with monthly mean ozone levels. The black curve shows ozone levels for the pre-ozone hole period (October 1979 in Antarctica and March 1979 for Arctic), and the red dotted curve shows ozone levels for the ozone hole period (October 1998 in Antarctica and March 1998 for Arctic). (b) Plot of the cosmic ray ionization rate with ozone loss as a function of altitude. The ozone loss represents springtime ozone depletion over Syowa, Antarctica, and was calculated by subtracting the ozone distribution curve for the pre-ozone hole period (1968–1980) from the ozone hole period (1991–1997). Reprinted from Ref. [59], Copyright (2010), with permission from Elsevier.

compared to early polar spring in both the Arctic and Antarctic lower stratosphere (16–20 km) [54]. Lu and Sanche conclude that the diminished CFC concentration by early spring is likely due to cosmic-ray-initiated dissociation of CFCs within PSCs (which are found predominately at lower stratospheric altitudes) during the wintertime when cosmic ray flux is at a maximum and solar radiation is absent [54].

Cosmic ray flux has an 11-year cycle, which is inversely correlated with the solar cycle [64]. At times of high solar activity, cosmic ray particle interactions with solar magnetic fields and material ejected from the sun reduce the number of cosmic rays reaching Earth's atmosphere. In 2001, Lu and Sanche demonstrated an inverse correlation of mean ozone content averaged over latitude with cosmic ray intensity variations that extended just over one 11-year cycle from 1979 to 1992 [54]. This data was more recently extended to 2008 with a continued observed inverse correlation (Fig. 7.6) [65].

The spatial and temporal links between cosmic ray flux and ozone depletion are suggestive of a significant CRE mechanism at work, and Lu contended that understanding cyclical patterns of

10. Lu, Q. B. (2015). *New Theories and Predictions on the Ozone Hole and Climate Change*. World Scientific, New Jersey.
11. Lu, Q.-B. (2013). Cosmic-ray-driven reaction and greenhouse effect of halogenated molecules: culpirts for atmospheric ozone depletion and global climate change. *Int. J. Mod. Phys. B*, **27**(17), p. 1350073.
12. Grooß, J.-U. and Müller, R. (2014). Comment on—Cosmic-ray-driven reaction and greenhouse effect of halogenated molecules: culprits for atmospheric ozone depletion and global climate change. *Int. J. Mod. Phys. B*, **28**(13), p. 1482001.
13. Rajasekhar, B. and Plane, J. M. C. (1993). An *ab initio* study of dissociative electron attachment to $NaHCO_3$ and $NaCO_3$, and the role of these reactions in the formation of sudden sodium layers. *Geophys. Res. Lett.*, **20**(1), pp. 21–24.
14. Marsh, N. D. and Svensmark, H. (2000). Low cloud properties influenced by cosmic rays. *Phys. Rev. Lett.*, **85**(23), pp. 5004–5007.
15. Vuitton, V., et al. (2009). Negative ion chemistry in Titan's upper atmosphere. *Planet. Space Sci.*, **57**(13), pp. 1558–1572.
16. Atreya, S. K., et al. (2006). Oxidant enhancement in martian dust devils and storms: Implications for life and habitability. *Astrobiology*, **6**(3), pp. 439–450.
17. Lutgens, F. K. and Tarbuck, E. J. (2013). *The Atmosphere: An Introduction to Meteorology*, 12th ed., Pearson, Boston, xxiii, 506 p.
18. Hanson, W. B. and Mantas, G. P. (1988). Viking electron temperature measurements: evidence for a magnetic field in the Martian ionosphere. *J. Geophys. Res.*, **93**(A7), p. 7538.
19. Kelley, M. (2009). *The Earth's Ionosphere*, Vol. 96, 2nd Ed., Academic Press.
20. Bone, N. (2007). *Aurora: Observing and Recording Nature's Spectacular Light Show*. Springer, New York.
21. Jones, D. B., et al. (2006). Electron-driven excitation of O2 under night-time auroral conditions: excited state densities and band emissions. *Planet. Space Sci.*, **54**(1), pp. 45–59.
22. Eastes, R. W. and Sharp, W. E. (1987). Rocket-borne spectroscopic measurements in the ultraviolet aurora: the Lyman-Birge-Hopfield bands. *J. Geophys. Res. A: Space Phys.*, **92**(A9), pp. 10095–10100.
23. Campbell, L. (2004). Infrared auroral emissions driven by resonant electron impact excitation of NO molecules. *Geophys. Res. Lett.*, **31**(10).
24. Campbell, L. and Brunger, M. J. (2007). Electron impact contribution to infrared NO emissions in auroral conditions. *Geophys. Res. Lett.*, **34**(22).

25. Verronen, P. T., et al. (2005). Diurnal variation of ozone depletion during the October-November 2003 solar proton events. *J. Geophys. Res. A: Space Phys.*, **110**(A9).

26. Gurevich, A. and Moskvitch, K. (2013). The lightning enigma. *New Sci.*, **219**(2929), pp. 28–29.

27. Scott, C. J., et al. (2014). Evidence for solar wind modulation of lightning. *Environ. Res. Lett.*, **9**(5), p. 055004.

28. Kolmašová, I., et al. (2016). Subionospheric propagation and peak currents of preliminary breakdown pulses before negative cloud-to-ground lightning discharges. *Geophys. Res. Lett.*, **43**(3), pp. 1382–1391.

29. Matejcik, S., et al. (1997). Electron attachment to molecules and clusters of atmospheric relevance: oxygen and ozone. *Plasma Sources Sci. Technol.*, **6**(2), pp. 140–146.

30. Fehsenfe, F. C., et al. (1967). Laboratory measurements of negative ion reactions of atmospheric interest. *Planet. Space Sci.*, **15**(2), pp. 373–379.

31. Kirkby, J., et al. (2011). Role of sulphuric acid, ammonia and galactic cosmic rays in atmospheric aerosol nucleation. *Nature*, **476**(7361), pp. 429–433.

32. Solanki, S. K., Krivova, N. A., and Haigh, J. D. (2013). Solar irradiance variability and climate, in *Annual Review of Astronomy and Astrophysics*, Vol. 51, Faber, S. M. and VanDishoeck, E., eds., Annual Reviews, Palo Alto, pp. 311–351.

33. Merikanto, J., et al. (2009). Impact of nucleation on global CCN. *Atmos. Chem. Phys.*, **9**(21), pp. 8601–8616.

34. Sloan, T. and Wolfendale, A. W. (2013). Cosmic rays, solar activity and the climate. *Environ. Res. Lett.*, **8**(4).

35. Svensmark, H., et al. (2007). Experimental evidence for the role of ions in particle nucleation under atmospheric conditions. *Proc. R. Soc. A: Math. Phys. Eng. Sci.*, **463**(2078), pp. 385–396.

36. Svensmark, H., Enghoff, M. B., and Pedersen, J. O. P. (2013). Response of cloud condensation nuclei (>50 nm) to changes in ion-nucleation. *Phys. Lett. A*, **377**(37), pp. 2343–2347.

37. Benestad, R. E. (2013). Are there persistent physical atmospheric responses to galactic cosmic rays? *Environ. Res. Lett.*, **8**(3), p. 035049.

38. Krissansen-Totton, J. and Davies, R. (2013). Investigation of cosmic ray-cloud connections using MISR. *Geophys. Res. Lett.*, **40**(19), pp. 5240–5245.

39. Almeida, J., et al. (2013). Molecular understanding of sulphuric acid-amine particle nucleation in the atmosphere. *Nature*, **502**(7471), pp. 359–363.
40. Dunne, E. M., et al. (2016). Global atmospheric particle formation from CERN CLOUD measurements. *Science*, **354**(6316), pp. 1119–1124.
41. Elkins, J. W. (1999). Chlorofluorocarbons (CFCs), in *The Chapman & Hall Encyclopedia of Environmental Science*, Fairbridge, R. W. and Alexander, D. E., eds., Kluwer Academic, Boston, MA, pp. 78–80.
42. Secretariat, O. (2006). *Handbook for the Montreal Protocol on Substances that Deplete the Ozone Layer*, 7th ed., United Nations Environment Programme, Nairobi, Kenya.
43. World Meteorological Organization (WMO) (2010). Scientific assessment of ozone depletion: 2010, Geneva, Switzerland, p. 517.
44. Doucet, J., Sauvageau, P., and Sandorfy, C. (1973). Vacuum ultraviolet and photoelectron spectra of fluoro-chloro derivatives of methane. *J. Chem. Phys.*, **58**(9), pp. 3708–3716.
45. Molina, M. J. and Rowland, F. S. (1974). Stratospheric sink for chlorofluoromethanes: chlorine atom-catalysed destruction of ozone. *Nature*, **249**(5460), pp. 810–812.
46. Toon, O. B. and Turco, R. P. (1991). Polar stratospheric clouds and ozone depletion. *Sci. Am.*, **264**, pp. 68–74.
47. National Oceanic & Atmospheric Administration (2012). South Pole total column ozone. Available from: http://www.esrl.noaa.gov/gmd/dv/spo_oz/.
48. Schoeberl, M. R. and Hartmann, D. (1991). The dynamics of the stratospheric polar vortex and its relation to springtime ozone depletions. *Science*, **251**, pp. 46–52.
49. Solomon, S. (1990). Progress towards a quantitative understanding of Antarctic ozone depletion. *Nature*, **347**, pp. 347–254.
50. Anderson, J. G., Toohey, D. W., and Brune, W. H. (1991). Free radicals within the Antarctic vortex: the role of CFCs in Antarctic ozone loss. *Science*, **251**, pp. 39–46.
51. Tolbert, M. A., et al. (1987). Reaction of chlorine nitrate with hydrogen chloride and water at Antarctic stratospheric temperatures. *Science*, **238**(4831), pp. 1258–1260.
52. Molina, M. J., et al. (1987). Antarctic stratospheric chemistry of chlorine nitrate, hydrogen chloride, and ice: release of active chlorine. *Science*, **238**(4831), pp. 1253–1257.

53. Leu, M.-T. (1988). Laboratory studies of sticking coefficients and heterogeneous reactions important in the Antarctic stratosphere. *Geophys. Res. Lett.*, **15**(1), pp. 17–20.

54. Lu, Q. B. and Sanche, L. (2001). Large enhancement in dissociative electron attachment to HCl adsorbed on H2O ice via transfer of presolvated electrons. *J. Chem. Phys.*, **115**(13), pp. 5711–5713.

55. Illenberger, E., Scheunemann, H.-U., and Baumgartel, H. (1979). Negative ion formation in CF_2Cl_2, CF_3Cl and $CFCl_3$ following low-energy (0–10eV) impact with near monoenergetic electrons. *Chem. Phys.*, **37**, pp. 21–31.

56. SPARC (2013). Lifetimes of Stratospheric Ozone-Depleting Substances, Their Replacements, and Related Species, Ko, M. K. W., Newman, P. A., Reimann, S., and Strahan, S. E., eds., SPARC Report No. 6, World Climate Research Program: Core Project of the WMO/ICSU/IOC.

57. Smith, D. and Adams, N. (1980). Elementary plasma reactions of environmental interest, in *Plasma Chemistry I*, Vepřek, S. and Venugopalan, M., eds., Springer Berlin Heidelberg, pp. 1–43.

58. Wayne, R. P. (1985). *Chemistry of Atmospheres*. Clarendon Press, Oxford.

59. Lu, Q.-B. (2010). Cosmic-ray-driven electron-induced reactions of halogenated molecules adsorbed on ice surfaces: Implications for atmospheric ozone depletion and global climate change. *Phys. Rep.*, **487**, pp. 141–167.

60. Eiben, K. and Taub, I. A. (1967). Solvated electron spectrum in irradiated ice. *Nature*, **216**(5117), pp. 782–783.

61. Lu, Q.-B. and Madey, T. E. (1999). Giant enhancement of electron-induced dissociation of chlorofluorocarbons coadsorbed with water or ammonia ices: Implications for atmospheric ozone depletion. *J. Chem. Phys.*, **111**(7), pp. 2861–2864.

62. Lu, Q. B. and Sanche, L. (2001). Enhanced dissociative electron attachment to CF_2Cl_2 by transfer of electrons in precursors to the solvated state in water and ammonia ice. *Phys. Rev. B*, **63**(15), p. 153403.

63. Lu, Q. B. and Sanche, L. (2003). Condensed-phase effects on absolute cross sections for dissociative electron attachment to CFCs and HCFCs adsorbed on Kr. *J. Chem. Phys.*, **119**(5), pp. 2658–2662.

64. Svensmark, H. and Friis-Christensen, E. (1997). Variation of cosmic ray flux and global cloud coverage: a missing link in solar-climate relationships. *J. Atmos. Sol. Terr. Phys.*, **59**(11), pp. 1225–1232.

65. Lu, Q. B. (2009). Correlation between cosmic rays and ozone depletion. *Phys. Rev. Lett.*, **102**(11), p. 118501.

66. Lu, Q.-B. (2010). Lu replies. *Phys. Rev. Lett.*, **105**(16), p. 169802.
67. Harris, N. R. P., Farman, J. C., and Fahey, D. W. (2002). Comment on—Effects of cosmic rays on atmospheric chlorofluorocarbon dissociation and ozone depletion. *Phys. Rev. Lett.*, **89**(21), p. 219801.
68. Patra, P. K. and Santhanam, M. S. (2002). Comment on—Effects of cosmic rays on atmospheric chlorofluorocarbon dissociation and ozone depletion. *Phys. Rev. Lett.*, **89**(21), p. 219803.
69. Müller, R. (2003). Impact of cosmic rays on stratospheric chlorine chemistry and ozone depletion. *Phys. Rev. Lett.*, **91**(5), p. 058502.
70. Müller, R. and Grooß, J.-U. (2009). Does cosmic-ray-induced heterogeneous chemistry influence stratospheric polar ozone loss? *Phys. Rev. Lett.*, **103**(22), p. 228501.
71. Alvarez-Madrigal, M. (2010). Comment on—Correlation between cosmic rays and ozone depletion. *Phys. Rev. Lett.*, **105**(16), p. 169801.
72. Coates, A. J. (2009). Interaction of Titan's ionosphere with Saturn's magnetosphere. *Philos. Trans. R. Soc. A: Math. Phys. Eng. Sci.*, **367**(1889), pp. 773–788.
73. Lavvas, P., et al. (2011). Energy deposition and primary chemical products in Titan's upper atmosphere. *Icarus*, **213**(1), pp. 233–251.
74. Coates, A. J., et al. (2007). Discovery of heavy negative ions in Titan's ionosphere. *Geophys. Res. Lett.*, **34**(22).
75. Hecht, M. H., et al. (2009). Detection of perchlorate and the soluble chemistry of martian soil at the Phoenix lander site. *Science*, **325**(5936), pp. 64–67.
76. Campbell, L., Allan, M., and Brunger, M. J. (2011). Electron impact vibrational excitation of carbon monoxide in the upper atmospheres of Mars and Venus. *J. Geophys. Res. A: Space Phys.*, **116**(A9), p. A09321.
77. Campbell, L. and Brunger, M. J. (2008). Electron cooling by carbon monoxide in the atmospheres of Mars and Venus. *PMC Phys. B*, **1**, p. 10.
78. Campbell, L., Brunger, M. J., and Rescigno, T. N. (2008). Carbon dioxide electron cooling rates in the atmospheres of Mars and Venus. *J. Geophys. Res. E: Planets*, **113**(E8), p. E08008.

Chapter 8

Electrons in Space: The Final (and First) Frontier

Nigel J. Mason[a,b] **and David Field**[c]

[a]*School of Physical Sciences, The Open University, Walton Hall, Milton Keynes, Bedfordshire, MK7 6AA, UK*
[b]*School of Physical Sciences, Ingram Building, University of Kent, Canterbury CT2 7NH, UK*
[c]*Department of Physics and Astronomy, Ny Munkegade 120 Building 1522, 8000 Aarhus C, Denmark*
n.j.mason@kent.ac.uk, dfield@phys.au.dk

In this chapter the role of low-energy electrons in our studies of space is reviewed. Electron collisions were amongst the earliest collision processes to occur after the Big Bang and played a crucial role in the development of the universe. Space and the stars within it are plasmas, and thus electrons are inherent in the physical and chemical processes that pervade them. Recently, rich (astro) chemistry prevalent in the universe has been discovered and the role of electron-induced collisions revealed; indeed electron collisions may underpin the formation of stars. To see the role of electrons in extraterrestrial phenomena we only have to look in our own solar system, with electrons driving the aurora and day glow of other planets and fashioning the chemistry on Titan, a macrocosm of our own planet's evolution, while electron-induced phenomena on comets may allow us to explore these strange new worlds.

Low-Energy Electrons: Fundamentals and Applications
Edited by Oddur Ingólfsson

ISBN 978-981-4800-00-6 (Hardback), 978-0-429-05882-0 (eBook)
www.panstanford.com

8.1 Introduction

Throughout this book it has been stated that electrons are ubiquitous, playing a role in many natural and technological processes, with many examples of electron-induced chemistry. In this chapter we will explore the full extent of the ubiquitous electron, revealing the role of electron collisions in the earliest history of the universe, their role in star and planet formation, and whether they may play a major role in the synthesis of molecules in the interstellar medium (ISM) that are the precursors of life. We will give examples of electron collisions in our own and other planetary atmospheres, leading to observations of aurorae. We will also discuss some recent results from Rosetta, the first mission to study, in situ, a comet, which reveal an unexpected role of electron collisions in local physical and chemical processes.

8.1.1 Electrons and the Big Bang

In the beginning, just one second after the Big Bang, the universe was filled with neutrons, protons, electrons, positrons, photons, and neutrinos. Within three minutes the first elements were formed during a process known as Big Bang nucleosynthesis, with protons and neutrons colliding to make deuterium, most of which combined to make helium, and trace amounts of lithium and beryllium (Fig. 8.1). Hydrogen and helium atoms then began to form as the density of the universe fell. This is thought to have occurred some 400,000 years after the Big Bang. Initially hydrogen and helium were ionized, but as the universe cooled down, electrons were captured by the ions in a process called electron recombination [1]. At the end of recombination, most of the protons in the universe were bound up in neutral atoms. Before such recombination occurred most of the photons (light) in the universe were interacting with electrons (and protons) through Thomson scattering, as the mean free path each photon could travel before encountering an electron was very short. Therefore, the universe was opaque, but once most of the protons in the universe were bound, the photons' mean free path became effectively infinite and the photons could now travel freely across the universe, which had now become transparent. This is

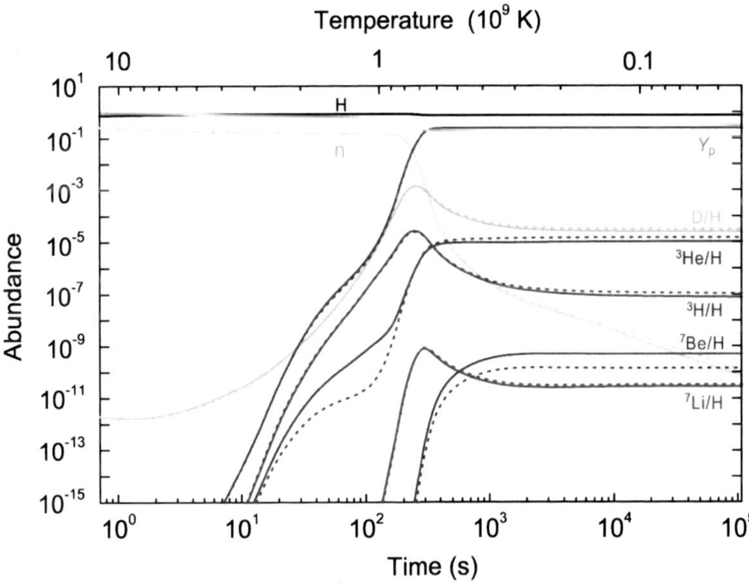

Figure 8.1 Nucleosynthesis immediately after the Big Bang, showing the concentration of different atomic species as a function of time after the Big Bang [2]. Courtesy: Prof. Jianjun He.

usually referred to as *decoupling*. The photons present at the time of decoupling are the photons in the cosmic microwave background (CMB), after being greatly cooled by the expansion of the universe.

Thus electron collisions are amongst the first and most important collisions in the universe. Indeed the earliest electron collision process is electron–proton and electron–He_2^+ recombination; before recombination, all of the electrons were free; after recombination, the electrons and protons were mostly bound to form atoms.

If the $p + e \rightarrow H(n) + h\upsilon$ collision takes place directly at the ground state ($n = 1$) of hydrogen, the emitted photon with $E = h\upsilon = 13.6$ eV can readily re-ionize a hydrogen atom, with no net effect. Radiative recombination to a higher energy level with subsequent decay to the ground state therefore does not achieve a lower ionization fraction, the emitted photons being quickly re-absorbed by H atoms, which are then promptly re-ionized, given the high density of photons with $E < 13.6$ eV. Therefore the process that lowers the ionization fraction (and hence makes the universe transparent) is two-photon emission from the 2S to the 1S ground state. This transition is highly forbidden, with a transition probability $<10^8$ times lower than the Lyman α line from 2P to 1S. To conserve energy and angular momentum, a pair of photons is emitted, neither of which is energetic enough to excite an atom from the ground state.

Evaluations of the electron–proton recombination cross sections have been extensive, with both complete and semi-analytical results being widely reported and reviewed for each final state of the product H atom (H(n)) from 0.1 to 10^4 eV [3, 4]. Similarly for electron–He^{2+}/He^+ interactions concise and compete data sets have been recorded and evaluated for

$$He^{2+} + e \rightarrow He^+ + h\upsilon,$$

$$He^+ + e \rightarrow He + h\upsilon.$$

A small amount of lithium is also created in the early stages of the universe via p + ^4He $\rightarrow$ ^5Li + $h\upsilon$, which requires electron interactions with Li cations as an initial step.

8.1.2 The Birth of the Molecular Universe

The first molecules to form in the universe are believed to be $D_2{}^+$, $H_3{}^+$, H_2D^+, D_2H^+, $D_3{}^+$, HeH^+, HeD^+, $He_2{}^+$, LiH^+, LiD^+, LiD, and LiH^+, most of which were formed in ion molecule collisions, which are barrier-less with rate constants that are often increasing at lower temperatures [4]. The first electron–molecule collisions were therefore once again recombination reactions such as

$$H_3{}^+ + e^- \rightarrow H + H + H \rightarrow H_2 + H$$

and

$$e^- + HeH^+ \rightarrow He + H.$$

Such dissociative recombination reactions have been the subject of considerable debate within the atomic and molecular community since experiments and theoretical estimates of the cross sections (rate constants) have revealed orders-of-magnitude differences [1]. Experiments performed in ion storage rings have, in part, resolved

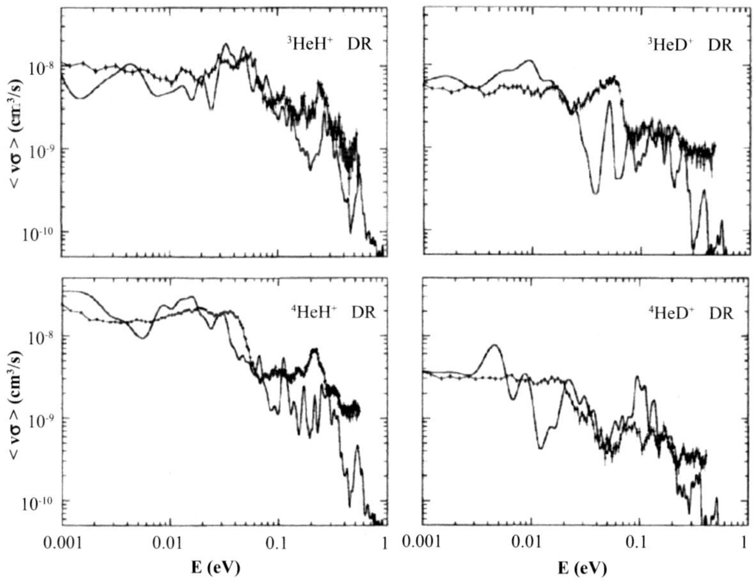

Figure 8.2 Comparison of the relative experimental differential recombination (DR) rate coefficients (dots) of helium-hydride isotope ions with the corresponding theoretical calculations. From Ref. [5].

such differences by revealing a strong dependence on the internal (ro-vibrational) energy of the target ion (Fig. 8.2). By cooling the ions in storage rings before the ions collided with electrons, the experimental data was found to be in better agreement with theoretical data, which assumed the target to be in its lowest state of excitation [6].

An additional electron process leading to the first anions in the universe is electron attachment

$$e^- + H \rightarrow H^- + h\upsilon,$$

which provided a route to the formation of molecular hydrogen

$$H^- + H \rightarrow H_2 + e^-$$

by associative detachment. Indeed this is the most probable route by which early H_2 was formed in the universe before the first stars and the subsequent production of dust, the surface of which is now believed to be the major source for molecular hydrogen production (see later).

Photodetachment of H^- plays a key role in the opacity of stars, a process first suggested by Wildt in 1939 [7]. Photodetachment is possible by virtue of energy conservation, a minimum-energy photon $h\upsilon$, equal to the electron affinity, is required to detach an electron from the negative ion obeying the law of conservation of energy, that is, $h\upsilon = E + \hbar^2 k^2/2m$, where $\hbar k$ is the linear momentum of the free electron and E is its detached energy. Once again semi-analytical formulae describing the cross section have been derived and provide a good fit to the (sparse) experimental data. This is in contrast to many other collisions involving anions (e.g., dissociative electron attachment [DEA], which plays an important role in planetary atmospheres; see later).

8.2 Electrons in the Interstellar Medium

It is now necessary to explore our current universe and review where the free electrons may be found that may influence its physics and chemistry. There are a great variety of environments in space, ranging from the fully ionized intergalactic medium with number densities of <1 particle m^{-3} and temperatures of several million Kelvin, through dense regions close to the surface of evolved stars with densities in excess of 10^{19} m^{-3}, weakly ionized, to regions of ionized hydrogen atoms around massive stars, so-called HII regions. In this chapter we will focus on those parts of the ISM where stars (and planets) may form in the future, that is, molecular clouds. There are two fundamental quantities that are important: the first is the electron energy and the second their abundance.

8.2.1 Electron Collisions and Energies in Molecular Clouds: The Gas Phase

The effect of electrons on any target depends very largely on their kinetic energy. For example, in addition to inelastic collisions causing the electronic excitation of atoms and the rotational, vibrational, and electronic excitation of molecules, electrons can cause chemistry. Electrons can collide with molecules or attach to molecules, causing them to break apart, either to form a negative ion and a neutral or to form two neutrals and a scattered electron:

$$e + AB \rightarrow AB^{-} \rightarrow A \quad + B \tag{8.1}$$

$$e + AB \rightarrow AB^{-} \rightarrow A + B + e \tag{8.2}$$

Equation 8.1 is called DEA and Eq. 8.2, electron impact dissociation. The cross section for DEA generally proceeds through a series of resonances, that is, temporary formation of the AB^{-} anion with lifetimes between picoseconds and femtoseconds. The likelihood of DEA depends critically on the electron energy. The second simple dissociative process turns out to be very difficult to study in the laboratory but is thought to proceed with cross sections of perhaps 1 $Å^2$ at most. In truth these two processes of DEA and ionizing dissociation are only mentioned here for completeness. Except for H_2 they probably play a negligible role, at least in the gas phase. This is discussed in more detail later. There is, however, a class of interactions that takes place close to zero energy to form a parent negative ion, which may be stabilized by infrared (IR) photon emission. This is very low energy attachment, within a few millielectron volt collision energy, in which virtual states may provide a gateway to attachment [8]. The importance of this process has new significance following the identification of negative molecular ions in space, the discovery of which is described in detail in an excellent recent review [9].

A further fundamentally important process is the reaction of electrons with positive ions to form neutral species. Indeed this is generally the final step in molecule formation in the ISM in the gas phase. Thus CO, the most common molecular species in the universe apart from H_2, is formed by dissociative recombination of electrons with HCO^{+}. This occurs with rate coefficients which may be typically 10^{-7} cm^3 s^{-1}, 2 orders of magnitude faster than ion–molecule reactions.

To determine the importance of particular electron processes it is necessary to know the electron energies in molecular clouds and the timescales on which their energies are achieved. For example, electrons in the ISM are cooled from their formation energy to the temperature of the ISM, 10 to 20K, on a timescale that is very rapid compared to any process of DEA. This is due to the fact that for DEA we need collisions between two particles, electrons and molecules, both of which are of very low abundance (see later). The exception is of course H_2, which is the most abundant molecule.

Consider a molecular cloud of density $n(H_2) > \sim 10^4$ cm^{-3}, in which the column density (density × scale) is sufficient to allow electrons to undergo many collisions. For example, the column density may be $>10^{22}$ cm^{-2}, given that a cloud is generally >1 pc (3×10^{18} cm) in scale. One may note that the latter figure of 1 pc would correspond to a cloud of $\sim$2000 solar masses. Dalgarno et al. have described in detail the slowing of electrons from initial energies of 1 keV down to 30 eV and below in the environment of a molecular cloud [10]. Electrons are formed by ionization of H_2 by cosmic rays, that is, mainly by high-energy protons. The initial average energy of electrons is $\sim$20 eV but can be up to 1 keV. Loss of energy by electronic excitation and ionization of He, H, and H_2 takes place with a combined cross section of ~ 1 Å^2. Evidently, given this cross section, these processes will occur within a radius of formation of the electron of a few times 10^{16} cm, that is, $\sim$0.01 pc, much smaller than the spatial extent of the clouds (see Fig. 8.3) in which these electrons are produced.

How do such 20 eV electrons, in their bath of gas and dust at 10 K, lose their energy in a molecular cloud? When electrons ionize H_2 they lose 15.4 eV. When they excite H_2 electronically, the minimum energy lost is $\sim$10 eV. As soon as the electron energy falls below this value, this channel becomes closed. At lower energies, DEA, vibrationally inelastic and rotationally inelastic collisions may lower the electron energy down to 60 meV (450 K), the threshold for $J = 0$–2 excitation. The formation of H^- by DEA is a weak, resonant process occurring between 3 and 15 eV with cross sections of only $\sim 10^{-4}$ Å^2 [11]. Vibrationally inelastic collisions $e^- + H_2$ (v = 0) $\rightarrow e^- + H_2$ (v = 1) occur with a loss of 0.555 eV and rotationally inelastic collisions $e^- + H_2$ ($J = 0$) $\rightarrow e^- + H_2$ ($J = 2$) with a loss of 60 meV.

Elastic collisions involve much less loss of energy per collision, but they are nevertheless very significant [12]. After n collisions, in a bath of cold H_2, the final/initial electron velocity, $\mathbf{v}_f/\mathbf{v}_i$ =

Figure 8.3 The molecular cloud, or "Bok globule," Barnard 68 of diameter 0.36 pc [13]. It is opaque in this image in visible light due to the presence of dust in the cloud (https://creativecommons.org/licenses/by/4.0/deed.en).

$(0.9994555)^n$. The energy change per collision is $1.088 \times 10^{-3} \times$ the electron energy in electron volts. Thus after 6340 collisions, the electron energy will be reduced from 10 eV to 0.1 eV, purely from elastic collisions. The elastic collision cross section has been measured to be typically 9 Å^2 at low energy around 100 meV, and using this figure we may estimate how quickly electrons are cooled by elastic collisions. The effective rate coefficient is that equivalent to 9 Å^2 and is therefore $\sim 1.7 \times 10^{-8}$ cm^3 s^{-1} using $k^{2\text{nd}}$ = Cross section × Electron velocity, at 100 meV. The time between collisions is therefore $5.9 \times 10^7/n(\text{H}_2)$ seconds. Thus in a cloud of density $n(\text{H}_2) = 10^4$ cm^{-3}, the electron energy is reduced form 10 eV to 0.1 eV in ~14 months, *purely by elastic collisions.*

Now compare the efficiency of rotationally inelastic collisions with elastic collisions in cooling electrons. For rotationally inelastic collisions at 1 eV, the effective rate coefficient = 5.4×10^{-9} cm^3 s^{-1}

using a cross section of 1 Å^2 [12]. The time between rotationally inelastic collisions therefore becomes $1/(5.4 \times 10^{-9} \times 10^4) = 1.85 \times 10^4$ s and therefore electrons lose 60 meV every 5 hours. The time between elastic collisions is $5.9 \times 10^7/n(H_2)$ s = 5900 s. Thus at 100 meV, an electron loses 0.1 meV every 1.6 hours, equivalent to 960 hours for 60 meV. Thus elastic collisions turn out to be 190 times less efficient at 1 eV than rotationally inelastic, but below 100 meV elastic collisions become more efficient in cooling electrons.

How do electrons further lose their kinetic energy and become thermalized to the ambient kinetic temperature of 1 meV? Are there any other mechanisms operative at very low energy, apart from the elastic collisions discussed earlier? The answer is yes there are, but they are of a very different nature. For electron energies <60 meV, another process may become important: rotationally inelastic collisions with polar molecules. The long-range electron–dipole attraction in such encounters leads to very high rotationally inelastic cross sections for the process e^- (cold) + molecule (J_{initial}) → e^- (colder) + molecule (J_{final}). Because of its high dipole moment, the molecule of most importance in this context is H_2O, so long as it is sufficiently abundant (see later). At a crude level of approximation, the cross sections for H_2O will be (dipole moment of H_2O/dipole moment of CO)^2 or ~250 times larger than for CO.

We now compare the electron energy transfer rate with H_2 and H_2O. Cross sections for rotationally inelastic collisions of electrons and H_2O may be extracted from experimental data [14]. Consider a 20 meV electron (240 K). For elastic collisions with H_2, the energy lost by the colliding electron, $\Delta E = 0.02$ meV, with a cross section of 7.5 Å^2, relevant to somewhat lower energies than 100 meV [12]. For a rotationally inelastic collision with H_2O, $\Delta E = 5$ meV with a cross-section = 1530 Å^2. Thus the relative efficiency of elastic versus rotationally inelastic collisions with H_2O at 20 meV is given by $7.5 \times 0.02/1530 \times 5 = 2 \times 10^{-5}$ and therefore H_2O is important in cooling electrons if the total population is $\geq 2 \times 10^{-5}$ H_2. Such large abundances are not encountered in quiescent molecular clouds but are relevant in situations in which a protostellar core warms up and evaporates ice mantles on grains (see later) and under the conditions around massive stars, where shocks evaporate ice mantles.

An important conclusion at this stage is that free electrons in the ISM are cooled rapidly to the ambient kinetic temperature before they can do anything of chemical interest, such as DEA or ice processing. To make the former clear, consider once more the case of DEA. To give DEA every chance of being active, let us say that we cool all electrons to 3 eV, which we may take as a typical energy at which DEA may be effective, and let us give DEA a cross section of 1 $Å^2$. This yields an effective rate coefficient k^{DA} of $\sim 10^{-8}$ cm^3 s^{-1}. If we assume that the target, M, for DEA is considerably more abundant than the electrons, that is the target is, say, 10^{-7} of the total number density of $H_2 = 10^4$ cm^{-3}, then $[M] = 10^{-3}$ cm^{-3}. The effective first-order collision rate of electrons with M, k^{DA} [M] resulting in a DEA process, becomes $10^{-8} \times 10^{-3} = 10^{-11}$ s^{-1}. This implies that every 10^{11} s, or $\sim$3200 years, a DEA process can take place. However, we have already seen above that in a cloud of density $n(H_2) = 10^4$ cm^{-3}, the electron energy is reduced from 10 eV to 0.1 eV in $\sim$14 months, purely by elastic collisions. Thus it is clear that DEA cannot compete. Active electrons, that is, any at the requisite energy for DEA of a few electron volts necessary for astrophysically interesting species, are whisked away to lower energy before they can contribute any significant chemical change. Nevertheless, low-energy electrons may, or even must, play a role in negative ion formation, as pointed out in Ref. [9], but what that role is remains unclear. In this connection, we note that negative ions can acquire a larger abundance than that of free electrons in some astronomical objects [9].

8.2.2 Electron Collisions and Energies in Molecular Clouds: The Dusty Universe

Once the first stars had been formed and died, a new ingredient became important in the interstellar makeup. This ingredient is particles of solid material, commonly known as dust. This dust is composed typically of iron magnesium silicates plus some graphitic component. The size of dust grains follows a lower law, such that their number is proportional to $a^{-3.5}$ where a is the grain radius. Observational data show that the properties of grains may be satisfactorily described using an average value of a, $<a> \sim 0.1$ μm.

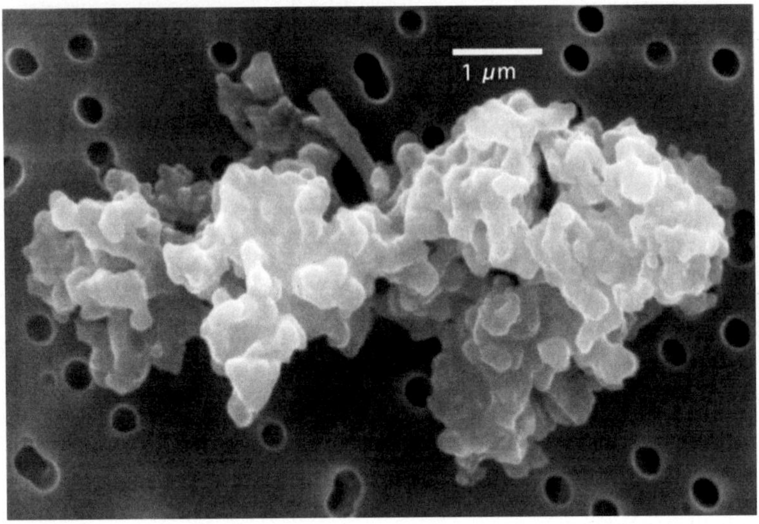

Figure 8.4 Interstellar dust grain. Note the irregular shape and size. Image credit: Donald E. Brownlee, University of Washington, Seattle, USA, and Elmar Jessberger, Institut für Planetologie, Münster, Germany (https://creativecommons.org/licenses/by/1.0/deed.en).

These grains may gradually accrete to form the primordial dust (Fig. 8.4), which is the building block of planets.

The cold surface of such dust grains allows H atoms to recombine to form molecular hydrogen. They also accrete other molecules made in the gas phase of the ISM to form an icy mantle upon and within which it is believed that much of the ISM chemistry occurs, yielding the rich array of nearly 200 molecular species that have now been reported in molecular clouds [15].

Dust grains are typically covered with a mantle of H_2O and CO ice. There will also be trace quantities of methanol and other ices, which may also be significant for subsequent electron-induced chemistry [16]. Such ice-covered grains are bombarded with ultraviolet (UV) light, X-rays, and cosmic rays that can penetrate the cloud (Fig. 8.5). We concentrate here upon electrons and how often they may hit dust grains. We also ask, what takes place when electrons encounter amorphous, probably porous, water ice at (say) 10 K?

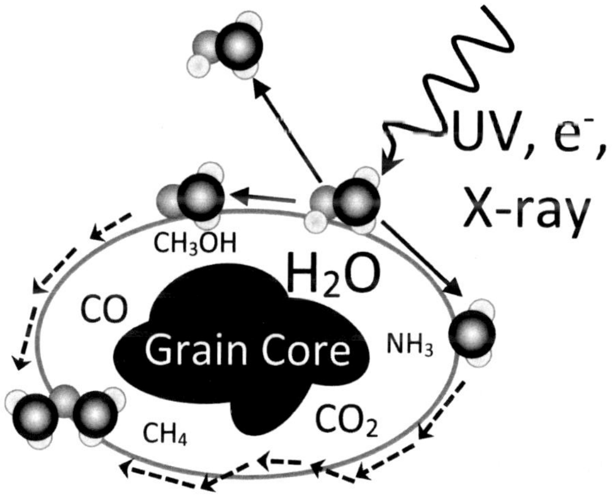

Figure 8.5 Schematic diagram of particle interactions with ice-covered dust grains in the interstellar medium, which can lead to molecular synthesis.

Electrons hit dust grains at the same rate as positive ions hit dust grains, satisfying the condition of zero net current, essentially as in the case of a floating potential associated with an insulator in a plasma. Using the number of grains $\sim 1.3 \times 10^{-12}\ n(\mathrm{H})\ \mathrm{cm}^{-3}$, grain size, $<a> = 0.1\ \mu\mathrm{m}$, the collision rate is $1.1 \times 10^{-17}\ n(\mathrm{H})\ \mathrm{s}^{-1}$ for 10 K H_3^+ ions, where n_H is the number of protons, that is $2n(H_2)$. This must also be the effective collision rate for electrons. Therefore the rate of electron collisions is once every 250,000 years in a typical dark cloud at $10^4\ H_2\ \mathrm{cm}^{-3}$. Because of the greater velocity of electrons and the requirement for zero net current to flow, grains therefore gain a negative charge, typically to the extent of $\sim$1 electron per grain. Thus in a star-forming region, with $n(\mathrm{H}) = 10^5\ \mathrm{cm}^{-3}$ an electron hits a grain every 25,000 years, and in such a region every 25,000 years there is the chance that an electron can do some interesting chemistry—or, at least, if not some chemistry, something of influence to the ISM. The age of dense clouds may be a few hundred thousand years to several million years and age of dark clouds 10^7–10^8 years. Thus $\sim$100 electrons from the ISM may strike a 0.1 μm grain in the lifetime of a dark cloud.

However, all regions of the spiral arms where molecular clouds are found are visited by shocks generated through supernova explosions every ~2 million years [17]. We may perhaps ignore this last complication, which requires that ices (of water, methanol, CO, etc.) be redeposited. However, with reference to a recent study of the role of low-energy electrons (<20 eV) in astrochemistry [16], it should be noted that since the energies of collisions of electrons with ices is thermal, in the astrophysical context higher-energy electron collisions, for example, at ≤ 20 eV, as studied in Ref. [16], do not take place in quiescent molecular clouds. Such events are of inherent interest but are likely to be of only limited importance in understanding molecular abundances. This is true in both ices themselves or in the gas phase, in environments in which ice processing is through evaporation of molecular material from grains, for example, protoplanetary disks. A related topic concerns cosmic ray induced photons in dense interstellar clouds involving electronic excitation of H_2 into the singlet B, C, D, and F states [10]. This results in the generation of a UV–vacuum ultraviolet (UV-VUV) photon field, which may be instrumental in inducing chemistry in ices on dust grains in molecular clouds. This is again discussed in Ref. [15] and elsewhere at some length. Probably the most important contribution would be the photodissociation of water ice, which absorbs in the UV below 160 nm. It should be noted that VUV absorption spectra in the ice phase may be considerably different from those in the gas phase. Indeed in water ice, the absorption band is blue-shifted by almost 1 eV [15]. Use of synchrotron radiation is revealing VUV spectra of ice that are important in astronomical and planetary observations [18–20].

So what takes place when electrons encounter amorphous water ice at (say) 10 K? Here our knowledge is still very limited. Three relevant studies may be found in "Cross Sections for Low-Energy (1–100 eV) Electron Elastic and Inelastic Scattering in Amorphous Ice" [21], "Dissociative Electron Attachment in Nanoscale Ice Films: Thickness and Charge Trapping Effects" [22], and "Transmission and Trapping of Cold Electrons in Water Ice" [23]. Only the last involves truly low-energy electrons, controlled below 100 meV, and it is found that porous amorphous ice and compact crystalline ice at 40 K show efficient trapping (Fig. 8.6), corroborating results in

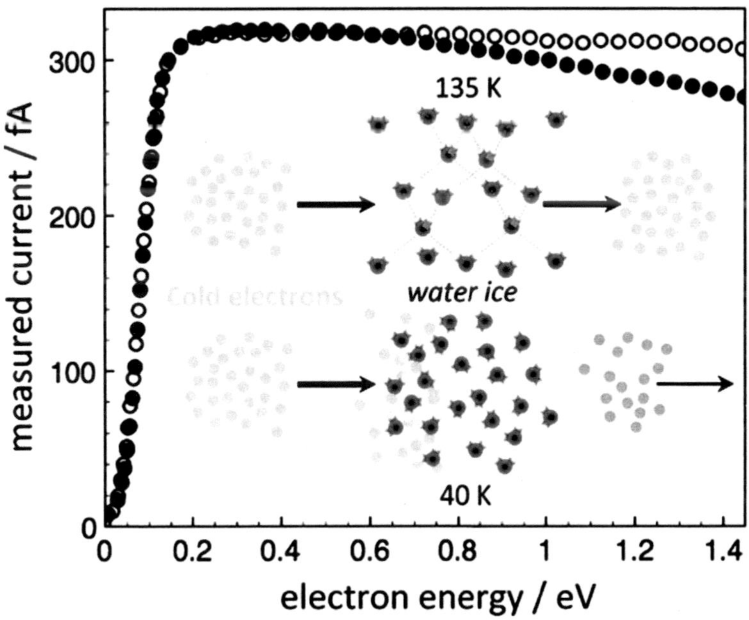

Figure 8.6 Low-energy ("cold") electrons pass unattenuated through crystalline ice at 135 K for energies between 0 and 650 meV. By contrast, both porous amorphous ice and compact crystalline ice at 40 K show efficient electron trapping. This may have profound consequences for star formation (see text). Reprinted with permission from Ref. [23]. Copyright (2011) American Chemical Society.

earlier work. Thus when electrons at very low energy encounter water ice on dust grains, they become trapped, causing the ice to become negatively charged. The events that may then transpire can be quite fundamental for our understanding of star formation, as we describe in the next section.

8.3 Electrons and the Formation of Stars

We turn now exclusively to the role of electrons in regions that are about to form stars, so-called cold or prestellar cores. The reason for this choice is that this is where a great deal of current interest lies—and is likely to continue to lie for many years hence. Cold cores form part of a major focus in astronomy: how do stars and their accompanying planets form? This is a rather recent question in the sense that it is driven, as is all astronomy, by new observations, especially those at high spatial resolution. The recognition that the Milky Way, is a dynamic object, continually changing, forming stars, with a supernova burst every 50 years [17], has become one that is very much of our time. Here we address just one question: how many electrons are there likely to be in one cubic meter of a cold core that may form a star? As we explain later, this is fundamental to understanding why stars can form at all or, at any rate, with an efficiency necessary to reproduce observations and to make galaxies.

8.3.1 Cold Cores and Star Formation

Charles Darwin spent 8 years studying barnacles and 20 studying earthworms to reach his conclusions. He taught us that only by the almost unending study of the smallest details can the bigger picture emerge. So it is with electrons in space; we require the application of painstakingly collected laboratory data. Above all, we need detailed observations coupled with exacting laboratory work and theory to unravel the mechanisms of star formation (Fig. 8.7) and the role of electrons in cold cores. The key to star formation is the role of magnetic fields. The magnetic field needs to leak away before a star can form. The fewer electrons there are in a cold core, the more rapidly this takes place. The cold core does not have unlimited time to form stars since, as mentioned before, in the Milky Way there are sufficient supernovae to churn up the material within $\sim$2 million years, inhibiting star formation [17, 24–26]. Therefore the number of electrons or, more generally, the degree of ionization of a cold core is fundamental to how stars form and ultimately planets. So the first question we must answer is, what is the fractional ionization of the gas in cold cores? We can then explore the effect electrons can have on the ability of the magnetic field to escape and thereby allow cold cores to collapse to form stars. These considerations allow some quite far-reaching general conclusions about the role of electrons and the degree of ionization in the earliest stages of star formation.

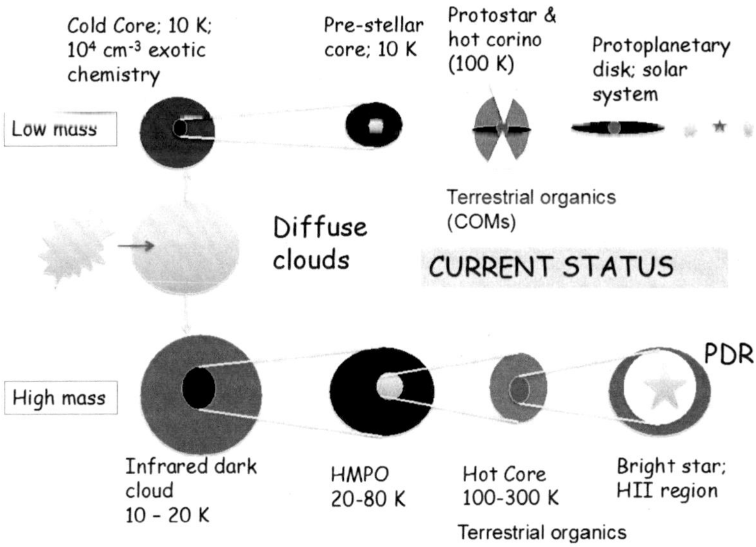

Figure 8.7 Schematic representation of star formation. Courtesy: Eric Herbst, Ohio State University.

First, what is a cold core? A cold core is an observed condensation of gas, formed by means that are still disputed, of density in excess of 10^4 cm^{-3} at its edge and extending to $>10^5$ cm^{-3} toward its centre. Cold cores are typically, though by no means always, embedded in an interstellar cloud whose density may be between 1 and 2 orders of magnitude lower. All but the very outer shell of gas in a cold core is shielded from ionizing radiation, through absorption of VUV by hydrogen molecules, which, with He, as always in the ISM form the major part of the gas, and through absorption by dust particles. Cold cores can be identified for example by the far-infrared (FIR) emission of cold dust at 1.3 mm or of CO molecules or of H_2O; see, for example, Refs. [27, 28] and [29], respectively, but cold cores may also contain complex organic species [30].

The steady-state concentration of electrons and ions is given by setting their rate of formation and their rate of removal to be equal. At this stage we need to recognize the great diversity that exists within the Milky Way (and in external galaxies), even among cold cores. It is therefore best at this stage to take a specific example, one that has been studied in great detail, and use this throughout our discussion. Therefore let us concentrate on the object B68 (Fig. 8.3) as the classic case of a prestellar cold core. We know a number of key parameters for B68, from numerous and very detailed observations, for example, Refs. [13, 31]. We use the conditions in B68, outlined later, to calculate the electron and positive ion concentrations. This may be achieved since the known parameters dictate the rate of ionization and of neutralization. However, several important questions remain unanswered, even in such a well-studied object such as B68. Do we also have negative ions, for example, polycyclic aromatic hydrocarbon ions (PAH^-) in the gas phase? We do not have an answer to this, but if such negative ions are formed, this may change our understanding. Nothing is immutable in modeling the ISM!

To compute the degree of ionization in B68, we need the following input. The material in B68 is composed of gas, 90% H_2, and 10% helium by number density, at a kinetic temperature of 10 K, and dust grains. The dust grains in B68 are coated with water ice and, in a layer on top of this water ice, CO ice. CO condenses on the grains at $\leq$26 K and therefore the major molecular constituent of the gas other than H_2, that is, CO, is frozen out, noting that water ice probably forms in situ on the grain rather than being frozen out from the gas phase. The gas density increases from the outside to the inside in B68, but let us initially take an average of $n\mathrm{H} \approx 10^{11}\ \mathrm{m}^{-3}$. The number of grains is given by $2.1\, n(\mathrm{H}_2) m(\mathrm{H})\, \xi / \pi a_\mathrm{b}^3 \delta\ \mathrm{m}^{-3}$, where $m(\mathrm{H})$ is the mass of the hydrogen atom, ξ is the grain/gas mass ratio (=0.013), a_b is a typical bare grain radius ($=10^{-7}$ m) or 0.1 μm, and δ is the density of grain material ($=2000\ \mathrm{kg\ m}^{-3}$) [26]. The grain radius with water and CO ice mantles becomes 1.15×10^{-7} m, given the abundance of water and CO in the gas phase before freeze-out. Later we will find out why we need to know the thickness of the CO mantle.

The degree of ionization in B68 is obtained through a steady-state balance between the rate of cosmic ray ionization per volume per second, $\zeta_\mathrm{c} n(\mathrm{H}_2)$, where ζ_c is typically $\sim 3 \times 10^{-17}\ \mathrm{s}^{-1}$ [23, 26, 28], and loss processes through ion–electron recombination both in the gas phase and at the grain surface. Note the inclusion of the grain surface effect. This is often erroneously omitted through an incorrect emphasis upon gas-phase electron–ion recombination events, a process mentioned earlier. These events are important in the chemistry, as noted before for the formation of CO, but turn out to be considerably less important in the charge balance than surface effects, as we now show.

In the gas phase, recombination processes active in removing electrons are very largely determined by $e^- + H_3{}^+$ interactions. In this connection, electron recombination with atomic ions is typically 3 to 4 orders of magnitude less efficient than with molecular ions. Thus the process $e^- + C^+ \rightarrow C^* \rightarrow C +$ photon takes place through radiative recombination, with an effective rate coefficient of ~3.6 × 10^{-11} cm^3 s^{-1}. With this in mind and taking account of the proportion of the positive ion budget that is composed of $H_3{}^+$, the effective rate coefficient, k^{2nd}, for recombination becomes 4 × 10^{-8} cm^3 s^{-1}. If we balance recombination with $\zeta_c n(H_2)$, the cosmic ray ionization rate, we obtain $[e^-] = [\text{ions}] = (\zeta_c\, n(H_2)/\, k^{2nd})^{1/2} \approx 8.7 \times 10^{-3}$ cm^{-3} or 8.7 × 10^3 m^{-3} for $n(H_2) = 10^{11}$ m^{-3}, say. This yields an ionization fraction of 8.7 × 10^{-8}, whereas the value in B68 is 5 × 10^{-9} relative to the abundance of H nuclei, a figure based on abundances of $H^{13}CO^+$ and DCO^+ deduced from observations [32]. Thus the gas-phase removal rate, k^{2nd} $[e^-][\text{ions}]$, is *too small* to account for the level of ionization in the medium. The discrepancy must be taken up by grain surface interactions.

The grain surface mechanism may operate as follows. First it is necessary to slightly rephrase the discussion of the charge balance between grains and gas in the ISM. The condition that must be maintained in the ISM is that there is no gradient in electrochemical potential between the electrons in the gas phase and those on the grains. This condition corresponds to the statement that an equilibrium charge is acquired by any grain, equivalent, as mentioned earlier, to the familiar floating potential of an insulator in a plasma. This is consistent with the observed experimental results for water ice [23], that as electrons at very low energy encounter water ice on dust grains they become trapped. Unfortunately there are no equivalent data for CO ices. However, the condition of zero gradient of electrochemical potential may be maintained by the presence of an approximately single negative charge on a grain, as discussed before.

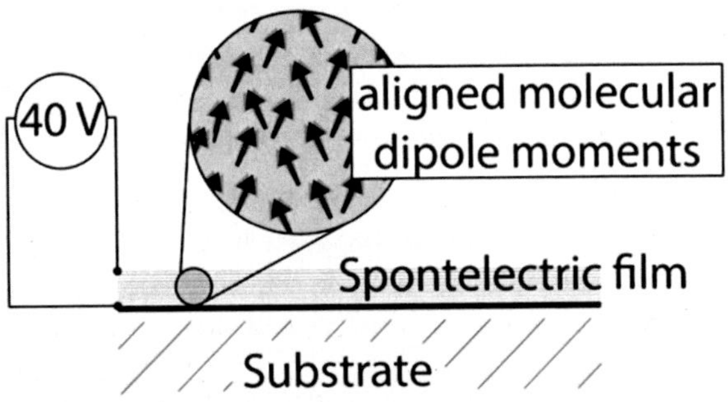

Figure 8.8 CO deposited on water ice on the surface of a dust grain exhibits the spontelectric effect. Reproduced with permission from Ref. [39]. Copyright (2017) by Springer Nature.

Due to local electric fields created through electron depletion, positive ions follow the electrons to grain surfaces, thus preserving electroneutrality in the plasma. On the grain surface, positive ions recombine efficiently with the mobile surface electrons [33–38]. This effect is considerably enhanced by the presence of solid CO on the surface and requires an explanation, provided below.

Experiments using reflection-absorption infrared spectroscopy (RAIRS) show that solid CO, deposited on water ice, exhibits the *spontelectric effect* [37–40]. The spontelectric effect gives rise to fields of $\sim 3.7 \times 10^7$ Vm^{-1} in solid CO, created through the spontaneous orientation of CO dipoles. This field is determined from spectral shifts, interpreted in terms of the vibrational Stark effect and the electronic Stark effect (Fig. 8.8). The field in solid CO creates a potential on the surface of the CO, which may be directly related to the polarization charge on the surface, given the thickness of the CO film. The conditions in B68, set out before, give this thickness as ~ 10 monolayers (ML). Note that this value is independent of grain radius. The charge that would be acquired is $q_p \approx +5.5$ charges, if the grain were left to its own devices; however, it is not.

As CO deposits, any grain surface becomes less negative due to the spontelectric effect and its electrochemical potential is effectively offset from that of the plasma. The electrochemical potential of the grain surface therefore no longer equals that of the gas-phase electrons, the system moving away from the nonspontelectric equilibrium. This causes a net flow of electrons toward the grain, down the electrochemical potential gradient. This in turn causes more positive ions to deposit on the dust grains. On the grain surface, the electrons and ions meet and recombine, thwarting the attempt to cause the grain surface to become positive. The scenario just described therefore increases the already dominant surface recombination rate. In the fully fledged CO layer of 10 ML, a new equilibrium between cosmic ray ionization and recombination on dust grains is now established through the requirement to maintain the equality of the electrochemical potential of gas-phase and grain surface electrons. This may be satisfied by a reduction in the degree of ionization in B68 through a factor of $q_p \approx 5.5$. The outcome is that the grain surface recombination mechanism is $\sim$40 times more efficient than the gas-phase recombination mechanism. In the absence of the spontelectric effect in CO, the factor is 7.5, with the grain surface mechanism still dominant, but a factor of q_p less efficient than in the presence of spontelectric CO. If we now balance the spontelectrically enhanced grain surface rate of removal with the cosmic ray ionization rate, we find a degree of ionization of $= 5.8 \times 10^{-9}$, a figure in agreement with an observationally based figure of 5×10^{-9} relative to H nuclei for B68 [32]. Our conclusion therefore is that the degree of ionization in those prestellar cores that are cold enough to freeze out CO, that is, cores containing grains at $T < 26$ K, will be rather low, lying around 5 to 6×10^{-9}. Observationally there are numerous cores, which show CO freeze-out [41, 42].

8.3.2 Electron Concentration and Star Formation

Why does electron concentration matter for the process of star formation? To address this question we need to step back a moment. What prevents a cold core from collapsing to form a protostar? The competition is between gravity pulling in and thermal, magnetic, and turbulent pressure pushing out. There are some very real puzzles in the earliest stages of star formation. These may be summarized as follows. Observations give lifetimes of prestellar cores of 0.25 to 1×10^6 years (Myr), while expulsion of the magnetic field necessary for gravitational collapse has been estimated to require ~3 Myr. However, almost the entire spiral arm star-forming volume of the galaxy is overrun by overlapping remnants of supernova explosions (SNEs) in 2 ± 0.5 Myr [14, 33–36]. The quiescent conditions necessary for gravitational collapse cannot therefore be met, since turbulence created by a supernova shock wave stirs up the medium. This creates a turbulent pressure, which inhibits gravitational collapse. These puzzles can all be simultaneously solved by a consideration of the electron concentrations in the cold cores, showing how electron collisions and interactions play a fundamental role in shaping the Galaxy as we find it.

If we recollect that pressure is energy per unit volume, then we may compare gas pressure with magnetic pressure, where the magnetic energy is given by $B_0^2/2\ \mu$ joules per unit volume, where μ is the permittivity of effectively free space and the subscript 0 refers to time zero. There exists a reasonably well-founded empirical relationship between density and magnetic field, $B_0 = 1.549 \times 10^{-15} <n(\mathrm{H})>^{0.65}$ Tesla [24]. If we take the radius of B68 to be 0.18 pc (Fig. 8.3) with a total mass of 3.1 solar masses [13], this corresponds to $<n(\mathrm{H})> = 4.46 \times 10^{10}\ \mathrm{m}^{-3}$ and thus $B_0 = 1.29 \times 10^{-8}$ Tesla, where we have taken a mean number density to simplify the discussion.

We can now use these figures to compare the magnetic pressure with the thermal pressure. This shows that initially the magnetic pressure is considerably the greater then the thermal pressure. In B68 the magnetic pressure is $B_0^2/2\ \mu = 6.7 \times 10^{-11}\ \mathrm{Jm}^{-3}$, whereas the thermal pressure, $<n(\mathrm{H})>\ kT = 6.2 \times 10^{-12}\ \mathrm{Jm}^{-3}$, that is, an order of magnitude lower. Indeed the magnetic pressure is considerably greater than that necessary to inhibit gravitational collapse. Here we just quote the result, whose derivation may be found in Chapter 41 of Ref. [24], which informs us that if the magnetic flux to mass ratio is above a critical value of ~0.15 T m^{-1} kg^{-1}, then the magnetic field will prevent gravitational collapse. In B68 the time zero flux to mass ratio is $\sim 2 \times 10^{-7}\ \mathrm{T\ m}^{-2}\ \mathrm{kg}^{-1}$, which is 13.4 times larger than the critical value. If a star is to form, there must therefore be a mechanism, and an associated timescale, through which the magnetic field is expelled from the cold core. This is the point at which electron concentrations become all-important and our knowledge of what controls these concentrations becomes a vital input into the astrophysics of star formation.

We now need to refine our model of B68, expressed in terms of some average number density, and recognize that it is more dense at the center than at the edges by a factor of 33 [13]. A greater magnetic pressure in the middle than the outside is therefore present, since $B_0^2 \propto <n(\mathrm{H})>^{0.65}$. The magnetic pressure will push out, flux will be lost, and, eventually, gravitational collapse will be able to take place. The rate at which the magnetic field can push out and be lost to B68 is proportional to the number of ions and electrons in the gas phase. The magnetic field captures the electrons and ions on the field lines, like "pearls on a string," and these electrons and ions, in turn, undergo collisions with passing H_2 molecules. The characteristic time for escape of the field contains a term $k^{\mathrm{m}}\chi n(\mathrm{H}_2)$, where k^{m} is the momentum transfer cross section between ions and H_2, $\sim 3 \times 10^{-9}$ cm^{3}s^{-1}, and χ is the degree of ionization of the medium, discussed before [42]. Through this drag between the gas and the charged particles, gravitational collapse is inhibited. Thus the degree of ionization of the medium determines how readily the magnetic field may "slip away" from the potentially star-forming core and thus make it able to collapse. A quantitative description of this may be found in Ref. [43], but suffice it to say that this description is based on *the knowledge of electron–molecule (and ion–molecule) cross sections* and the ability of grains to trap cold electrons. Moreover, we should recollect that the presence of cold electrons is itself determined from *a knowledge of electron molecule collisions in the gas phase.*

Let us now return to the puzzles in the earliest stages of star formation set out before. The chief of these was that the timescale for the expulsion of the magnetic field relative to the time between disruptions by supernovae. The lifetime to expulsion of the field was comparable with or greater than the time between supernova disruptions. As explained in detail in Ref. [43], we find a lifetime for expulsion of the field of ~0.6 Myr for B68, based essentially on a computation of the degree of ionization in the medium. This shows that B68, and doubtless many other such cold cores, will indeed form a star in due course: we find that B68 is, in fact, at present, teetering on the brink of collapse and the star formation process is very unlikely to be interrupted by the intervention of supernova remnants. This understanding has been achieved due to a combination of remarkable observational data and our experimentally and theoretically acquired knowledge of how electrons behave in space.

8.4 Electrons and Planetary Atmospheres

8.4.1 Planetary Aurorae

Let us now look inward toward our own solar system and planet. One of the first applications of our understanding of electron collisions was to model Earth's ionosphere. The terrestrial ionosphere was first characterized by the development of the radar in the 1930s, whereupon the plasma nature of Earth's upper atmosphere was identified. The ionosphere of any planet is formed by interaction of the terrestrial or planetary atmosphere with the solar wind, and its persistence and spatial properties are defined by the planet's magnetic field. Figure 8.9 shows the structure of Earth's upper atmosphere and typical electron densities.

The "capture" of solar wind protons and electrons by the planetary magnetosphere may lead to their penetrating deeper into the planetary atmosphere where they may produce an avalanche of secondary electrons that collide with the molecules comprising that atmosphere. The most dramatic example of such electron interactions with terrestrial molecules is the bright-color curtains often seen in northern and southern polar regions, where the electron density is highest. These are the aurorae manifest as "curtains" of light (Fig. 8.10).

Very high in the ionosphere (above 300 km), oxygen is the most common atom and excitation collisions $e^-(\varepsilon) + O\ (^3P) \rightarrow e^-(\varepsilon - \Delta\varepsilon) + O\ (^1D)$ occur, the decay of which $O\ (^1D) \rightarrow O\ (^3P)$ produces transitions at 630 and 634 nm that create a red aurora.

At lower altitudes, between 100 and 300 km the collision $e^-(\varepsilon) + O\ (^3P) \rightarrow e^-(\varepsilon - 4.17\ \text{eV}) + O\ (^1S)$ may occur, the decay of which $O\ (^1S) \rightarrow O\ (^3P)$ gives the strong yellow-to-green light at 557.7 nm that is most commonly seen in aurorae. $O\ (^1S)$ atoms can

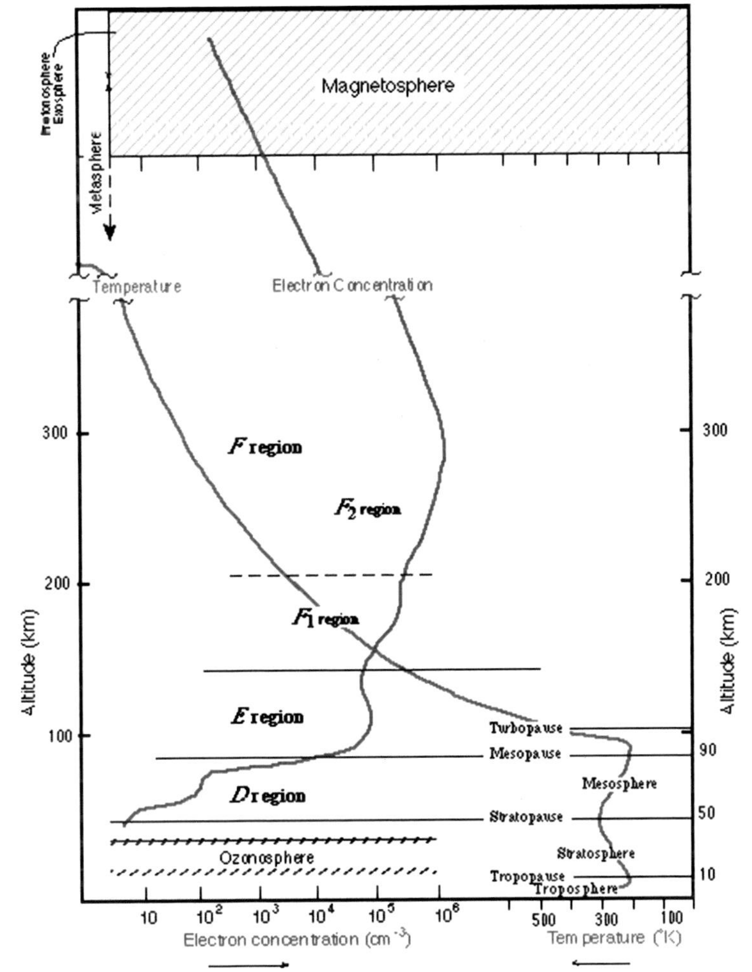

Figure 8.9 Electron and temperature profiles of the terrestrial ionosphere, showing its different regions. See the text for details.

also be produced by energy transfer from the molecular nitrogen metastable $N_2\ A^3\Sigma_u^+$. $N_2\ A^3\Sigma_u^+ + O \rightarrow N_2 + O\ (^1S)$ and dissociative recombination $e^- + O_2^+ \rightarrow O\ (^1S\ ^1D$ and $^3P)$.

Around 100 km, nitrogen molecules produce a crimson/purple light 427.8 nm that often seems to form the lower fringes on auroral curtains; this is due to emission from the molecular cation $N_2^+\ B^2\Sigma_u^+$

Figure 8.10 Terrestrial aurorae, a magnificent example of electrons (and protons) is space interacting with Earth's atmosphere.

to the first vibrational level of the electronic ground state of the molecular nitrogen ion ${N_2}^+ X^2\Sigma_u^+$ [44].

Aurorae are not limited to the visible but can also be seen in the X-ray and the UV spectrum; indeed excitation of the nitrogen molecule into the $a_1\pi$ electronic state and its subsequent decay (the so-called Lyman–Birge–Hopfield band) yields the brightest aurorae, albeit in the UV.

Similarly auroral emissions are seen in the IR, and an excellent work by Campbell and Brunger [45] discusses the role of electron interactions with NO to determine the electron impact contribution to infrared NO emissions in the aurora.

Most of the solar system planets are subject to the same intense bursts from the sun during the solar cycle, and all are subject to photoionization processes by solar irradiations. It is therefore not surprising that aurorae were both predicted and then observed on other planets with magnetospheres, such as the gas giant planets (Jupiter, Saturn, Uranus, and Neptune). Figure 8.11 illustrates an auroral oval observed on Jupiter, which is much larger in its extent than those on Earth and caused by emission from a more exotic molecular

Figure 8.11 Auroral oval on Jupiter caused by electron interactions with H_3^+. Photograph: NASA/ESA/PA.

species H_3^+ and hydrocarbon cations [46]. Aurorae have also been observed on Saturn in both the UV due to H Lyman-α and H_2 Lyman- and Werner-band transitions and in the infrared arising from H_3^+. However the large gas giants (Jupiter and Saturn) may also have an "internal" aurora created by ejection of material from their moons, which, when ionized by solar radiation, penetrate into the planet's atmosphere causing local aurorae, indeed such a process derived from the moon Io leads to permanent UV and X-ray aurora seen on Jupiter, while Enceladus with its water plumes injects electrons and ions 240,000 km along magnetic field lines into Saturn's north pole, inducing the observed aurorae.

Several planetary moons may have their own aurorae. For example, the Jovian moon Ganymede, which is also embedded in Jupiter's large magnetosphere, has its own aurorae arising from its own magnetic field believed to be induced by a saltwater ocean under an ice-covered surface. Aurorae have also been observed on planets with no magnetospheres, for example, on Mars, where the solar wind strikes the planet's atmosphere directly and high-energy electrons penetrate deep into the planet's atmosphere, exciting carbon dioxide to produce UV aurorae. Weak emissions observed in the atmosphere of Venus have also been ascribed to auroral phenomena.

Thus the study of planetary aurorae requires an extensive knowledge of electron impact with many molecular systems from simple hydrogen to hydrocarbon species, for example, methane, acetylene, ethylene, and ethane in the Jovian atmosphere [47].

8.4.2 Planetary Ionospheres

The study of planetary aurorae is intrinsically linked to the physical and chemical structures of the planetary ionosphere. Detailed models of the Jovian ionospheres have been developed to interpret mission data [48, 49] and a Saturn thermosphere-ionosphere model (STIM) was developed to support and interpret Cassini data [50]. Such models contain extensive electron-scattering cross sections over the necessary wide energy range.

In the terrestrial atmosphere, one region, the so-called E-region, has a high concentration of anions, an observation ascribed to the formation of O^- anions by DEA to molecular oxygen, $e^- + O_2 \rightarrow O^- + O$. Until recently this was the only known case of anion formation in any planetary atmosphere, but in 2004 the Cassini–Huygens mission reached the Saturnian system and explored the atmosphere of the Saturnian moon Titan. Titan is the only solar system moon to have a thick atmosphere, composed of 98.4% nitrogen and 1.4% methane, with a surface pressure in excess of that of Earth at 1.5 bars. The electron spectrometer (ELS), one of the sensors making up the Cassini Plasma Spectrometer (CAPS), revealed the existence of copious amounts of negative ions in Titan's upper atmosphere [51]. The observations at closest approach ($\sim$1000 km) showed evidence of negatively charged ions up to $\sim$10,000 amu/q, as well as two distinct peaks at 22 ± 4 and 44 ± 8 amu/q with evidence for a third one at 82 ± 14 amu/q (Fig. 8.12). The formation of these anions may be attributed to DEA to a set of molecules known as tholins, a class of heteropolymer molecules formed by solar UV irradiation of methane and nitrogen, the simplest building blocks

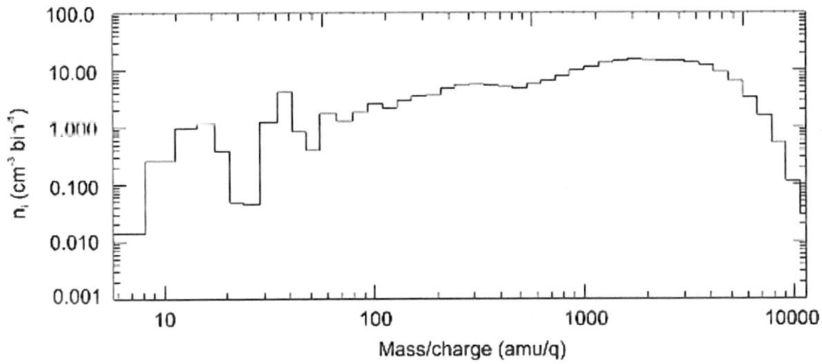

Figure 8.12 Anion spectrum of Titan's atmosphere recorded by the Cassini–Huygens mission at an altitude of 953 km. Adapted from Ref. [51] with permission. Copyright (2007) by John Wiley and Sons.

of which are HCN and CH_3N. The negative charge may then be transferred to more acidic molecules such as HC_3N, HC_5N, or C_4H_2. The three low mass peaks observed by the ELS are then attributed to CN^-, C_3N^-, and C_5N^- [52], which can then act as intermediates in the formation of the even larger negative ions observed by ELS, which are, in turn, precursors to the aerosols observed at lower altitudes.

Chemical and physical models of Titan's ionosphere have been developed by Vuitton and coworkers [53] to interpret this data and require further data on electron interactions with nitriles and tholin compounds. This research is given further emphasis by the discovery of Tholins on Pluto and its moon Charon by the New Horizons mission [54]. Whether anions are present in the tenuous atmosphere of Pluto and Charon is unknown, and more detailed models of the solar wind interaction with Pluto's atmosphere (and the electron interactions therein) are needed. However, exospheric neutral atoms and molecules escape from Pluto, depositing material onto Charon, while traveling into interplanetary space for millions of kilometres [55]. Eventually, the neutrals are ionized by solar extreme ultraviolet (EUV) photons and/or by collisions with solar wind electrons. Producing a comet-like interaction of the solar wind with Pluto, cometary tails are known to have anions (see the next section).

8.4.3 Comets and New Evidence for Electron Interactions from the Rosetta Mission

One of the unexpected yet fascinating discoveries of the recent ESA Rosetta mission to Comet 67P/Churyumov–Gerasimenko has been the recognition that electrons, not photons as previously assumed, are responsible for the dissociation of water and carbon dioxide erupting from the comet surface as it approached the sun [56–58]. Analyzing emission spectra recorded by the OSIRIS camera and the ALICE spectrometer in the UV it was found that the observed OH emissions and [O I] 630 nm emission in the inner coma could not be matched to any photon-induced model. However, since electron impact of water produces 100 times more OH than photodissociation and electron impact of CO_2 40 times more O I yield than photodissociation, these observations may be attributed to electron dissociative excitation of H_2O and CO_2. The electrons are produced by UV photons from the sun ionizing water molecules in the comet's coma; these photoelectrons then dissociate water and CO_2 in the coma, forming the excited oxygen and OH observed by Alice and Osiris.

Similarly the emissions detected in the NH filter of the Osiris camera are most likely the result of electron dissociative excitation that produces OH^+ ions from H_2O molecules directly in an excited state, while data from the CN filter suggest that further emissions are the result $CO_2{}^+$ ions produced by electron impact dissociative excitation and of fluorescent emission by CN radicals, most likely from the molecule HCN. However, the lack of experimental cross sections hampers the interpretation of these observations and *absolute* cross sections for the production of O I ^1D from CO_2 and H_2O and of CN from HCN by electron impact are urgently needed.

Electron collisions may also explain physical phenomena of the comet. The intensity of the emission in the inner coma was seen to decrease, despite increasing gas production rates, as the comet approached the sun. The most likely explanation is that because the gas production rates increased much faster than the ionizing solar radiation, collisions between electrons and neutral water molecules lowered electron temperatures in the inner coma

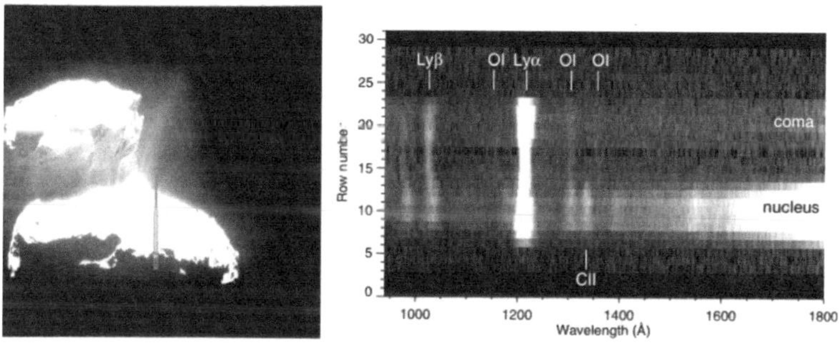

Figure 8.13 Emission spectra (right) from the inner comma of Comet 67P studied by the Rosetta spacecraft provide clear evidence of the role of electrons in space. Credit: ESA/Rosetta/MPS for OSIRIS Team MPS/UPD/LAM/IAA/SSO/INTA/UPM/DASP/IDA). See the text for details.

below the activation threshold of the dissociative impact excitation reactions [58].

These observations provide the first evidence for discrete electron collisions in an astronomical object and are leading to new programs to explore electron-driven processes in other main-body comets from ground-based observations. For example, electron impact excitation may also explain large-scale jet-like structures around comets [59]. New databases will need to be assembled to develop models of cometary emissions that include photoionization (the source of active electrons), electron-induced dissociation yielding fluorescent atomic, molecular, and ionic fragments, and electron-driven reactions induced on molecular positive ions, including dissociative recombination reactions.

The rich inventory of species observed on Comet 67P further complicates the problem, since not only water and CO_2 and their fragments need to be included but many hydrocarbons and nitriles must be included. Key species may be radicals such as OH and CN for which experiments are impractical. Comet 67P data also suggests the range of electron energies necessary to be studied is large, from the cold component observed below 0.1 eV to a 5–10 eV electron population, being interpreted as electrons retaining the energy they obtained when released in the ionization process to energies as high as 100–200 eV.

It should also be noted that anions have also been detected in cometary atmospheres. In studies of Halley's comet the dominant anion production mechanisms are found to be polar photodissociation of water and radiative electron attachment to carbon chains in the inner coma, followed by proton transfer from C_2H_2 and HCN to produce C_2H^- and CN^-, respectively. However, farther out in the coma, beyond 2000 km, where electron temperatures reach in the order of 10^3–10^5 K. DEA of C_2H_2 and HCN may dominate the production of C_2H^- and CN^- [60, 61]. Anion studies of Comet 67P have yet to be reported but maybe in archived data.

8.5 Conclusions

In this chapter we reviewed the role of electrons in space. Electrons are integral to the characteristics of planetary atmospheres and the observation of auroral phenomena on Earth and other planets in our solar system. Electrons may provide some of the first insights into the characteristics of planets around other stars, since exoplanetary aurorae could define the habitability of such planets, a strong magnetosphere protecting a planet's surface from the solar wind being one condition for life to develop [62]. The recent observation of aurorae on a brown dwarf [63] demonstrates that the technology for observation of exoplanetary aurorae is developing rapidly and we may be able to resolve such phenomena around exoplanets in the next decade.

Recent results from Rosetta reveal that in environments where it has been accepted that photon processes would dominate, secondary electrons produced by photoionization may, in contrast, dominate.

In the ISM the primary source of electrons in dark clouds, in which stars form, is through cosmic ray ionization. We have seen in this connection that electrons may play a critical role in star formation through their control of rate of escape of magnetic fields. Electron collisions also played a key role in shaping the early universe after the Big Bang.

Electrons in space therefore play an essential function in shaping the universe, and further studies have the prospect of revealing their part in creation of planets and the synthesis of prebiotic molecules, which are the harbinger of life itself.

Acknowledgments

The authors wish to recognize support from the UK STFC and generous support from the EU Framework programs (FP6 and 7 and H2020), which has allowed them to develop many collaborations within and across the astrochemistry and astrobiology communities, in particular FP7 Lassie ITN training program Grant Agreement 238258, Europlanet FP7 grant agreement 228319, and Europlanet 2020 RI, which has received funding from the European Union's Horizon 2020 research and innovation programme under Grant Agreement 654208. We also acknowledge support of COST Action CM0805 The Chemical Cosmos and COST Action TD1308 Origins and Evolution of Life on Earth and in the Universe (ORIGINS). NJM also recognizes support of UK STFC, EPSRC, The Royal Society, and Leverhulme Trust for their support of his research on electron physics and chemistry over past decade.

References

1. Laarson, M. and Orel, A. (2008). *Dissociative Recombination of Molecular Ions*. Cambridge University Press, Cambridge.
2. Hou, S. Q., He, J. J., Parikh, A., Kahl, D., Bertulani, C. A., Kajino, T., Mathews, G. J., and Zhao, G. (2017). *Astrophys. J.*, **834**, p. 165.
3. Janev, R. K., Langer, W. D., Evans, K. Jr., and Post, D. E. Jr. (2012). *Elementary Processes in Hydrogen-Helium Plasmas: Cross Sections and Reactions Rate Coefficients*. Springer-Verlag Berlin Heidelberg.
4. Lepp, S., Stancil, P. C., and Dalgarno, A. (2002). *J. Phys. B: At. Mol. Opt. Phys.*, **35**, pp. R57–R80.
5. Tanabe, T., Katayama, I., Ono, S., Chida, K., Watanabe, T., Arakaki, Y., Haruyama, Y., Saito, M., Odagiri, T., Hosono, K. (1999). *J. Phys. B: At. Mol. Opt. Phys.*, **32**, p. 5221.
6. Larsson, M. (2012). *Phil. Trans. R. Soc. A*, **370**, pp. 5118–5129.
7. Wildt, R. (1939). *Astrophys. J.*, **89**, p. 295.

8. Jones, N. C., Field, D., Ziesel, J. P., and Field, T. A. (2005). *J. Chem. Phys.*, **122**, p. 074301.
9. Millar, T. J., Walsh, C., and Field, T. A. (2017). *Chem. Rev.*, **117**, pp. 1765–1795.
10. Dalgarno, A., Yan, M., and Liu, W. (1999). *ApJ. Supp. Ser.*, **125**, p. 237.
11. Krishnakumar, E., Denifl, S., Cadez, I., Markelj, S., and Mason, N. J. (2011). *Phys. Rev. Lett.*, **106**, p. 243201.
12. Randell, J., Lunt, S. L., Mrotzek, G., Ziesel, J. P., and Field, D. (1994). *J. Phys. B: At. Mol. Opt. Phys.*, **27**, p. 2369.
13. Nielbock, M., Launhardt, R., Steinacker, J., Stutz, A. M., Balog, Z., Beuther, H., et al. (2012). *Astron. Astrophys.*, **547**, p. A11.
14. Curik, R., Ziesel, J. P., Jones, N. C., Field, T. A., and Field, D. (2006). *Phys. Rev. Lett.*, **97**, p. 123202.
15. https://www.astro.uni-koeln.de/cdms/molecules
16. Boyer, M. C., Rivas, N., Tran, A. A., Verish, C. A., and Arumainayagam, C. R. (2016). *Surf. Sci.*, **652**, p. 26.
17. Diehl, R., Halloin, H., Kretschmer, K., Lichti, G. G., Schoenfelder, V., et al. (2006). *Nature*, **439**, p. 45.
18. Mason, N. J., Dawes, A., Holtom, P. D., Mukerji, R. J., Davis, M. P., Sivaraman, B., Kaiser, R. I., Hoffmann, S. V., and Shaw, D. A. (2006). *Faraday Discuss.*, **133**, pp. 311–329.
19. Dawes, A., Pascual, N., Hoffmann, S. V., Jones, N. C., and Mason, N. J. (2017). *Phys. Chem. Chem. Phys.*, **19**, p. 27544.
20. Sivaraman, B., Pavithraaa, S., Lob, J.-I., Raja Sekhar, B. N., Hill, H., Cheng, B. M., and Mason, N. J. (2016). *Astrophys. J.*, **85**, p. 141.
21. Michaud, M., Wen, A., and Sanche, L. (2003). *Radiat. Res.*, **159**, p. 2.
22. Simpson, W. C., Orlando, T. M., Parenteau, L., Nagesha, K., and Sanche, L. (1998). *J. Chem. Phys.*, **108**, pp. 5027–5034.
23. Balog, R., Cicman, P., Field, D., Jones, N. C., Feketeova, L., Hoydalsvik, K., Field, T. A., and Ziesel, J. P. (2011). *J. Phys. Chem. A*, **115**, pp. 6820–6824.
24. Draine, B. T. (2011). *Physics of the Interstellar and Intergalactic Medium*. Princeton University Press.
25. Korolev, V. V., Vasiliev, E. O., Kovalenko, I. G., and Shchekinov, Yu. A. (2015). arXiv:1506.09038v1.
26. Tielens, A. G. G. M. (2006). *The Physics and Chemistry of the Interstellar Medium*, Cambridge University Press.

27. Juvela, M., Mattila, K., Lehtinen, K., Lemke, D., Laureijs, R., and Prusti, T. (2002). *Astron. Astrophys.*, **382**, p. 583.

28. Jones, P., Cunningham, M., Tóth, L., Liu, T., and Lopez, J. (2015). *J. Proc. Int. Astron. Union*, **11**(A29B), p. 60. doi:10.1017/S1743921316004440 and references therein.

29. Bergin, E. A. and Snell, R. L. (2002). *ApJ. L*, **581**, p. L105.

30. Vastel, C., Ceccarelli, C., Lefloch, B., and Bachiller, R. (2014). *ApJ. L*, **795**, p. L2.

31. Roy, A., André, Ph., Palmeirim, P., Attard, M., Könyves, V., Schneider, N., et al. (2014). *Astron. Astrophys.*, **562**, p. A138.

32. Maret, S. and Bergin, E. A. (2007). *Astrophys. J.*, **664**, pp. 956–963.

33. Umebayahsi, T. and Nakano, T. (1980). *Publ. Astron. Soc. Jpn.*, **32**, p. 405.

34. Draine, B. T. and Sutin, B. (1987). *Astrophys. J.*, **320**, p. 803.

35. Field, D. (2000). *Astron. Astrophys.*, **362**, p. 774.

36. Rosu-Finsen, A., Lasne, J., Cassidy, A., McCoustra, M. R. S., and Field, D. (2016). *Phys. Chem. Chem. Phys.*, **18**, pp. 5159–5171.

37. Balog, R., Cicman, P., Jones, N., and Field, D. (2009). *Phys. Rev. Lett.*, **102**, p. 073003.

38. Field, D., Plekan, O., Cassidy, A., Balog, R., Jones, N., and Dunger, J. (2013). *Int. Rev. Phys. Chem.*, **32**, p. 345.

39. Plekan, O., Rosu-Finsen, A., Cassidy, A. M., Lasne, J., McCoustra, M. R. S., and Field, D. (2017). *Eur. Phys. J. D*, **71**, p. 162.

40. Chen, Y.-J., Munoz Caro, G. M., Aparicio, S., Jimenez-Escobar, A., Lasne, J., Rosu-Finsen, A., McCoustra, M. R. S., Cassidy, A. M., and Field, D. (2017). *Phys. Rev. Lett.*, **119**, p. 157703.

41. Walmsley, C. M. and Flower, D. R. (2004). Pineau des Forês, complete depletion in prestellar cores. *Astron. Astrophys.*, **418**, p. 1035.

42. Flower, D. R. (2003). Pineau des Forêts. *Mon. Not. R. Astron. Soc.*, **341**, p. 1272.

43. Rosu-Finsen, A., Lasne, J., Cassidy, A., McCoustra, M. R. S., and Field, D. (2016). *Astrophys. J.*, **832**, p. 1.

44. Ashrafi, M., Lanchester, B. S., Lummerzheim, D., Ivchenko, N., and Jokiaho, O. (2009). *Ann. Geophys.*, **27**, p. 2545.

45. Campbell, L. and Brunger, M. J. (2007). *Geophys. Res. Lett.*, **34**, p. L22102.

46. Gérard, J. C., Bonfond, B., Grodent, D., Radioti, A., Clarke, J. T., Gladstone, G. R., Waite, J. H., Bisikalo, D., and Shematovich, V. I. (2014). *J. Geophys. Res.*, **119**, p. 9072.

47. Perry, J. J., Kim, Y. H., Fox, J. L., and Porter, H. S. (1999). *J. Geophys. Res.*, **104**, p. 16541.
48. Grodent, D., Waite, J. H., and Gérard, J. C. (2001). *J. Geophys. Res.*, **106**, p. 12933.
49. Millward, G., Miller, S., Stallard, T., Aylward, A. D., and Achilleos, N. (2002). Icarus, **160**, pp. 95–107.
50. Galand, M., Moore, L., Mueller-Wodarg, I., Mendillo, M., and Miller, S. (2011). *J. Geophys. Res.*, **116**, p. A09306.
51. Coates, A. J., Crary, F. J., Lewis, G. R., Young, D. T., Waite Jr., J. H., and Sittler Jr., E. C. (2007). *Geophys. Res. Lett.*, **34**, p. L22103.
52. Horvath, G., Aranda-Gonzalvo, Y., Mason, N. J., Zahoran, M., and Matejcik, S. (2009). *EPJ Appl. Phys.*, **49**, p. 13105.
53. Vuitton, V., Lavvas, P., Yelle, R. V., Galand, M., Wellbrock, A., Lewis, G. R., Coates, A. J., and Wahlund, J.-E. (2009). *Planet. Space Sci.*, **57**, p. 1558.
54. Grundy, W. M., Binzel, R. P., Buratti, B. J., Cook, J. C., Cruikshank, D. P., et al. (2016). *Science*, **351**, p. 9189.
55. Cravens, T. E. and Strobel, D. F. (2015). *Icarus*, **246**, pp. 303–309.
56. Feldman, P. D., A'Hearn, M. F., Bertaux, J.-L., Feaga, L. M., Wm Parker, J., Schindhelm, E., Steffl, A. J., Stern, S. A., Weaver, H. A., Sierks, H., and Vincent, J.-B. (2015). *Astron. Astrophys.*, **583**, p. A8.
57. Feldman, P. D., A'Hearn, M. F., Feaga, L. M., Bertaux, J.-L., Noonan, J., Wm Parker, J., Schindhelm, E., Steffl, A. J., Stern, S. A., and Weaver, H. A. (2016). *Astrophys. J. Lett.*, **825**, p. L8.
58. Bodewits, D., Lara, L. M., A'Hearn, M. F., et al. (2016). *Astron. J.*, **152**, p. 130.
59. A'Hearn, M. F., Hoban, S., Birch, P. V., Bowers, C., Martin, R., and Klinglesmith, A. (1986). *Nature*, **324**, p. 649.
60. Cordiner, M. A. and Charnley, S. B. (2014). *Meteorit. Planet. Sci.*, **49**, p. 21.
61. Rodgers, S. D. and Charnley, S. B. (2002). *Mon. Not. R. Astron. Soc.*, **330**, p. 660.
62. Smith, D. S., Scalo, J., and Wheeler, J. C. (2004). *Origins Life Evol. Biosphere*, **34**, p. 513.
63. Hallinan, G., Littlefair, S. P., Cotter, G., Bourke, S., Harding, L. K., Pineda, J. S., Butler, R. P., Golden, A., Basri, G., Doyle, J. G., Kao, M. M., Berdyugina, S. V., Kuznetsov, A., Rupen, M. P., and Antonova, A. (2015). *Nature*, **523**, pp. 568–571.

Index